T0181402

Advances in Imaging and Sensing

Devices, Circuits, and Systems

Series Editor
Krzysztof Iniewski
Emerging Technologies CMOS Inc.
Vancouver, British Columbia, Canada

PUBLISHED TITLES:

Advances in Imaging and Sensing
Shuo Tang and Daryoosh Saeedkia

Analog Electronics for Radiation Detection
Renato Turchetta

Atomic Nanoscale Technology in the Nuclear Industry
Taeho Woo

Biological and Medical Sensor Technologies
Krzysztof Iniewski

Building Sensor Networks: From Design to Applications
Ioanis Nikolaidis and Krzysztof Iniewski

**Cell and Material Interface: Advances in Tissue Engineering,
Biosensor, Implant, and Imaging Technologies**
Nihal Engin Vrana

Circuits and Systems for Security and Privacy
Farhana Sheikh and Leonel Sousa

Circuits at the Nanoscale: Communications, Imaging, and Sensing
Krzysztof Iniewski

CMOS: Front-End Electronics for Radiation Sensors
Angelo Rivetti

**CMOS Time-Mode Circuits and Systems: Fundamentals
and Applications**
Fei Yuan

Design of 3D Integrated Circuits and Systems
Rohit Sharma

Electrical Solitons: Theory, Design, and Applications
David Ricketts and Donhee Ham

Electronics for Radiation Detection
Krzysztof Iniewski

Electrostatic Discharge Protection: Advances and Applications
Juin J. Liou

PUBLISHED TITLES:

Embedded and Networking Systems:
Design, Software, and Implementation
Gul N. Khan and Krzysztof Iniewski

Energy Harvesting with Functional Materials and Microsystems
Madhu Bhaskaran, Sharath Sriram, and Krzysztof Iniewski

Gallium Nitride (GaN): Physics, Devices, and Technology
Farid Medjdoub

Graphene, Carbon Nanotubes, and Nanostuctures:
Techniques and Applications
James E. Morris and Krzysztof Iniewski

High-Speed Devices and Circuits with THz Applications
Jung Han Choi

High-Speed Photonics Interconnects
Lukas Chrostowski and Krzysztof Iniewski

High Frequency Communication and Sensing:
Traveling-Wave Techniques
Ahmet Tekin and Ahmed Emira

High Performance CMOS Range Imaging:
Device Technology and Systems Considerations
Andreas Süss

Integrated Microsystems: Electronics, Photonics, and Biotechnology
Krzysztof Iniewski

Integrated Power Devices and TCAD Simulation
Yue Fu, Zhanming Li, Wai Tung Ng, and Johnny K.O. Sin

Internet Networks: Wired, Wireless, and Optical Technologies
Krzysztof Iniewski

Ionizing Radiation Effects in Electronics: From Memories to Imagers
Marta Bagatin and Simone Gerardin

Labs on Chip: Principles, Design, and Technology
Eugenio Iannone

Laser-Based Optical Detection of Explosives
Paul M. Pellegrino, Ellen L. Holthoff, and Mikella E. Farrell

Low Power Emerging Wireless Technologies
Reza Mahmoudi and Krzysztof Iniewski

Medical Imaging: Technology and Applications
Troy Farncombe and Krzysztof Iniewski

Metallic Spintronic Devices
Xiaobin Wang

PUBLISHED TITLES:

MEMS: Fundamental Technology and Applications
Vikas Choudhary and Krzysztof Iniewski

Micro- and Nanoelectronics: Emerging Device Challenges and Solutions
Tomasz Brozek

Microfluidics and Nanotechnology: Biosensing to the Single Molecule Limit
Eric Lagally

MIMO Power Line Communications: Narrow and Broadband Standards, EMC, and Advanced Processing
Lars Torsten Berger, Andreas Schwager, Pascal Pagani, and Daniel Schneider

Mixed-Signal Circuits
Thomas Noulis

Mobile Point-of-Care Monitors and Diagnostic Device Design
Walter Karlen

Multisensor Attitude Estimation: Fundamental Concepts and Applications
Hassen Fourati and Djamel Eddine Chouaib Belkhiat

Multisensor Data Fusion: From Algorithm and Architecture Design to Applications
Hassen Fourati

MRI: Physics, Image Reconstruction, and Analysis
Angshul Majumdar and Rabab Ward

Nano-Semiconductors: Devices and Technology
Krzysztof Iniewski

Nanoelectronic Device Applications Handbook
James E. Morris and Krzysztof Iniewski

Nanomaterials: A Guide to Fabrication and Applications
Sivashankar Krishnamoorthy

Nanopatterning and Nanoscale Devices for Biological Applications
Šeila Selimovic´

Nanoplasmonics: Advanced Device Applications
James W. M. Chon and Krzysztof Iniewski

Nanoscale Semiconductor Memories: Technology and Applications
Santosh K. Kurinec and Krzysztof Iniewski

Novel Advances in Microsystems Technologies and Their Applications
Laurent A. Francis and Krzysztof Iniewski

Optical, Acoustic, Magnetic, and Mechanical Sensor Technologies
Krzysztof Iniewski

Optical Fiber Sensors: Advanced Techniques and Applications
Ginu Rajan

PUBLISHED TITLES:

Optical Imaging Devices: New Technologies and Applications
Ajit Khosla and Dongsoo Kim

Organic Solar Cells: Materials, Devices, Interfaces, and Modeling
Qiquan Qiao

Physical Design for 3D Integrated Circuits
Aida Todri-Sanial and Chuan Seng Tan

Power Management Integrated Circuits and Technologies
Mona M. Hella and Patrick Mercier

Radiation Detectors for Medical Imaging
Jan S. Iwanczyk

Radiation Effects in Semiconductors
Krzysztof Iniewski

Reconfigurable Logic: Architecture, Tools, and Applications
Pierre-Emmanuel Gaillardon

Semiconductor Radiation Detection Systems
Krzysztof Iniewski

Smart Grids: Clouds, Communications, Open Source, and Automation
David Bakken

Smart Sensors for Industrial Applications
Krzysztof Iniewski

Soft Errors: From Particles to Circuits
Jean-Luc Autran and Daniela Munteanu

Solid-State Radiation Detectors: Technology and Applications
Salah Awadalla

Structural Health Monitoring of Composite Structures Using Fiber
Optic Methods
Ginu Rajan and Gangadhara Prusty

Technologies for Smart Sensors and Sensor Fusion
Kevin Yallup and Krzysztof Iniewski

Telecommunication Networks
Eugenio Iannone

Testing for Small-Delay Defects in Nanoscale CMOS Integrated Circuits
Sandeep K. Goel and Krishnendu Chakrabarty

Tunable RF Components and Circuits: Applications in Mobile Handsets
Jeffrey L. Hilbert

VLSI: Circuits for Emerging Applications
Tomasz Wojcicki

PUBLISHED TITLES:

Wireless Medical Systems and Algorithms: Design and Applications
Pietro Salvo and Miguel Hernandez-Silveira

Wireless Technologies: Circuits, Systems, and Devices
Krzysztof Iniewski

Wireless Transceiver Circuits: System Perspectives and Design Aspects
Woogeun Rhee

FORTHCOMING TITLES:

Introduction to Smart eHealth and eCare Technologies
Sari Merilampi, Krzysztof Iniewski, and Andrew Sirkka

Magnetic Sensors: Technologies and Applications
Laurent A. Francis and Kirill Poletkin

Nanoelectronics: Devices, Circuits, and Systems
Nikos Konofaos

Noise Coupling in System-on-Chip
Thomas Noulis

Radio Frequency Integrated Circuit Design
Sebastian Magierowski

Semiconductor Devices in Harsh Conditions
Kirsten Weide-Zaage and Malgorzata Chrzanowska-Jeske

X-Ray Diffraction Imaging: Technology and Applications
Joel Greenberg and Krzysztof Iniewski

Advances in Imaging and Sensing

Edited by
Shuo Tang
University of British Columbia
Vancouver, British Columbia, Canada

Daryoosh Saeedkia
TeTechS Inc.
Waterloo, Ontario, Canada

Managing Editor
Krzysztof Iniewski
Emerging Technologies CMOS Services Inc.
Vancouver, British Columbia, Canada

CRC Press
Taylor & Francis Group
Boca Raton London New York

CRC Press is an imprint of the
Taylor & Francis Group, an **informa** business

MATLAB® is a trademark of The MathWorks, Inc. and is used with permission. The MathWorks does not warrant the accuracy of the text or exercises in this book. This book's use or discussion of MAT-LAB® software or related products does not constitute endorsement or sponsorship by The MathWorks of a particular pedagogical approach or particular use of the MATLAB® software.

CRC Press
Taylor & Francis Group
6000 Broken Sound Parkway NW, Suite 300
Boca Raton, FL 33487-2742

First issued in paperback 2022

ISBN-13: 978-1-498-71475-4 (hbk)
ISBN-13: 978-1-03-233990-0 (pbk)
DOI: 10.1201/9781315371283

Visit the Taylor & Francis Web site at
http://www.taylorandfrancis.com

and the CRC Press Web site at
http://www.crcpress.com

Contents

Preface..xi
Series Editor... xiii
Editor xv
Contributors xvii

SECTION I Technology and Devices

Chapter 1 Theory and Experiments on THz Devices on Graphene 3

Taiichi Otsuji and Victor Ryzhii

Chapter 2 Microscopic Many-Body Theory for Terahertz Quantum
Cascade Lasers 37

Qi Jie Wang and Tao Liu

Chapter 3 Photonic Engineering of Terahertz Quantum Cascade Lasers 61

Qi Jie Wang and Guozhen Liang

SECTION II Imaging Sensors and Systems

Chapter 4 Intense Few-Cycle Terahertz Pulses and Nonlinear Terahertz
Spectroscopy 85

Ibraheem Al-Naib and Tsuneyuki Ozaki

Chapter 5 Applications of Terahertz Technology for Plastic Industry 107

Daniel M. Hailu and Daryoosh Saeedkia

Chapter 6 Cross-Sectional Velocity Distribution Measurement Based on
Fiber-Optic Differential Laser Doppler Velocimetry....................... 121

Koichi Maru

Chapter 7   Integrated CMOS Optical Biosensor Systems 141

Sameer Sonkusale and Jian Guo

Chapter 8   Design of CMOS Microsystems for Time-Correlated Single-
Photon Counting Fluorescence Lifetime Analysis........................... 169

Liping Wei and Derek Ho

SECTION III   *Biomedical Applications*

Chapter 9   Fiber Delivery of Femtosecond Pulses for Multiphoton Endoscopy 195

Shuo Tang and Jiali Yu

Chapter 10 *In Vivo* Flow Cytometer: A Powerful Tool to Study Cancer
Metastasis 213

Xun-Bin Wei, Dan Wei, Ping Yang, Zhen-Yu Niu, and Huamao Ye

Chapter 11 Transition Metal Luminophores for Cell Imaging 227

Christopher S. Burke, Aisling Byrne, and Tia E. Keyes

Chapter 12 Three-Dimensional Electrical Impedance Tomography for
Pulmonary Embolism: A Simulation Study of a 16-Electrode
System 255

*D. Trang Nguyen, M.A. Barry, R. Kosobrodov,
A. Thiagalingam, and Alistair L. McEwan*

Index 275

Preface

Imaging and sensing technologies are essential for industrial and biomedical applications. Compared to x-ray, computerized tomography (CT), magnetic resonance imaging (MRI), and ultrasound, optical, and terahertz imaging and sensing have the advantages of noninvasive, nonionizing, noncontact, and high resolution. By utilizing the different wavelength ranges of the electromagnetic spectrum, optical and terahertz technologies can probe the structure and properties of organic and inorganic materials, including biological tissues. Optical waves cover ultraviolet, visible, and infrared spectrum from 100 nm to 100 μm. Terahertz waves have longer wavelength than optical waves and range from 100 μm to 1 mm. The wavelength is directly related to the photon energy carried by the electromagnetic wave, which determines the properties of the wave–material interaction. Utilizing such properties, a variety of optical and terahertz imaging and sensing techniques and applications have been developed. This book includes the state-of-the-art advancement in optical and terahertz imaging and sensing on device design, system development, and applications from experts in the field.

The advancement of optical imaging and sensing has been empowered by the invention of new laser sources, optical fiber components, and optoelectronic devices. For example, CMOS technology has enabled the integration of optical detection with electronic signal processing, which has made the systems more compact and portable. New contrast agents have been developed to not only enhance the signal intensity but also specifically target cancer cells. Due to its unique properties, optical imaging and sensing has found essential applications in the biomedical field. It has high resolution in the submicron region, which can image cells and extracellular matrix. It can image tissue structures below the surface noninvasively for detecting cancer in the early stage. From the rich signals of light absorption, scattering, fluorescence, and luminescence, not only the morphology but also functional properties, such as tissue constituents, oxygenation, and flow velocity, can all be detected by optical technology. The advancement of optical imaging and sensing is moving toward portable systems for *in vivo* detection in real time and noninvasively, which will have a powerful impact on the diagnosis and treatment of diseases.

Meanwhile, the real-world applications for terahertz technology have largely been unknown until recent years. The elusive properties of these waves have long posed challenges to researchers who worked to find an affordable way to generate terahertz to be bright enough for day-to-day use. Today, terahertz technology is an ever-evolving and penetrating force that is making its mark across research labs and industries alike. The unique characteristics of terahertz waves render them a viable solution for challenges facing multiple markets. On one hand, terahertz has the proven potential to largely shift the way current scientific research is being done in multiple fields. On a larger, more industrial scale, the use of terahertz technology presents notable opportunities to create a shift in traditional processes in a vast array of areas ranging from security to manufacturing. Due to its ability to see through common materials that current modalities such as x-ray and infrared cannot, terahertz can be utilized to identify and conduct precise measurements of materials such as plastics, rubber, and glass. This enables terahertz to be applied to areas such as plastic bottle manufacturing, plastic recycling, material identification, void detection, and so on. With a wide array of applications already known, the use of terahertz in varying industries will undoubtedly continue to grow until the once puzzling gap between microwaves and infrared is at last closed.

The book is grouped into three sections: Section I includes three chapters on devices for generating and detecting terahertz waves; Section II includes five chapters on system developments for imaging and sensing; Section III includes four chapters on applications in the biomedical field. These chapters highlight the state of the art of optical and terahertz imaging and sensing. They can provide the readers with a good understanding about the principles of the technology and how to develop systems and applications using the technology.

The editors would like to express gratitude to all the contributing authors for their significant and indispensable efforts that resulted in this book.

Shuo Tang
Daryoosh Saeedkia
Krzysztof Iniewski

MATLAB® is a registered trademark of The MathWorks, Inc. For product information, please contact:

The MathWorks, Inc.
3 Apple Hill Drive
Natick, MA 01760-2098 USA
Tel: 508-647-7000
Fax: 508-647-7001
E-mail: info@mathworks.com
Web: www.mathworks.com

Series Editor

Krzysztof (Kris) Iniewski is managing R&D at Redlen Technologies Inc., a start-up company in Vancouver, Canada. Redlen's revolutionary production process for advanced semiconductor materials enables a new generation of more accurate, all-digital, radiation-based imaging solutions. Kris is also a founder of Emerging Technologies CMOS Inc. (www.etcmos.com), an organization of high-tech events covering communications, microsystems, optoelectronics, and sensors. In his career, Dr. Iniewski held numerous faculty and management positions at the University of Toronto, University of Alberta, SFU, and PMC-Sierra Inc. He has published more than 100 research papers in international journals and conferences. He holds 18 international patents granted in the USA, Canada, France, Germany, and Japan. He is a frequent invited speaker and has consulted for multiple organizations internationally. He has written and edited several books for CRC Press, Cambridge University Press, IEEE Press, Wiley, McGraw-Hill, Artech House, and Springer. His personal goal is to contribute to healthy living and sustainability through innovative engineering solutions. In his leisurely time, Kris can be found hiking, sailing, skiing, or biking in beautiful British Columbia. He can be reached at kris.iniewski@gmail.com.

Editor

Shuo Tang received her BS and MS in electronics from Peking University, China, in 1992 and 1997, respectively. She received her PhD in electrical engineering from the University of California, Los Angeles, in 2003. From 2003 to 2006, she was a post-doctoral researcher at the Beckman Laser Institute at the University of California, Irvine. She began as an assistant professor at the Department of Electrical and Computer Engineering at the University of British Columbia in 2007, where she is currently an associate professor since 2013.

Prof. Tang's research interests are in the broad areas of biomedical optical imaging systems and devices, including multiphoton microscopy, optical coherence tomography, photoacoustic imaging, and micro-endoscopy systems for biomedical applications. She is the author or coauthor of over 90 journal articles, conference papers, and book chapters.

Daryoosh Saeedkia received his PhD in electrical and computer engineering from the University of Waterloo in 2005. In 2010, he founded TeTechS Inc, a leading innovator and provider of advanced terahertz measurement solutions for scientific and industrial applications worldwide. He has more than 12 years of industrial and academic experience in terahertz photonics technology, has delivered 12 invited talks, and has published more than 55 scientific articles, a book, and a book chapter in the area of terahertz technology. Dr. Saeedkia is the recipient of the 2008 Douglas R. Colton Medal for Research Excellence in Canada and the 2006 NSERC Innovation Challenge Award.

Contributors

Ibraheem Al-Naib
University of Dammam
Dammam, Saudi Arabia

M.A. Barry
University of Sydney
Sydney, New South Wales, Australia

Christopher S. Burke
School of Chemical Sciences
and
Dublin City University
Dublin, Ireland

Aisling Byrne
School of Chemical Sciences
and
Dublin City University
Dublin, Ireland

Jian Guo
Department of Electrical and Computer
    Engineering
Tufts University
Medford Massachusetts

Daniel M. Hailu
TeTechS Inc.
Waterloo, Ontario, Canada

Derek Ho
Department of Physics and Materials
    Science
City University of Hong Kong
Kowloon, Hong Kong

Tia E. Keyes
School of Chemical Sciences
and
Dublin City University
Dublin, Ireland

R. Kosobrodov
University of Sydney
Sydney, New South Wales, Australia

Guozhen Liang
School of Electrical and Electronic
    Engineering
Nanyang Technological University
Singapore

Tao Liu
School of Electrical and Electronic
    Engineering
and
School of Physical and Mathematical
    Sciences
Nanyang Technological University
Singapore

Koichi Maru
Department of Electronics and
    Information Engineering
Kagawa University
Takamatsu, Japan

Alistair L. McEwan
University of Sydney
Sydney, New South Wales, Australia

D. Trang Nguyen
University of Sydney
Sydney, New South Wales, Australia

Zhen-Yu Niu
Med-X Research Institute
and
School of Biomedical Engineering
Shanghai Jiao Tong University
Xuhui, Shanghai, People's Republic of
    China

Taiichi Otsuji
Research Institute of Electrical
    Communication
Tohoku University
Sendai, Japan

Tsuneyuki Ozaki
Institut National de la Recherche
    Scientifique
Centre Énergie Matériaux
    Télécommunications
Montréal, Québec, Canada

Victor Ryzhii
Research Institute of Electrical
    Communication
Tohoku University
Sendai, Japan

Daryoosh Saeedkia
TeTechS Inc.
Waterloo, Ontario, Canada

Sameer Sonkusale
Department of Electrical and Computer
    Engineering
Tufts University
Medford Massachusetts

Shuo Tang
Department of Electrical and Computer
    Engineering
University of British Columbia
Vancouver, British Columbia, Canada

A. Thiagalingam
University of Sydney
Staff SpecialistWestmead Hospital
Sydney, New South Wales, Australia

Qi Jie Wang
School of Electrical and Electronic
    Engineering
and
School of Physical and Mathematical
    Sciences
Nanyang Technological University
Singapore

Dan Wei
Med-X Research Institute
and
School of Biomedical Engineering
Shanghai Jiao Tong University
Xuhui, Shanghai, People's Republic of
    China

Liping Wei
Department of Physics and Materials
    Science
City University of Hong Kong
Kowloon, Hong Kong

Xun-Bin Wei
Med-X Research Institute
and
School of Biomedical Engineering
Shanghai Jiao Tong University
Xuhui, Shanghai, People's Republic of
    China

and

Institutes of Biomedical Sciences
and
Department of Chemistry
Fudan University
Yangpu, Shanghai, People's Republic of
    China

Ping Yang
Med-X Research Institute
and
School of Biomedical Engineering
Shanghai Jiao Tong University
Xuhui, Shanghai, People's Republic of
    China

Huamao Ye
Department of Urology
Shanghai Changhai Hospital
Yangpu, Shanghai, People's Republic of
China

Jiali Yu
Department of Electrical and Computer
Engineering
University of British Columbia
Vancouver, British Columbia, Canada

Section I

Technology and Devices

1 Theory and Experiments on THz Devices on Graphene

Taiichi Otsuji and Victor Ryzhii

CONTENTS

1.1 Introduction 3
1.2 Optoelectronic Properties of Graphene 4
    1.2.1 Optical Conductivity in Doped and Pumped Graphene 4
    1.2.2 Ultrafast Carrier Dynamics and THz Gain in Pumped Graphene 6
    1.2.3 Current-Injection Pumping 9
1.3 Plasmonic Properties of Graphene 11
    1.3.1 Dispersions and Modes in Graphene Plasmons 11
    1.3.2 Giant THz Gain via Excitation of Surface Plasmon Polaritons in
         Pumped Graphene 11
    1.3.3 Superradiant THz Plasmonic Lasing in Graphene Metasurfaces 16
    1.3.4 Plasmonic THz Photodetection and Photomixing 18
    1.3.5 Plasmonic THz Intensity Modulation 18
1.4 Double Graphene–Layered van der Waals Heterostructures 19
    1.4.1 Intergraphene Layer Resonant Tunneling 19
    1.4.2 Photon-Assisted Resonant Tunneling 22
    1.4.3 Plasmon-Assisted Resonant Tunneling 24
    1.4.4 THz Photomixers 28
    1.4.5 THz Modulators 28
1.5 Conclusion 30
Acknowledgments 31
References 31

1.1 INTRODUCTION

This chapter reviews recent advances in theory and experiments on graphene-based terahertz (THz) devices. The groundbreaking discovery of graphene [1–3] has stimulated the study on graphene-based THz-active devices [4–10]. Electrons and holes in graphene behave as relativistic charged particles of massless Dirac fermions holding enormously ultrafast transport properties, which can break through the speed and frequency limit on conventional electronic and optoelectronic devices like transistors, photodetectors, lasers, and light-emitting diodes. In particular, the creation of graphene-based THz lasers (still to be realized) is one of the hottest topics [11–18]. Massless Dirac fermions of electrons and holes in gapless and linear symmetric band structures in graphene support a weak gain in a wide THz frequency range under optical or electrical pumping [11–15]. The fundamental physics behind the lasing operation in graphene-based heterostructures is based on the carrier population inversion resulting from the imbalance between the ultrafast energy relaxation of the excited carriers via the optical phonon emission and their relatively slow recombination [11–15].

The physics of plasmons and surface plasmon polaritons (SPPs) in graphene is an important aspect to be exploited in the THz device implementation. The slow-wave nature of the graphene SPPs can dramatically enhance the interaction between THz photons and graphene [6–8,19–21]. As a consequence, the excitation of the SPPs in doped graphene can exhibit a giant absorption at around the plasmon mode frequencies [6,8], whereas the excitation of the SPPs in carrier-population-inverted graphene can produce a giant gain enhancement effect and the superradiant plasmonic lasing in metasurface structures [22–26].

Graphene-based van der Waals heterostructures have recently attracted considerable attention [27–29]. The double-graphene layer (DGL) heterostructures consisting of a core shell (in which a thin tunnel barrier layer is sandwiched by the two graphene layers [GLs] being independently connected with the side contacts), and the outer gate stack have been studied recently [30–32]. The DGL core shell works as a nanocapacitor [33,34] exhibiting inter-GL resonant tunneling (RT) when the band offset between the two GLs is aligned [30]. The RT produces a strong nonlinearity with a negative differential conductance (NDC) in the DGL current–voltage characteristics [31,35]. The excitation of the graphene plasmons in the GLs strongly modulates the inter-GL RT [36,37] so that it could dramatically enhance the quantum efficiency of various device operations like detection [36], self-oscillation [36], frequency multiplication [36], intensity modulation [38], photomixing [39], and lasing [40] in the THz device implementations. When the band offset is aligned to the THz photon energy, the DGL structure can mediate photon-assisted RT, resulting in the resonant emission or detection of the THz radiation [41–43]. The cooperative double-resonant excitation of the graphene plasmons and the inter-GL RT enables enormous enhancement of the efficiency of the device functional operation.

1.2   OPTOELECTRONIC PROPERTIES OF GRAPHENE

1.2.1   Optical Conductivity in Doped and Pumped Graphene

Optical conductivity in graphene is derived from the Kubo formula [44,45]:

$$\sigma(\omega) = \sigma_{inter}(\omega) + \sigma_{intra}(\omega)$$

$$= \frac{ie^2\omega}{\pi}\int_0^\infty \frac{f(\varepsilon - \varepsilon_F) - f(-\varepsilon - \varepsilon_F)}{(2\varepsilon)^2 - (\varepsilon + i\delta)^2}\, d\varepsilon + \frac{ie^2\varepsilon_F}{\pi\hbar^2(\omega + i/\tau_m)}, \tag{1.1}$$

where

    e is the elementary charge

    $f(\varepsilon)$ is the Fermi–Dirac distribution function

    ε_F is the Fermi energy

    $\hbar$ is the reduced Planck's constant

    δ is the broadening parameter for the interband transition

    τ_m is the momentum relaxation time

The first term is the conductivity driven by the interband carrier transitions, whereas the second term is the conductivity driven by the intraband carrier transport. This gives the important aspects of the frequency dependence of the conductivity profiles described as follows.

Due to the gapless and linear energy spectra of electrons and holes in graphene, interband carrier transition exhibits the flat response with a universal absorption coefficient $\alpha = \pi e^2/(c\hbar) \approx 1/137 \approx 2.3\%$ over broadband photon energies from sub-millimeter waves to near ultraviolet when the photon energies exceed twice the Fermi energy $2\varepsilon_F$ where c is the speed of light in vacuum [46]. When the photon energy $\hbar\omega$ is smaller than $2\varepsilon_F$, the Pauli-state blocking prohibits the optical absorption. Thus, in general, the real part of the interband optical conductivity $\text{Re}(\sigma_{\hbar\omega}^{inter})$ at photon energy $\hbar\omega$ given by the universal quantum conductivity $e^2/(4\hbar)$ and the carrier distribution function $f(\hbar\omega)$:

$$\text{Re}\left(\sigma_{\hbar\omega}^{inter}\right) = \frac{e^2}{4\hbar}\left(f\left(\frac{-\hbar\omega}{2}\right) - f\left(\frac{\hbar\omega}{2}\right)\right), \tag{1.2}$$

where $f(-\hbar\omega/2)$ and $f(\hbar\omega/2)$ are the electron distribution functions at the initial and the final states of the transition, respectively.

On the other hand, intraband conductivity takes the Drude-like dependence with parameters $\omega\tau_m$; the real part of the intraband conductivity $\text{Re}\left(\sigma_{\hbar\omega}^{intra}\right)$ is given by

$$\text{Re}\left(\sigma_{\hbar\omega}^{intra}\right) \approx \frac{(\ln 2 + \varepsilon_F/2k_BT)e^2}{\pi\hbar}\frac{k_BT\tau_m}{\hbar\left(1+\omega^2\tau_m^2\right)}, \tag{1.3}$$

where

    k_B is the Boltzmann constant

    T is the temperature

Figure 1.1 plots the real part of the conductivity in (a) doped and (b) pumped graphene. The THz frequency regime is an intersection region where intraband (interband) conductivity decreases (increases) with increasing frequency [47]. When graphene is doped, intraband Drude conductivity shifts upward, whereas the Pauli-state-blocking limited roll-off frequency of the interband conductivity shifts upward. Thus, THz conductivity can be substantially modulated by doping carriers,

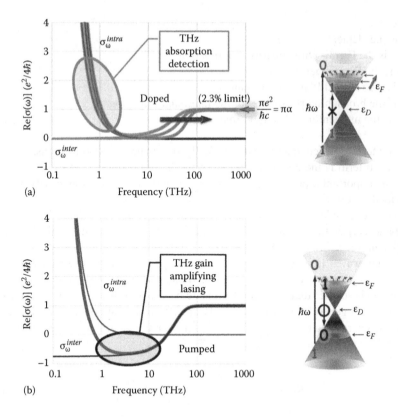

FIGURE 1.1 Real part of the conductivity in (a) doped graphene and (b) pumped graphene. The numeric number "0" and "1" in the band diagrams denote the values of the electron distribution function $f(\hbar\omega/2)$ in the conduction band and $f(-\hbar\omega/2)$ in the valence band.

which could be performed by the electrostatic gating. This, in turn, makes graphene active THz functional devices such as photodetectors [48–55], intensity modulators [47,56,57], and filters [58].

According to Equation 1.2, when graphene is optically pumped, $\text{Re}(\sigma_{\hbar\omega}^{inter})$ could take a negative value when $f(\hbar\omega/2)>f(-\hbar\omega/2)$. Generally, optical pumping is performed with photon energies $\hbar\omega$ larger than $2\varepsilon_F$. Therefore, the condition $f(\hbar\omega/2)>f(-\hbar\omega/2)$ means that the carrier population is inverted [11]. This effect and application to THz lasers will be discussed in detail later.

1.2.2 Ultrafast Carrier Dynamics and THz Gain in Pumped Graphene

The honeycomb lattice of graphene is formed with strong in-plane covalent σ-bonds of carbons having a low atomic mass. This gives rise to extremely high optical phonon energies, ~198 meV (1600 cm^{-1}) at the Brillouin zone center and ~163 meV (1340 cm^{-1}) at the Brillouin zone edges [59,60]. Due to such an extremely high optical phonon energy, interaction with optical phonons for excited carriers makes ultrafast

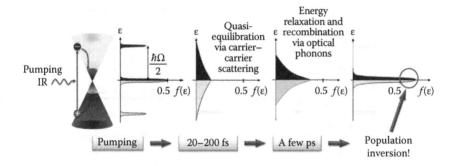

FIGURE 1.2 Carrier relaxation and recombination dynamics in optically pumped graphene.

energy relaxations through intravalley, intervalley, intraband, and interband scattering [61–65]. Figure 1.2 illustrates the carrier relaxation and recombination dynamics in optically pumped graphene. As soon as the electrons and holes are photogenerated, collective excitations due to carrier–carrier scattering quasi-equilibrate the total energy distributions to Fermi-like distributions within a timescale of 10–100 fs [61–63]. Then, carriers in the high-energy tails of their distributions emit optical phonons, cooling the entire carriers and accumulating around the Dirac points. Because of the fast intraband relaxation (picosecond or less), relatively slow interband relaxation via optical phonon emissions (picoseconds) and very slow interband recombination (≫1 ps) of photoelectrons/photoholes, a large excess of electrons and holes builds up just above and below the Dirac points (in a wide THz frequency range). As a consequence, with sufficiently intense excitation, a population inversion can be obtained [65,66], leading to spontaneous emission of THz radiation [25,26].

Interband carrier–carrier scatterings like Auger-type recombination and impact-ionization-type carrier multiplication processes are the major recombination/generation processes in narrow-gap semiconductors. They have been thought to be theoretically forbidden in graphene [67]. They may, however, take place due to higher-order many-body effects and/or imperfections of crystal qualities [68–73], which could also extremely modify the carrier relaxation dynamics, limiting the population inversion and, hence, preventing the negative conductivity. Recent experimental works demonstrate Auger-like carrier relaxation dynamics [69–73] particularly in the case of defect-originated symmetry breaking and/or lattice deformation caused by extremely high-field excitation. One should take care for synthesizing and processing high quality of graphene to obtain the population inversion.

The nonequilibrium relaxation dynamics of both the electrons and the holes can be characterized by the Fermi–Dirac distribution function with carrier temperature T_C and quasi-Fermi energy ε_F [65]. The time evolutions of T_C and ε_F after pumping at $t = 0$ with an optical pulse are numerically calculated [65]. Figure 1.3 plots the typical results for time evolutions of T_C and ε_F in intrinsic graphene when it is pumped with the center photon energy of 0.8 eV, the pulse width of 80 fs in full width at half maxima, and the pulse intensity of 10^8 W/cm². Instantly after the pumping ($t = 0$ ps), ε_F instantly falls down due to the carrier–carrier scattering. Then, ε_F is rapidly elevated by carrier cooling via emission of optical phonons at high-energy tails. When the

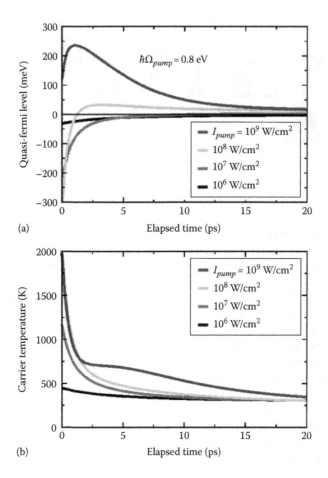

FIGURE 1.3 Numerically simulated time evolution of the quasi-Fermi level ε_F (a) and carrier temperature T_C (b) of monolayer intrinsic graphene after impulsive pumping. (After Otsuji, T. et al., *IEEE J. Sel. Top. Quant. Electron.*, 19, 8400209, 2013.)

pumping intensity exceeds a certain threshold level, ε_F becomes positive; hence, the population is inverted. This result proves the occurrence of the population inversion. After that, the recombination process follows more slowly (~10 ps).

By using the calculated values of T_C and ε_F, time evolution of THz conductivity is calculated as shown in Figure 1.4 [17,66]. At sufficiently strong excitation, the interband-stimulated emission of photons can prevail over the intraband (Drude) absorption. In this case, the real part of the dynamic conductivity of graphene, $\mathrm{Re}(\sigma_\omega)$, can be negative in THz range due to the gapless energy spectrum. This effect can be exploited in graphene-based THz lasers with optical or current-injection pumping [11,12,14]. Graphene photonic lasers with the Fabry–Pérot resonators based on dielectric or slot-line waveguides were proposed for lasing the THz photons [13,74,75]. Stimulated emissions of THz [25,26] and near-infrared photons [76] from population-inverted graphene were observed experimentally.

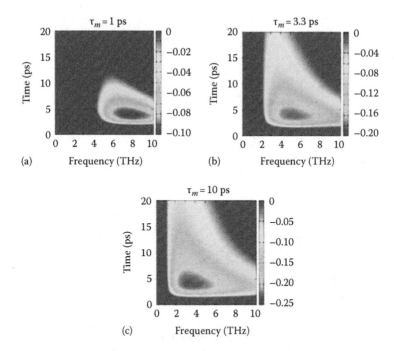

FIGURE 1.4 Numerically simulated temporal evolution of the real part of the dynamic conductivity Re[$\sigma(\omega)$] (normalized to the minimum quantum conductivity $e^2/4\hbar$) in intrinsic monolayer graphene photoexcited with 0.8 eV, 80 fs pump fluence 8 μJ/cm^2 (average intensity 1×10^8 cm^2) at the time of 0 ps having different momentum relaxation times τ_m of (a) 1, (b) 3.3, and (c) 10 ps. Any positive values of Re[$\sigma(\omega)$] are clipped to the zero level (shown with red color) to focus on the negative-valued region. (After Otsuji, T. et al., *J. Phys. D: Appl. Phys.*, 45, 303001, 2012.)

1.2.3 CURRENT-INJECTION PUMPING

As is described in Section 1.2.2 with Figure 1.3(b), the carrier temperature is elevated by optical pumping. By increasing the pumping photon energy, the carrier temperature increases, stimulating the carrier–carrier scattering. Such a carrier heating, in turn, suppresses the formation of population inversion as is seen in Figure 1.3 [14,77]. In the case of optical pumping with sufficiently low photon energy, however, the electron–hole plasma can be cooled down [14,66,77]. Another important parameter is the optical phonon decay time [14], which stands for the thermal conductivity of graphene. A longer phonon decay time (like suspended free-standing graphene) suppresses the carrier cooling, preventing from population inversion and thus from the negative conductivity [14]. Current-injection pumping can substantially reduce the pumping threshold because electrical pumping can easily reduce the pumping energy to the levels below the order of "meV" when a p–i–n junction is formed like semiconductor laser diodes. A dual-gate structure can make a p–i–n junction in the graphene channel, as shown in Figure 1.5a [12,14]. Gate biasing controls the injection level, whereas drain biasing controls the separation of the quasi-Fermi levels of

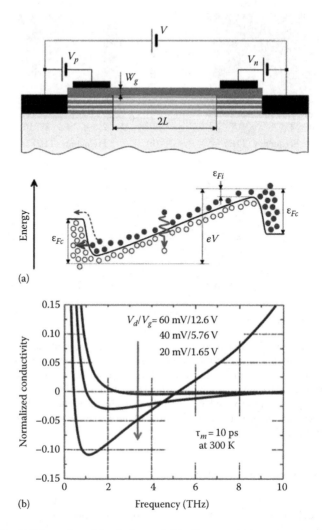

FIGURE 1.5   (a) Schematics of the dual-gate field-effect transistor structure and corresponding band diagram for electrically pumping graphene to create current-injection (terahertz)
THz lasers. Opposite biases on the two gates, G1 and G2, create effective p–i–n lateral junc-
tions. A small bias between drain D and source S injects electrons and holes from under the
gates into the intrinsic region between the gates. Sufficient injection will create population
inversion in this region and thus the possibility of optical gain for photons of energy $\hbar\omega_0$. ε_{Fi} is
the quasi-Fermi energy in the intrinsic section given by the drain–source bias, whereas ε_{Fc} is
the Fermi energy of the p-type and n-type region given by the dual-gate biases. (b) Simulated
real parts of the THz conductivities (negative values corresponding to gain) of monolayer
graphene at 300 K. The axis unit is normalized to the quantum conductivity $e^2/4\hbar$. (After
Ryzhii, M. and Ryzhii, V., *Jpn. J. Appl. Phys.*, 46, L151, 2007; Ryzhii, V. et al., *J. Appl. Phys.*,
113, 244505, 2013.)

electrons and holes and thus the lasing gain profiles (photon energy and gain). To minimize undesired tunneling current that lowers the injection efficiency, the distance between the dual-gate electrodes must be sufficiently long. Typical calculated conductivities are plotted in Figure 1.5b [14]. Compared to the results for optical pumping shown in Figure 1.4, the lower cutoff frequency of the gain spectra (the spectra of the negative conductivity) can be extended to below 1 THz so that the advantage of current injection is clearly seen. Waveguiding the THz-emitted waves with less attenuation is another key issue.

1.3 PLASMONIC PROPERTIES OF GRAPHENE

1.3.1 DISPERSIONS AND MODES IN GRAPHENE PLASMONS

The dispersion relations for plasmons have been formulated in various graphene structures [19,21,78–85]. In particular, gated plasmons are of major concerns for use in practical frequency-tuned device applications. The dispersion relation and the damping rate for gated graphene plasmons are theoretically revealed; it holds a superlinear dispersion in which the gate-to-graphene distance d and the momentum relaxation rate (collision frequency) deform the linear dispersion relations (Figure 1.6a) [21,25,27]. The plasmon phase velocity is proportional not to the 1/2 power but to the 1/4 power of the gate bias and of d, which is quite different from those in conventional semiconductors.

Semiclassical Boltzmann equations derive the hydrodynamic kinetics of electron–hole plasma wave dynamics [20]. Intrinsic graphene and photoexcited graphene hold equidensity of electrons and holes so that bipolar e–h modes of plasmons are excited. Due to the freedoms of e–h motions, e–h plasma waves and charge-neutral sound-like waves are excited (Figure 1.6b). The e–h plasmons are strongly damped, whereas the e–h sound-like waves are survived. On the other hand, doped graphene holds a large fraction of the majority and minority carriers. The plasma waves originating from the minority carriers are strongly damped, resulting in unipolar modes of plasmons that originate from the majority carriers.

1.3.2 GIANT THz GAIN VIA EXCITATION OF SURFACE PLASMON POLARITONS IN PUMPED GRAPHENE

It is noted that the negative THz conductivity in optically pumped monolayer graphene is limited to or below the universal quantum conductivity as seen in Equation 1.2. This is because the absorption of THz photons that can contribute to the stimulated photon emission is only made possible via interband transition processes with absorbance limited to $\pi e^2/\hbar c \approx 2.3\%$ [46]. To overcome this limit, using the plasmons in graphene is very promising. There are several factors to exploit the graphene plasmons: (1) the excitation and propagation of the plasmons along population-inverted graphene [21,22], (2) the resonant plasmon absorption in structured graphene like microribbon arrays [23] and microdisk arrays [86], and (3) the superradiant THz emission mediated by the plasmons [23,24].

As compared with the lasing associated with the stimulated emission of photons, the stimulated emission of plasmons by the interband transitions in population-inverted

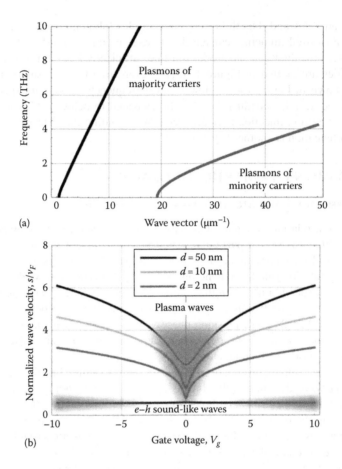

FIGURE 1.6 (a) Dispersion relation of gated plasmons in doped graphene. (b) Plasma wave velocity versus gate bias for undoped graphene with different gate-to-graphene distances d. Black line corresponds to the electron–hole sound-like waves in the vicinity of the neutrality point. Regions of strong damping are filled. (After Svintsov, D. et al., *J. Appl. Phys.*, 111, 083715, 2012.)

graphene can be a much stronger emission process [21–24]. Nonequilibrium plasmons in graphene can be coupled with the transverse magnetic (TM) modes of electromagnetic waves, resulting in the formation and propagation of the SPPs [22]. The SPP gain in pumped graphene can be very high due to small group velocity of the plasmons and strong confinement of the plasmon field in the vicinity of the GL. The effective refractive index ρ of the graphene SPP along the z coordinate derived from Maxwell's equations is

$$\sqrt{n^2 - \rho^2} + n^2\sqrt{1 - \rho^2} + \frac{4\pi}{c}\sigma_\omega\sqrt{1 - \rho^2}\sqrt{n^2 - \rho^2} = 0, \qquad (1.4)$$

where

n is the substrate refractive index

c is the speed of light in vacuum

σ_ω is the graphene conductivity [22]

When $n = 1$, ρ becomes

$$\rho = \sqrt{1 - \frac{c^2}{4\pi^2\sigma_\omega^2}}. \tag{1.5}$$

Thus, the absorption coefficient α is obtained as the imaginary part of the wave vector along the z coordinate: $\alpha = \mathrm{Im}(q_z) = 2\mathrm{Im}(\rho \cdot \omega/c)$. Figure 1.7 plots calculated value of α for monolayer graphene on a SiO$_2$/Si substrate ($\mathrm{Im}(n) \sim 3 \times 10^{-4}$) at 300 K.

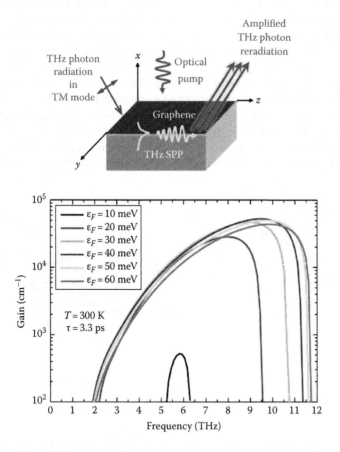

FIGURE 1.7 Simulated gain coefficient for monolayer graphene on a SiO$_2$/Si substrate ($\mathrm{Im}(n) \sim 3 \times 10^{-4}$) at 300 K. To drive graphene in the population inversion with a negative dynamic conductivity, quasi-Fermi energies are parameterized at $\varepsilon_F = 10, 20, 30, 40, 50, 60$ meV and a carrier momentum relaxation time $\tau_m = 3.3$ ps is assumed. (After Watanabe, T. et al., *New J. Phys.*, 15, 075003, 2013.)

To drive graphene in the population inversion with a negative dynamic conductivity, quasi-Fermi energies are parameterized at ε_F = 10, 20, 30, 40, 50, 60 meV and a carrier momentum relaxation time τ_m = 3.3 ps is assumed [26]. The results demonstrate a giant THz gain (negative values of absorption) on the order of 10^4 cm^{-1}. An increase in the substrate refractive index and, consequently, stronger localization of the surface plasmon electric and magnetic fields results in markedly larger gain, that is, negatively larger absorption coefficient. Waveguiding the THz-emitted waves with less attenuation is a critical key issue to create a graphene THz laser.

We conducted optical-pump, THz-probe, and optical-prove measurement at room temperature for intrinsic monolayer undoped graphene on a SiO$_2$/Si substrate [25,26]. The experimental setup is shown in Figure 1.8. An 80 fs (in FWHM), 20 MHz repetition, 1550 nm fiber laser was utilized as the optical pump and probe pulse source. The pumping laser beam, being linearly polarized, was focused with a beam diameter of about 120 μm onto the sample and the CdTe from the backside, while the probing beam is cross-polarized and focused from the topside. The CdTe can rectify the optical pump pulse to emit the envelope THz probe pulse. The emitted primary THz beam grows along the Cherenkov angle to be detected at the CdTe top surface as the primary pulse (marked with "①" in Figure 1.8) and then reflects to the graphene sample. When the substrate of the sample is conductive, the THz probe pulse transmitting through graphene again reflects back to the CdTe top surface, which is electro-optically detected as a THz photon echo signal (marked with "②" in Figure 1.8). Figure 1.9a shows the temporal responses measured for different pumping pulse intensities up to 3 × 10^7 W/cm^2 (pump fluence 2.4 μJ/cm^2, almost one order below the Pauli blocking) [26]. It is clearly seen that the peak obtained

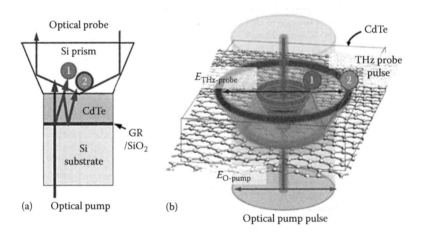

FIGURE 1.8 Experimental setup of the time-resolved optical-pump, THz-probe, and optical-probe measurement based on a near-field reflective electro-optic sampling. (a) Cross-sectional image of the pump/probe geometry and (b) bird's view showing the trajectories of the optical-pump and THz-probe beams. The polarization of the optical pump and THz probe pulse is depicted with red and dark blue arrows, respectively. (After Watanabe, T. et al., *New J. Phys.*, 15, 075003, 2013.)

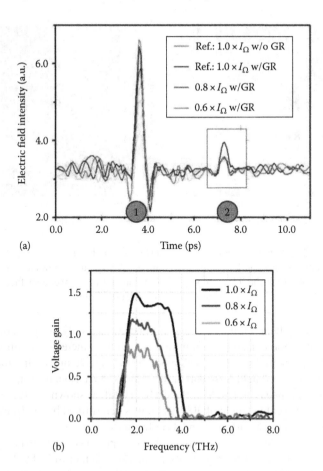

FIGURE 1.9 (a) Measured temporal responses of the THz photon-echo probe pulse (designated with "②") for different pumping intensities I_Ω (3×10^7 W/cm²), $0.8 \times I_\Omega$, and $0.6 \times I_\Omega$. (b) Corresponding voltage gain spectra of graphene obtained by Fourier transforming the temporal responses of the secondary pulse measured at graphene flake, normalized to that at the position without graphene. (After Watanabe, T. et al., *New J. Phys.*, 15, 075003, 2013.)

on graphene is more intense than that obtained on the substrate without graphene, and that the obtained gain factor exceeds the theoretical limit given by the quantum conductance by more than one order of magnitude as was measured in Reference 25. The Fourier-transformed gain spectra (Figure 1.9b) are well reproduced and showed similar pumping intensity dependence with the results in Reference 25.

We observed the spatial distribution of the THz probe pulse under the linearly polarized optical-pump- and THz-probe-pulse conditions at the maximum pumping intensity [26]. The optical probe pulse position was changed step by step to measure the in-plane spatial distributions of the THz probe pulse radiation. Observed field distributions for the primary and secondary pulse intensity are shown in Figure 1.10 [26]. The primary pulse field is situated along the circumference with diameter ~50 μm concentric to the center of optical pumping position. The secondary pulse

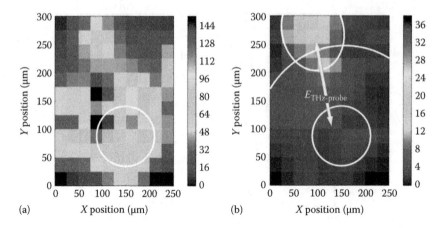

FIGURE 1.10 Spatial field distribution of the THz probe pulse intensities. (a) The primary pulse shows nonpolar distribution, whereas (b) the secondary pulse shows a strong localization to the area in which the THz probe pulse is impinged to graphene surface in the TM modes. (After Watanabe, T. et al., *New J. Phys.*, 15, 075003, 2013.)

(THz photon echo) field, on the contrary, is concentrated only at the restricted spot area in and out of the concentric circumference with diameter ~150 μm where the incoming THz probe pulse takes a TM mode being capable of exciting the SPPs in graphene. The observed field distribution reproduces the reasonable trajectory of the THz echo pulse propagation inside of the CdTe crystal, as shown in Figure 1.6 when we assume the Cherenkov angle of 30°, which was determined by the fraction of the refractive indices between infrared and THz frequencies.

How to couple the incoming/outgoing THz pulse photons to the surface plasmons in graphene is considered. One possibility of the excitation of SPPs by the incoming THz probe pulse is the spatial charge-density modulation at the area of photoexcitation by optical pumping. The pump beam having a Gaussian profile with diameter ~120 μm may define the continuum SPP modes in a certain THz frequency range as seen in various SPPs waveguide structures [87]. After the short propagation on the order of ~10 μm, the SPPs approach the edge boundary of illuminated and dark areas so that they could mediate the THz electromagnetic emission. The plasmon group velocity in graphene (exceeding the Fermi velocity) and propagation distance gives propagation time on the order of 100 fs. According to the calculated gain spectra shown in Figure 1.9b, the gain enhancement factor could reach or exceed ~10 at the gain peak frequency of 4 THz, which is dominated in the optically probed secondary pulse signals.

1.3.3 SUPERRADIANT THz PLASMONIC LASING IN GRAPHENE METASURFACES

The amplification of THz waves by stimulated generation of resonant plasmons in a planar periodic array of graphene plasmonic microcavities (PA-GPMC), as shown in Figure 1.11a, has been theoretically derived [23,24]. Graphene microcavities are confined beneath the metal-grating fingers located on a flat surface of a dielectric substrate.

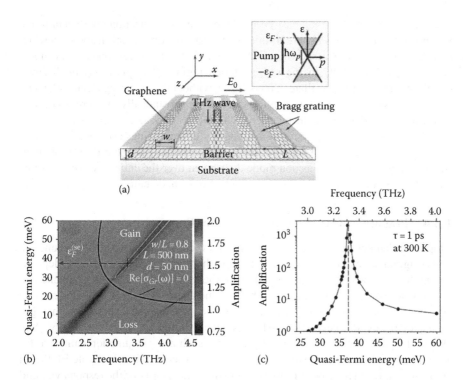

FIGURE 1.11 (a) Schematic view of the array of graphene micro-/nanocavities. The incoming electromagnetic wave is incident from the top at normal direction to the structure plane with the polarization of the electric field across the metal-grating contacts. The energy band structure of pumped graphene is shown schematically in the inset. (b) Contour map of the absorbance as a function of the quasi-Fermi energy and the frequency of incoming THz wave for the array of graphene microcavities with period $L = 500$ nm and the length of a graphene microcavity $w = 400$ nm. The electron scattering time in graphene is $\tau_m = 10^{-12}$ s. Blue and red arrows mark the quasi-Fermi energies for the maximal absorption and for the plasmonic lasing regime, respectively, at the fundamental plasmon resonance. (c) The variation of the power amplification coefficient along the fundamental plasmon resonance lobes is shown in panel (b) around the self-excitation regime marked by vertical dashed lines. Bullets are the simulation results and solid lines are guides for the eye. (After Popov, V.V. et al., *J. Opt.*, 15, 114009, 2013.)

Suppose that the graphene is optically or electrically pumped. Figure 1.11b shows the contour map of the calculated absorbance as a function of the quasi-Fermi energy (which corresponds to the pumping strength and, hence, the population inversion) and the THz wave frequency for the PA-GPMC with a period $L = 500$ nm and the length of each microcavity $w = 400$ nm [24]. With increasing ε_F, the energy gain can balance the energy loss so that the net energy loss becomes zero, $\text{Re}[\sigma(\omega)] = 0$, with corresponding graphene transparency. Above the graphene transparency line, the THz wave amplification at the plasmon resonance frequency is several orders of magnitude stronger than away from the resonances (the latter corresponding to the photon amplification in population-inverted graphene [11–15]).

At a certain value of the quasi-Fermi energy, the amplification coefficient at the plasmon resonance tends toward infinity with corresponding amplification linewidth shrinking down to zero as shown in Figure 1.11c [24]. This corresponds to plasmonic lasing in the PA-GPMC in the self-excitation regime. The plasmons in different graphene microcavities oscillate in phase (even without the incoming electromagnetic wave) because the metal-grating fingers act as electrostatically synchronizing elements between adjacent graphene microcavities (by applying a mechanical analogy, one may think of rigid crossbars connecting oscillating springs arranged in a chain) as shown in Figure 1.11b. Therefore, the plasma oscillations in the PA-GPMC constitute a single collective plasmon mode distributed over the entire area of the array, which leads to the enhanced *superradiant* electromagnetic emission from the array.

1.3.4 PLASMONIC THz PHOTODETECTION AND PHOTOMIXING

Graphene-channel field-effect transistors (FETs) can form a lateral p–i–n junction in its original state at the source and/or drain ohmic metal contact portions depending on the fractions of the work functions of the metal electrodes and graphene and on the gate biasing condition. Thus, they can be exploited as p–i–n photodetectors [88,89]. At optical frequencies, excellent photodetection performance with high-frequency bandwidth beyond 40 GHz was demonstrated by the group of IBM [88]. Due to the gapless energy spectra of graphene, such a p–i–n junction can work for detection of THz radiation. Various novel device structures like dual-gate FETs and graphene sandwich have been proposed, and their superiority of the responsivity and high-frequency performances were analytically studied [48–52].

On the other hand, the THz detectors using the resonant excitation of plasma oscillations in the standard heterostructures like high-electron-mobility transistor with a gated 2D electron (or hole) plasma in its channel, proposed by Dyakonov and Shur [90], were also extensively studied theoretically and successfully fabricated and analyzed [91–95]. The operation of these detectors is associated with the rectified component of the current due to the nonlinearity of the 2D electron plasma oscillations. At the plasma resonances, when the THz signal frequency is approaching the resonant plasma frequency and its harmonics, the responsivity of such detectors becomes fairly high. Plasma resonant phenomena in the gated GL structures can also be used in the THz detection, providing the enhanced performance. Experimental demonstrations of the THz detection using graphene FETs have been reported recently and improvement on detection responsivity has been progressing [9,53–55].

1.3.5 PLASMONIC THz INTENSITY MODULATION

Modulation of optical conductivity can work for modulation of transmittance/reflectance of incoming photons as mentioned in Section 1.2. First, demonstration of graphene optical modulator was performed in the infrared optical frequency region by using interband absorption in a double-layer graphene capacitor structure [96]. On the other hand, THz intensity modulators based on intraband absorption in graphene have been proposed and experimentally demonstrated by the group of Jena, which is probably the first real "THz" application of graphene [55,57]. The device exploits the electrical tuning of *intraband* Drude conductivity for making transmittance modulation to the THz electromagnetic radiation. They experimentally demonstrated a fairly high modulation index up to ~70% by utilizing a simple structure of a back-gate biased planar graphene capacitor in which the back-gate electrode acts as a THz reflector [57]. It is also theoretically demonstrated that by introducing plasmonic effects with a patterned graphene ribbon arrays, the modulation index could become 100% [86].

1.4 DOUBLE GRAPHENE–LAYERED VAN DER WAALS HETEROSTRUCTURES

1.4.1 INTERGRAPHENE LAYER RESONANT TUNNELING

We developed a device model describing the nonlinear electrical properties of the DGL structure [36]. We assume the GL to be originally undoped intrinsic graphene. Figure 1.12 depicts the cross-sectional schematic of the DGL in (a) and corresponding energy band diagrams under different bias conditions in (b) and (c), respectively [36]. Responding to a dc bias V_0 applied to the DGL, electrons and holes are injected from the negatively and positively biased GL, respectively. Due to the linear dependence of the density of states on energy in graphene, a band offset Δ appears as shown in Figure 1.12b and varies with V_0. In this situation, only nonadiabatic, non-RT takes place due to the hot carriers thermally excited by the acoustic and/or

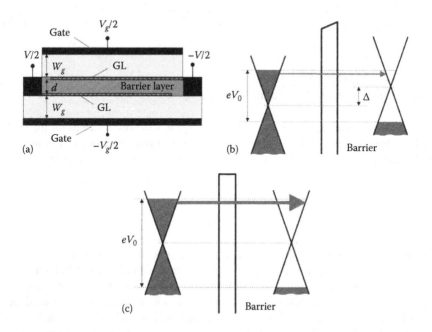

FIGURE 1.12 (a) Schematic cross section of the double-graphene layer. (b) Band diagram having a band offset. (c) Band diagram when the band offset is aligned. (After Ryzhii, V. et al., *J. Phys. D: Appl. Phys.*, 46, 315107, 2013.)

optical phonons. The band offset Δ can be aligned by applying a pertinent gate bias V_g as shown in Figure 1.12c. In this case, all the excess electrons in the n-type GL can resonantly tunnel through the barrier layer while preserving their energy and momentum. This results in a drastic increase of the tunneling current.

The static carrier density Σ_0 in the GL is expressed as a function of V_g and V_0:

$$\Sigma_0 = \Sigma_i + \Sigma_g = \frac{\kappa_g V_g}{4\pi e W_g} + \frac{\kappa V_0}{4\pi e d}, \tag{1.6}$$

where

κ_g and W_g are the permittivity and thickness of the outer gate stack insulator, respectively

κ and d are the permittivity and thickness of the tunnel barrier layer, respectively

e is the elementary charge [36]

The corresponding static band offset Δ_0 is given by

$$\Delta_0 = 2\varepsilon_F - eV_0, \tag{1.7}$$

where ε_F is the Fermi energy of electrons and holes in the pertinent GLs [36]. When the electron and hole systems in GLs are degenerated,

$$\varepsilon_F = \hbar v_F \sqrt{\pi \left(\Sigma_i + \Sigma_g\right)}, \tag{1.8}$$

where

$\hbar$ is the reduced Planck constant

v_F is the electron and hole Fermi velocity [36]

Figure 1.13 plots the Δ_0 values as a function of V_g for different V_0 values for the DGL having a few-layered h-BN ($\kappa = 4.0$, $d = 4.5$ nm) as the tunnel barrier and a 5 μm thick h-BN gate insulation layer. The band offset alignment voltage V_t is given by

$$V_t = \frac{2\varepsilon_F}{e} = \frac{2\hbar v_F \sqrt{\pi \Sigma_i}}{e}. \tag{1.9}$$

Under such conditions, the DGL current–voltage characteristics exhibit a strong non-linearity with the NDC as was shown by the numerical calculations in Reference 35. Figure 1.14 reproduces the calculated results of Reference 35. The parameter $\bar{\beta}_{eh}$ expresses the electron–hole damping factor over the GL area reflecting the areal factor ($\propto L^2$), the carrier scattering, the carrier recombination, and the carrier diffusivity [35]. The $\bar{\beta}_{eh}$ values set in Figure 1.14 assume micrometer to tens micrometers range for the values of L and room temperature environments. The NDC can be exploited to device an oscillator similar to other NDC devices, such as the RT diodes and

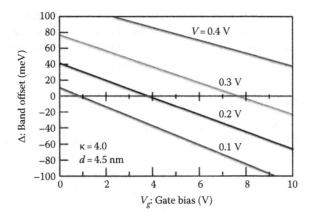

FIGURE 1.13 Band offset versus gate bias for a double-graphene layer having a few-layered h-BN tunnel barrier (κ = 4.0, d = 4.5 nm) and a 5 μm thick h-BN gate insulation layer.

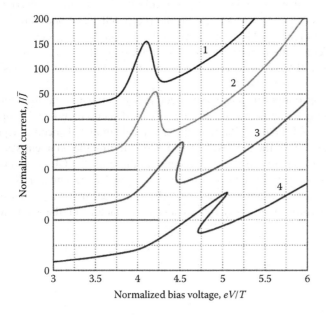

FIGURE 1.14 Current–voltage characteristics for different values of parameters $\overline{\beta}_{eh}$. 1: $\overline{\beta}_{eh} = 0.001$; 2: $\overline{\beta}_{eh} = 0.002$; 3: $\overline{\beta}_{eh} = 0.005$; 4: $\overline{\beta}_{eh} = 0.01$. The negative differential conductance curves exhibit a transition from N-shape to Z-shape. (After Ryzhii, M. et al., *J. Appl. Phys.*, 115, 024506, 2014.)

Gunn diodes. The carrier transit time across the RT should be fast to support THz oscillations. The scaling rule for the oscillation frequencies versus the DGL dimensions follows those for conventional NDC devices. The cutoff frequency is mainly limited by the DGL capacitance and the contact resistance time constant. Because of the lack of the substitutional impurity doping technique in graphene, the reduction of the contact resistance is yet to be achieved. The NDC curves exhibit a transition from the N-shape to Z-shape with decreasing the $\bar{\beta}_{eh}$ values, thus increasing the DGL length L. The Z-shape NDC means a hysteresis, multiple-state transition, giving rise to higher functionality in the device implementations.

1.4.2 Photon-Assisted Resonant Tunneling

As is described in Section 1.2, the DGL supports the inter-GL RT when the band offset is aligned. In this section, we consider the cases in which the band offset alignment is performed via interacting with photons. As depicted in Figure 1.15, when the energy level of the final states is lower (higher) than those of the initial states, emission (absorption) of photons whose energy $\hbar\omega$ can compensate for the band offset Δ and the depolarization shift occurs, resulting in inter-GL RT. In this case, only the TM photon modes (the electric field intensity is directed perpendicular to the GL plane) can assist the RT due to the selection rule of the quantum mechanical RT. In comparison with emission/detection of THz photon radiation via the interband transitions in intrinsic graphene [15], not only the Fermi-surface carriers but also all the excess electrons with respect to the p-type GL can contribute to the emission or detection of the THz radiation, resulting in highly efficient emitter/detector device implementations. The photoemission-assisted RT gives rise to spontaneous THz emission. If the DGL is installed in a pertinent THz cavity, THz-lasing operation could be feasible [41,42].

We developed a pertinent device model and substantiated the superior quantum efficiency (gain for emission and responsivity for detection) of the DGL operation in References 41–43. Figure 1.16a reproduces typical simulated spectral dependencies of the modal gain of the proposed DGL lasers utilizing the photoemission-assisted inter-GL RT (solid curves) in comparison with those for the GL lasers utilizing the interband intra-GL transitions (dashed lines) originally presented in Reference 42. The DGL active-area dimension L and the injected carrier densities (correlated with Δ) determine the conductivities and areal factors of the gain overlapping

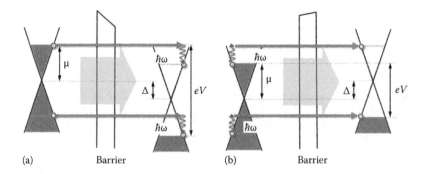

(a)   Barrier   (b)   Barrier

FIGURE 1.15 Band diagrams of THz emitter/detector structures with (a) photoemission-assisted inter-graphene layer (GL) and (b) photoabsorption-assisted inter-GL radiative transitions. Wavy arrows indicate the inter-GL radiative transitions. The inter-GL transitions in the structures (a) and (b) work for the TM-mode THz photon radiations.

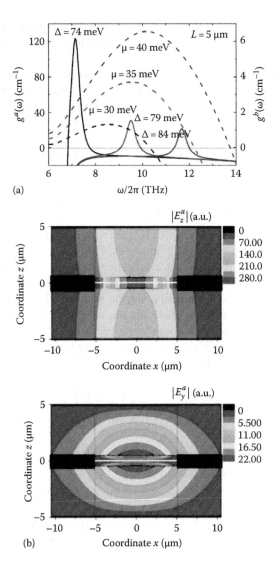

FIGURE 1.16 (a) Simulated frequency dependence of the THz gain $g^a(\omega)$ for the double-graphene layer (DGL) inter-graphene layer (GL) resonant tunneling (RT) laser (solid lines) for different band offset energies between the Dirac points Δ and of the THz gain $g^b(\omega)$ for the GL intra-GL transition laser (dashed lines) for different Fermi energies in GL. $L = 5$ μm and $W = 5$ μm. (b) Simulated spatial distributions of the photon electric field components $|E_z^a|$ (upper) and $|E_y^a|$ (lower) in the TM mode in the DGL inter-GL RT laser. (After Ryzhii, V. et al., *Opt. Express*, 21, 31569, 2013.)

in the GLs. Figure 1.16b shows a typical electric field distribution. Note that the TM-mode coupling in the proposed structure prevents the Drude absorption in GLs and contributes to an increased gain. Very recently, we succeeded in experimental observation of THz photon emission in a fabricated DGL [96]. The excess THz photon emission beyond the background blackbody radiation was observed only under the DGL bias conditions in which the energy level of the Dirac point of the p-type GL is lower than that of the n-type GL, giving rise to a clear manifestation of the occurrence of the photoemission-assisted RT.

Figure 1.17 reproduces the simulated dependence of the detector photoresponsivities on (a) THz photon energy, (b) bias voltage, and (c) injected carrier densities for the DGL structures with different thicknesses of the barrier layer, with d originally presented in Reference 43. The DGL responsivity exhibits fairly sharp peaks associated with the photoabsorption-assisted inter-GL RT transitions. The peak values of the responsivity tuned by the bias-voltage-dependent carrier densities are rather high in a wide spectral range.

1.4.3 Plasmon-Assisted Resonant Tunneling

We consider the DGL interacting with THz photons via graphene plasmons. The in-plane electric field components in TE modes or TM modes of photons can directly excite the GL surface plasmons. Such a collective excitation of the graphene carrier densities produces a local ac field among the GLs, which dynamically modifies the static V_0, so that the inter-GL tunneling may be dramatically enhanced or damped depending on in-phase or out-of-phase plasma excitation. Here we represent the essence of the device modeling studies of References 36 and 97.

The local value of the inter-GL RT current density is expressed as a function of the local value of the inter-GL voltage V as follows:

$$j_t = j_t^{\max} \exp\left[-\left(\frac{V - V_t}{\Delta V_t}\right)^2\right], \tag{1.10}$$

where
$j_t^{\max}$ is the peak value of the current density
$\Delta V_t \simeq 2\sqrt{2\pi}\hbar v_F/el$ determines the peak width, l is the coherence length [36]

Using Equation 1.10 and assuming small-signal variations of the local potential difference between the GLs $(\delta\varphi_+ - \delta\varphi_-)$, the variation of the RT current density can be expressed as

$$\delta j_t \simeq \sigma_t(\delta\varphi_+ - \delta\varphi_-) + \beta_t(\delta\varphi_+ - \delta\varphi_-)^2, \tag{1.11}$$

where
$\sigma_t = (dj_t/dV)|_{V=V_0}$ is the differential tunneling conductivity
$\beta_t = (1/2)(d^2 j_t/dV^2)|_{V=V_0}$ is the second derivative factor expressing the nonlinearity of the RT current–voltage characteristics [36]

The spatial distributions of $\delta\varphi_+(x)$ and $\delta\varphi_-(x)$ in the GL plane can be found from linearized hydrodynamic equations coupled with the Poisson equation in the gradual-channel approximation. Searching for the ac potential in the form

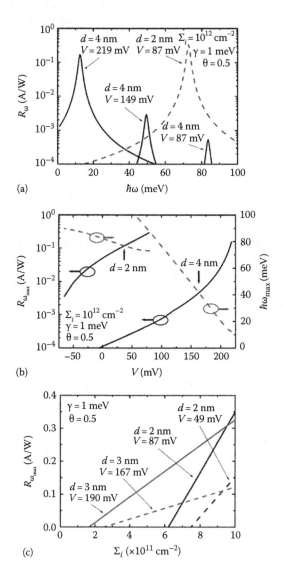

FIGURE 1.17 Simulated (a) double-graphene layer–photodiode (DGL-PD) responsivity R_ω versus photon energy $\hbar\omega$ for the inter-graphene layer barrier layer thickness $d = 4$ nm (solid lines) and $d = 2$ nm (dashed line) at different applied voltages V, (b) DGL-PD responsivity maximum $R_{\omega max}$ (solid lines) and corresponding photon energy $\hbar\omega_{max}$ (dashed lines) versus bias voltages V for different thicknesses d, and (c) $R_{\omega max}$ versus electron and hole densities $\Sigma_i \propto V_g/W_g$ at different bias voltages V and thicknesses d. (After Ryzhii, V. et al., *Appl. Phys. Lett.*, 104, 163505, 2014.)

of $\delta\varphi_{\pm} = \delta\varphi_{\pm}(x)exp(-i\omega t)$, where ω is the complex signal frequency, the second derivative term in Equation 1.11 can be neglected for simplicity. As a consequence, the dynamic system of equations in question can be reduced as follows:

$$\frac{d^2\delta\varphi_+}{dx^2} + \frac{(\omega + iv_t)(\omega + iv)}{s^2}(\delta\varphi_+ - \delta\varphi_-) = 0 \tag{1.12}$$

and

$$\frac{d^2\delta\varphi_-}{dx^2} + \frac{(\omega + iv_t)(\omega + iv)}{s^2}(\delta\varphi_- - \delta\varphi_+) = 0, \tag{1.13}$$

where
   v is the collision frequency (inverse momentum relaxation time) of electrons and
      holes in the GLs
   v_t is the characteristic frequency of the inter-GL tunneling
   s ($\propto \Sigma_0^{1/4}$) is the characteristic velocity of the electron–hole plasma waves [36]

Assuming that the total voltage between the contacts to the GLs is equal to $V = V_0 + \delta V_\omega$, where δV_ω is the small-signal *ac* voltage component, it then leads to the following boundary conditions:

$$\delta\varphi_{\pm}|_{x=\mp L} = \pm\frac{\delta V_\omega}{2}\exp(-i\omega t), \quad \frac{d\delta\varphi_{\pm}}{dx}\bigg|_{x=\mp L} = 0. \tag{1.14}$$

Solving Equations 1.12 and 1.13 under the boundary conditions in Equation 1.14, we obtain

$$\delta\varphi_+ - \delta\varphi_- = \delta V_\omega\left(\frac{((\cos\gamma_\omega x)/(\gamma_\omega\sin\gamma_\omega L))}{((\cos\gamma_\omega x)/(\gamma_\omega\sin\gamma_\omega L)) - L}\right), \tag{1.15}$$

where $\gamma_\omega = \sqrt{2(\omega + iv_t)(\omega + iv)}/s$ [36].

   The net *ac* current δJ_ω including the displacement current is calculated as

$$\delta J_\omega = H\left(-i\frac{\kappa\omega}{4\pi d} + \sigma_t\right)\int_{-L}^{+} dx(\delta\varphi_+ - \delta\varphi_-), \tag{1.16}$$

where H is the width of the DGL in the direction perpendicular to the current [36]. Using Equations 1.15 and 1.16, the small-signal admittance $Y_\omega = \delta J_\omega/\delta V_\omega$ is obtained as

$$Y_\omega = -i\left(\frac{\kappa HL}{2\pi d}\right)\sqrt{\left(\frac{\omega + iv_t}{\omega + iv}\right)}\frac{\omega_p}{(\cot\gamma_\omega - \gamma_\omega L)}, \tag{1.17}$$

where $\omega_p = \pi s/(\sqrt{2}L)$ is the characteristic plasma frequency [36].

Figure 1.18 reproduces the real and imaginary part of the Y_ω for different plasma quality factors Q ($\equiv \omega_p/\nu$) calculated in Reference 36. As seen in Figure 1.18, Re(Y_ω) is negative in a narrow range of small ω, which is due to NDC at $(V_0 - V_t) \gtrsim \Delta V_t$ associated with RT. Since Im(Y_ω) does not change its sign when Re(Y_ω) < 0, electron–hole plasma in this DGL system is stable. Equation 1.17 gives the following dispersion equation for plasma oscillations [36]:

$$\cot \gamma_\omega - \gamma_\omega L. \tag{1.18}$$

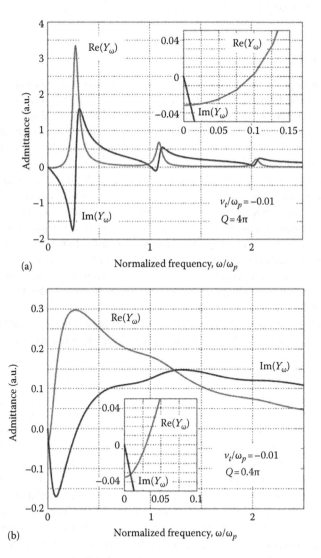

(a)

(b)

FIGURE 1.18 Real and imaginary part of the admittance Re(Y_ω) and Im(Y_ω) as functions of frequency normalized to ω_p for different Q values. (a) $Q = 4\pi$; (b) $Q = 0.4\pi$. (After Ryzhii, V. et al., *J. Phys. D: Appl. Phys.*, 46, 315107, 2013.)

Equation 1.18 determines the complex frequency $\omega = \omega' + i\omega''$; ω' is the real plasma oscillation frequency and ω'' is their damping/growth rage. For practical cases $\omega_p \gg \nu, |\nu_t|$, Equation 1.18 yields $\omega' = \omega_n, (n = 0,1,2,3,...)$ and $\omega'' = \Gamma$, where

$$\omega_0 \simeq \frac{0.86}{\pi}\omega_p, \quad \omega_n \simeq n\omega_p + \frac{\omega_p}{\pi^2 n}, \quad \Gamma \simeq -\frac{\nu + \nu_t}{2}. \tag{1.19}$$

According to Equation 1.18, $\mathrm{Re}(Y_\omega)$ exhibits aperiodic resonant peaks clearly as seen in Figure 1.18a when the plasmon damping is weak (thus, the plasmon quality factor Q is rather high).

1.4.4 THz Photomixers

We now consider the case of the incoming optical radiation of the photomixed dual-CW laser beam whose frequencies, Ω_1 and Ω_2, are different from ω_{PM} on the order of THz ($|\Omega_1 - \Omega_2| = \omega_{PM}$). The electric field intensity of the photomixed signal has the ω_{PM} component exciting the GLs plasmons and perturbing the local ac field and potential ($\delta\varphi_+, \delta\varphi_-$) at the frequency ω_{PM}. This situation is a similar to that of the DGL optical modulator discussed earlier. One can imagine that if ω_{PM} matches the plasmon-mode frequency ω_p^n, the inter-GL tunneling current may be resonantly enhanced. If the side contacts are introduced to a pertinent THz antenna, the ω_{PM} current component could be radiated so that the DGL can work for the THz photomixers (PMs). We developed a device model for the DGL PM and calculated its characteristics and evaluated the performance in comparison with the existing other devices [39]. The resonant response of DGL PMs associated with the excitation of plasma oscillation provides significant advantages over the existing THz PMs based on the low-temperature-grown GaAs photoconductive switch [98], p–i–n photodiodes [99], and uni-travelling-carrier photodiodes (UTC-PDs) [100]. Relatively strong absorption in GLs and ultrashort transit time across the inter-GL barrier can result in the DGL-PM performance exceeding that of UTC-PDs even with the integrated plasma cavity [101]. A detailed quantitative discussion can be found in Reference 96.

1.4.5 THz Modulators

Figure 1.19 depicts a schematic and corresponding band diagram of the DGL core shell working as an optical modulator [38]. The DGL side contacts work for an optical slot-line waveguide. The side contacts can also be connected to or be a part of THz antenna, which converts the incoming THz radiation into the modulation voltage. The electrons and holes cannot respond to the optical frequencies but can absorb the photon energies. The DGL modulator utilizes the variation of absorption due to the electrically controlled Pauli-state blocking effect. Our model accounts for the interband and intraband absorption and the plasma effects in GLs determining the spatiotemporal distributions of the electron and hole densities and the absorption coefficient. The developed model yields the dependence of the modulation depth on control voltage for strong but slow modulation signals and for small-signal modulation in a wide range of frequencies.

The modulation characteristics were calculated in Reference 38. Figure 1.20 reproduces the calculated frequency dependence of the normalized modulation depth

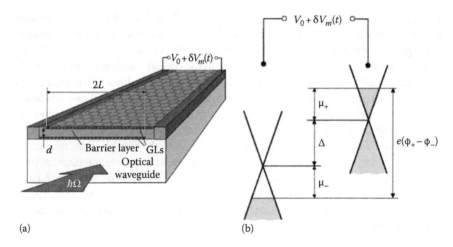

(a)   (b)

FIGURE 1.19 (a) Schematic view of a double-graphene layer (DGL) core shell working for an optical intensity modulator coupled with an incident light having a photon energy in the slot-line waveguide mode. The electric field intensity vector should be directed to the DGL length axis. (b) Band diagram of the DGL under a dc bias V_0 and a modulation voltage $\delta V_m(t)$ applied condition. The shaded areas indicate the states occupied by the electrons. (After Ryzhii, V. et al., *J. Appl. Phys.*, 102, 104507, 2012.)

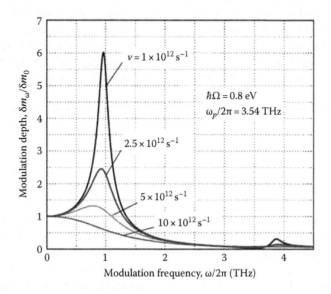

FIGURE 1.20 Normalized modulation depth $\delta m_\omega / \delta m_0$ versus modulation frequency $\omega/2\pi$ for different electron and hole collision frequencies ν. (After Ryzhii, V. et al., *J. Appl. Phys.*, 102, 104507, 2012.)

$\delta m_\omega/\delta m_0$ when incident photon energy is $\hbar\Omega = 0.8$ eV and the characteristic plasma frequency is $\omega_p/2\pi = 3.54$ THz for different collision frequencies ν. For the characteristic plasma velocities $s = (3{:}75{\sim}7{:}5) \times 10^8$ cm/s, this frequency corresponds to the length of the GL (size of the waveguide) close to the optical wavelength under consideration ($2L = 0.75{-}1.5$ µm). When the collision frequency is lower than the plasma characteristic frequency ($\nu < \omega_p$), the plasma quality factor Q ($\equiv \omega_p/\nu$) exceeds unity. Then the plasmon resonance becomes pronounced so that the photon absorption is resonantly enhanced, causing a sharp increase in the modulation depth at aperiodic plasmon-mode frequencies given by Equation 1.19. At the 0th plasmon mode, the peak value of $\delta m_\omega/\delta m_0$ is ~6 at $\nu = 1 \times 10^{12}$ Hz, that is, at momentum relaxation time $\tau = 1$ ps. Recent works demonstrate an excellent quality of graphene synthesis having τ values larger than 1 ps at room temperature. We can expect high $\delta m_\omega/\delta m_0$ values of the order of 10 to be achieved, enabling applications to the optical communication network systems.

1.5 CONCLUSION

We overviewed recent advances in theoretical and experimental studies of graphene-based THz devices. The gapless energy spectrum of the Dirac fermions (electrons and holes) in graphene and their linear dispersion enable unique features of graphene-based device characteristics, particularly in the THz range of frequencies. The characteristics of the devices with the gated graphene structure can effectively be modified by the applied voltages. This can lead to a substantial extension of the functionality of various devices such as photodetectors, modulators, mixers, and filters. Ultrafast energy relaxation and specifics of the recombination dynamics of electrons and holes in optically or electrically pumped graphene structures give rise to the interband population inversion, which, in turn, can provide negative dynamic conductivity in a wide THz range. This opens up new opportunities to create solid-state THz emitters, in particular, THz lasers. However, the realization of such THz devices needs to overcome the strong THz radiation losses inherent in different materials.

Due to relatively strong interaction of 2D plasmons (plasmon polaritons) in various graphene structures, their use can dramatically enhance the light–matter interaction at THz frequencies, drastically improving the quantum efficiency, providing highly sensitive detection of THz radiation, highly effective amplification, and powerful lasing. Plasmonic THz gain in graphene can exceed the values of 10^4 cm^{-1} in a wide THz spectral range. The latter is four orders of magnitude higher than the THz photonic gain in graphene. The gain enhancement effect of the SPPs on the THz-stimulated emission from optically pumped graphene has recently been experimentally discovered. Graphene nanoribbon arrays placed between gratings with metallic fingers can enable strong stimulated emission associated with the transformation of the cooperative plasmon modes. Due to a strong confinement of the plasmon modes in graphene-based microcavities and the superradiant nature of electromagnetic emission from such plasmonic microcavity arrays, the amplification of THz radiation with the frequencies close to the plasmon resonance frequencies is enhanced by several orders of magnitude. This provides the superradiant THz lasing if the plasmon gain in graphene balances the net damping of plasmon modes (due to both the dissipative and radiative damping) in the graphene microcavity arrays.

The DGL heterostructure with a core–shell structure consisting of a pair of the GLs separated by the tunnel barrier layer is sandwiched between the outer gate stack layers at both sides, which was recently proposed as a key element for a variety of novel devices. The strong nonlinearity of the current–voltage characteristics of DGL-based heterostructure associated, in particular, with the inter-GL RT can be exploited in different functional THz devices. The excitation of plasmons in DGL heterostructures strongly modulates the inter-GL RT current. This might dramatically enhance the efficiency of operation of various devices, in particular, THz emitters including frequency multipliers, PM, and lasers, as well as THz and fast optical modulators. As shown, the plasmon-assisted inter-GL RT can also be a new mechanism for resonant emission and detection of THz radiation. The cooperative double-resonant excitation of the plasmons in DGL structures combined with the inter-GL RT will open a new paradigm in highly efficient functional device implementation.

ACKNOWLEDGMENTS

Thanks to Stephane Boubanga Tombet, Akira Satou, Takayuki Watanabe, Maxim Ryzhii, Vyacheslav V. Popov, Alexander Dubinov, Vladimir Ya. Aleshkin, Vladimir Mitin, Dmitry Svintsov, Vladimir Vyurkov, Michael S. Shur, Eiichi Sano, Tetsuya Suemitsu, Hirokazu Fukidome, and Maki Suemitsu for their extensive contributions. Part of the works contributed by the authors were financially supported by JST-CREST, Japan, and JSPS Grant-in-Aid for Specially Promoting Research #23000008, Japan.

REFERENCES

1. K.S. Novoselov, A.K. Geim, S.V Morozov, D. Jiang, Y. Zhang, S.V. Dubonos, I.V. Grigorieva, and A.A. Frisov, Electric field effect in atomically thin carbon films, *Science* 306, 666 (2004).
2. K.S. Novoselov, A.K. Geim, S.V. Morozov, D. Jiang, M.I. Katsnelson, I.V. Grigorieva, S.V. Dubonos, and A. Firsov, Two-dimensional gas of massless Dirac fermions in graphene, *Nature* 438, 197 (2005).
3. Y. Zhang, Y.-W. Tan, H.L. Stormer, P. Kim, Experimental observation of the quantum Hall effect and Berry's phase in graphene, *Nature* 438, 201 (2005).
4. A.K. Geim and K.S. Novoselov, The rise of graphene, *Nat. Mater.* 6, 183 (2007).
5. F. Schwierz, Graphene transistors, *Nat. Nanotechnol.* 5, 487–496 (2010).
6. F. Bonaccorso, Z. Sun, T. Hasan, and A.C. Ferrari, Graphene photonics and optoelectronics, *Nat. Photon.* 4, 611–622 (2010).
7. Q. Bao and K.P. Loh, Graphene photonics plasmonics and broadband optoelectronic devices, *ACS Nano* 6, 3677–3694 (2012).
8. A.N. Grigorenko, M. Polini, and K.S. Novoselov, Graphene plasmonics, Graphene plasmonics, *Nat. Photon.* 6, 749–759, 2012.
9. A. Tredicucci and M.S. Vitiello, Device concepts for graphene-based terahertz photonics, *IEEE J. Sel. Top. Quant. Electron.* 20, 8500109 (2014).
10. R.R. Hartmann, J. Kono, and M.E. Portnoi, Terahertz science and technology of carbon nanomaterials, *Nanotechnology* 25, 322001 (2014).

11. V. Ryzhii, M. Ryzhii, and T. Otsuji, Negative dynamic conductivity in optically pumped graphene, *J. Appl. Phys.* 101, 083114 (2007).
12. M. Ryzhii and V. Ryzhii, Injection and population inversion in electrically induced p-n junction in graphene with split gates, *Jpn. J. Appl. Phys.* 46, L151 (2007).
13. V. Ryzhii, A. Dubinov, T. Otsuji, V. Mitin, and M.S. Shur, Terahertz lasers based on optically pumped multiple graphene structures with slot-line and dielectric waveguides, *J. Appl. Phys.* 107, 054505 (2010).
14. V. Ryzhii, M. Ryzhii, V. Mitin, and T. Otsuji, Toward the creation of terahertz graphene injection laser, *J. Appl. Phys.* 110, 094503 (2011).
15. V. Ryzhii, I. Semenikhin, M. Ryzhii, D. Svintsov, V. Vyurkov, A. Satou, and T. Otsuji, Double injection in graphene p–i–n structures, *J. Appl. Phys.* 113, 244505 (2013).
16. T. Otsuji, S.A. Boubanga Tombet, A. Satou, H. Fukidome, M. Suemitsu, E. Sano, V. Popov, M. Ryzhii, and V. Ryzhii, Graphene materials and devices in terahertz science and technology, *MRS Bull.* 37, 1235–1243 (2012).
17. T. Otsuji, S.A. Boubanga Tombet, A. Satou, H. Fukidome, M. Suemitsu, E. Sano, V. Popov, M. Ryzhii, and V. Ryzhii, Graphene-based devices in terahertz science and technology, *J. Phys. D: Appl. Phys.* 45, 303001 (2012).
18. T. Otsuji, S. Boubanga Tombet, A. Satou, M. Ryzhii, and V. Ryzhii, Terahertz-wave generation using graphene-toward new types of terahertz lasers, *IEEE J. Sel. Top. Quant. Electron.* 19, 8400209 (2013).
19. V. Ryzhii, A. Satou, and T. Otsuji, Plasma waves in two-dimensional electron-hole system in gated graphene heterostructures, *J. Appl. Phys.* 101, 024509 (2007).
20. D. Svintsov, V. Vyurkov, S. Yurchenko, T. Otsuji, and V. Ryzhii, Hydrodynamic model for electron-hole plasma in graphene, *J. Appl. Phys.* 111, 083715 (2012).
21. F. Rana, Graphene Terahertz Plasmon Oscillators, *IEEE Trans. Nanotechnol.* 7, 91–99 (2008).
22. A.A. Dubinov, Y.V. Aleshkin, V. Mitin, T. Otsuji, and V. Ryzhii, Terahertz surface plasmons in optically pumped graphene structures, *J. Phys. Condens. Mater.* 23, 145302 (2011).
23. V.V. Popov, O.V. Polischuk, A.R. Davoyan, V. Ryzhii, T. Otsuji, and M.S. Shur, Plasmonic terahertz lasing in an array of graphene nanocavities, *Phys. Rev. B* 86, 195437 (2012).
24. V.V. Popov, O.V. Polischuk, S.A. Nikitov, V. Ryzhii, T. Otsuji, and M.S. Shur, Amplification and lasing of terahertz radiation by plasmons in graphene with a planar distributed Bragg resonator, *J. Opt.* 15, 114009 (2013).
25. S. Boubanga-Tombet, S. Chan, T. Watanabe, A. Satou, V. Ryzhii, and T. Otsuji, Ultrafast carrier dynamics and terahertz emission in optically pumped graphene at room temperature, *Phys. Rev. B* 85, 035443 (2012).
26. T. Watanabe, T. Fukushima, Y. Yabe, S.A. Boubanga Tombet, A. Satou, A.A. Dubinov, V.Ya. Aleshkin, V. Mitin, V. Ryzhii, and T. Otsuji, The gain enhancement effect of surface plasmon polaritons on terahertz stimulated emission in optically pumped monolayer graphene, *New J. Phys.* 15, 075003 (2013).
27. K.S. Novoselov and A.H. Castro Neto, Two-dimensional crystals-based heterostructures: Materials with tailored properties, *Phys. Scr.* T146, 014006 (2012).
28. T. Georgiou, R. Jalil, B.D. Bellee, L. Britnell, R.V. Gorbachev, S.V. Morozov, Y.-J. Kim et al., Vertical field-effect transistor based on graphene-WS2 heterostructures for flexible and transparent electronics, *Nat. Nanotechnol.* 8, 100–103 (2013).
29. M.C. Lemme, L.-J. Li, T. Palacios, and F. Schwierz, Two-dimensional materials for electronic applications, *MRS Bull.* 39, 711–718 (2014).
30. L. Britnell, R.V. Gorbachev, R. Jalil, B.D. Belle, F. Shedin, A. Mishchenko, T. Georgiou et al., Field-effect tunneling transistor based on vertical graphene heterostructures, *Science* 335, 947–950 (2012).

31. L. Britnell, R.V. Gorbachev, A.K. Geim, L.A. Ponomarenko, A. Mishenko, M.T. Greenaway, T.M. Fromhold, K.S. Novoselov, and L. Eaves, Resonant tunneling and negative differential conductance in graphene transistors, *Nat. Commun.* 4, 1794–1799 (2013).
32. P. Zhao, R.M. Feenstra, G. Gu, and D. Jena, SymFET: A proposed symmetric graphene tunneling field-effect transistor, *IEEE Trans. Electron. Dev.* 60, 951–957 (2013).
33. G. Shi, Y. Hanlumyuang, Z. Liu, Y. Gong, W. Gao, B. Li, J. Kono, J. Lou, R. Bajtai, P. Sharma, and P.M. Ajayan, Boron-nitride-graphene nanocapacitor and the origins of anomalous size-dependent increase of capacitance, *Nano Lett.* 14, 1739–1744 (2014).
34. M. Liu, X. Yun, and X. Zhang, Double-layer graphene optical modulator, *Nano Lett.* 12, 1482–1485 (2012).
35. M. Ryzhii, V. Ryhii, T. Otsuji, P.P. Maltsev, V.G. Leiman, N. Ryabova, and V. Mitin, Double injection, resonant-tunneling recombination, and current-voltage characteristics in double-graphene-layer structures, *J. Appl. Phys.* 115, 024506 (2014).
36. V. Ryzhii, A. Satou, T. Otsuji, M. Ryzhii, V. Mitin, and M.S. Shur, Dynamic effects in double graphene-layer structures with inter-layer resonant-tunneling negative conductivity, *J. Phys. D: Appl. Phys.* 46, 315107 (2013).
37. D. Svintsov, V. Vyurkov, V. Ryzhii, and T. Otsuji, Voltage-controlled surface plasmon-polaritons in double graphene layer structures, *J. Appl. Phys.* 113, 053701 (2013).
38. V. Ryzhii, T. Otsuji, M. Ryzhii, V.G. Leiman, S.O. Yurchenko, V. Mitin, and M.S. Shur, Effect of plasma resonances on dynamic characteristics of double graphene-layer optical modulator, *J. Appl. Phys.* 102, 104507 (2012).
39. V. Ryzhii, M. Ryzhii, V. Mitin, M.S. Shur, A. Satou, and T. Otsuji, Terahertz photomixing using plasma resonances in double-graphene layer structures, *J. Appl. Phys.* 113, 174506 (2013).
40. A.A. Dubinov, V.Ya. Aleshkin, V. Ryzhii, M.S. Shur, and T. Otsuji, Surface-plasmons lasing in double-graphene-layer structures, *J. Appl. Phys.* 115, 044511 (2014).
41. V. Ryzhii, A.A. Dubinov, V.Ya. Aleshkin, M. Ryzhii, and T. Otsuji, Injection terahertz laser using the resonant inter-layer radiative transitions in double-graphene-layer structure, *Appl. Phys. Lett.* 103, 163507 (2013).
42. V. Ryzhii, A.A. Dubinov, T. Otsuji, V.Ya. Aleshkin, M. Ryzhii, and M. Shur, Double-graphene-layer terahertz laser: Concept, characteristics, and comparison, *Opt. Express* 21, 31569–31579 (2013).
43. V. Ryzhii, T. Otsuji, V.Ya. Aleshkin, A.A. Dubinov, M. Ryzhii, V. Mitin, and M.S. Shur, Voltage-tunable terahertz and infrared photodetectors based on double-graphene-layer structures, *Appl. Phys. Lett.* 104, 163505 (2014).
44. L. Falkovsky and A. Varlamov, Space-time dispersion of graphene conductivity, *Eur. Phys. J. B* 56, 281 (2007).
45. T. Ando, Y. Zheng, and H. Suzuura, Dynamical conductivity and zero-mode anomaly in honeycomb lattices, *J. Phys. Soc. Jpn.* 71, 1318–1324 (2002).
46. R.R. Nair, P. Blake, A.N. Grigorenko, K.S. Novoselov, T.J. Booth, T. Stauber, N.M.R. Peres, and A.K. Geim, Fine Structure Constant Defines Visual Transparency of Graphene, *Science* 320, 1308 (2008).
47. B. Sensale-Rodriguez, R. Yan, L. Liu, D. Jena, and H.G. Xing, Graphene for reconfigurable terahertz optoelectronics, *Proc. IEEE* 101, 1705–1716 (2013).
48. V. Ryzhii, N. Ryabova, M. Ryzhii, N.V. Baryshnikov, V.E. Karasik, V. Mitin, and T. Otsuji, Terahertz and infrared photodetectors based on multiple graphene layer and nanoribbon structures, *Opto-Electron. Rev.* 20, 15–25 (2012).
49. V. Ryzhii, V. Mitin, M. Ryzhii, N. Ryabova, and T. Otsuji, Device model for graphene nanoribbon phototransistor, *Appl. Phys. Express* 1, 063002 (2008).
50. V. Ryzhii and M. Ryzhii, Graphene bilayer field-effect phototransistor for terahertz and infrared detection, *Phys. Rev. B* 79, 245311 (2009).

51. V. Ryzhii, M. Ryzhii, V. Mitin, and T. Otsuji, Terahertz and infrared photodetection using p–i–n multiple-graphene structures, *J. Appl. Phys.* 106, 084512 (2009).

52. M. Ryzhii, T. Otsuji, V. Mitin, and V. Ryzhii, Characteristics of p–i–n terahertz and infrared photodiodes based on multiple graphene layer structures, *Jpn. J. Appl. Phys.* 50, 070117-1–070117-6 (2011).

53. L. Vicarelli, M.S. Vitiello, D. Coquillat, A. Lombardo, A.C. Ferrari, W. Knap, M. Polini, V. Pellegrini, and A. Tredicucci, Graphene field effect transistors as room-temperature terahertz detectors, *Nat. Mater.* 11, 865–871 (2012).

54. W. Knap, S. Rumyantsev, M.S. Vitiello, D. Coquillat, S. Blin, N. Dyakonova, M. Shur, F. Teppe, A. Tredicucci, and T. Nagatsuma, Nanometer size field effect transistors for THz detectors, *Nanotechnology* 24, 214002 (2013).

55. D. Spirito, D. Coquillat, S.L. De Bonis, A. Lombardo, M. Bruna, A.C. Ferrari, V. Pellegrini, A. Tredicucci, W. Knap, and M.S. Vitello, High performance bilayer-graphene terahertz detectors, *Appl. Phys. Lett.* 104, 061111 (2014).

56. B. Sensale-Rodriguez, R. Yan, M.M. Kelly, T. Fang, K. Tahy, W.S. Hwang, D. Jena, L. Liu, and H.G. Xing, Broadband graphene terahertz modulators enabled by intraband transitions, *Nat. Commun.* 3, 780 (2012).

57. B. Sensale-Rodriguez, R. Yan, S. Rafique, M. Zhu, W. Li, X. Liang, D. Gundlach et al., Extraordinary control of terahertz beam reflectance in graphene electro-absorption modulators, *Nano Lett.* 12, 4518–4522 (2012).

58. L. Ju, B. Geng, J. Horng, C. Girit, M. Martin, Z. Hao, H.A. Bechtel et al., Graphene plasmonics for tunable terahertz metamaterials, *Nat. Nanotechnol.* 6, 630–634 (2011).

59. J. Maultzsch, Double-resonant Raman scattering in graphite: Interference effects, selection rules, and phonon dispersion, *Phys. Rev. B* 70, 155403 (2004).

60. H. Suzuura and T. Ando, Zone-boundary phonon in graphene and nanotube, *J. Phys. Soc. Jpn.* 77, 044703 (2008).

61. J.M. Dawlaty, S. Shivaraman, M. Chandrashekhar, F. Rana, and M.G. Spencer, Measurement of ultrafast carrier dynamics in epitaxial graphene, *Appl. Phys. Lett.* 92, 042116 (2008).

62. P.A. George, J. Strait, J. Dawlaty, S. Shivaraman, M. Chandrashekhar, F. Rana, and M.G. Spencer, Ultrafast optical-pump terahertz-probe spectroscopy of the carrier relaxation and recombination dynamics in epitaxial graphene, *Nano Lett.* 8, 4248 (2008).

63. M. Breusing, C. Ropers, and T. Elsaesser, Ultrafast carrier dynamics in graphite, *Phys. Rev. Lett.* 102, 086809 (2009).

64. F. Rana, P.A. George, J.H. Strait, J. Dawlaty, S. Shivaraman, M. Chandrashekhar, and M.G. Spencer, Carrier recombination and generation rates for intravalley and intervalley phonon scattering in graphene, *Phys. Rev. B* 79, 115447 (2009).

65. A. Satou, T. Otsuji, and V. Ryzhii, Theoretical study of population inversion in graphene under pulse excitation, *Jpn. J. Appl. Phys.* 50, 070116 (2011).

66. A. Satou, V. Ryzhii, Y. Kurita, and T. Otsuji, Threshold of terahertz population inversion and negative dynamic conductivity in graphene under pulse photoexcitation, *J. Appl. Phys.* 113, 143108 (2013).

67. M.S. Foster and I.L. Aleiner, Slow imbalance relaxation and thermoelectric transport in graphene, *Phys. Rev. B* 79, 085415 (2009).

68. F. Rana, Electron-hole generation and recombination rates for Coulomb scattering in graphene, *Phys. Rev. B* 76, 155431 (2007).

69. J.H. Strait, H. Wang, S. Shivaraman, V. Shields, M. Spencer, and F. Rana, Very slow cooling dynamics of photoexcited carriers in graphene observed by optical-pump terahertz-probe spectroscopy, *Nano Lett.* 11, 4902–4906 (2011).

70. T. Winzer, A. Knorr, and E. Malic, Carrier multiplication in graphene, *Nano Lett.* 10, 4839–4843 (2010).

71. P.A. Obraztsov, M.G. Rybin, A.V. Tyurnina, S.V. Garnov, E.D. Obraztsova, A.N. Obraztsov, and Y.P. Svirko, Broadband light-induced absorbance change in multilayer graphene, *Nano Lett.* 11, 1540–1545 (2011).
72. T. Winzer and E. Malic, Impact of Auger processes on carrier dynamics in graphene, *Phys. Rev. B* 85, 241404(R)-1–241404(R)-5 (2012).
73. S. Tani, F. Blanchard, and K. Tanaka, Ultrafast carrier dynamics in graphene under a high electric field, *Phys. Rev. Lett.* 99, 166603-1–166603-5 (2012).
74. A.A. Dubinov, V.Y. Aleshkin, M. Ryzhii, T. Otsuji, and V. Ryzhii, Terahertz laser with optically pumped graphene layers and Fabry–Perot resonator, *Appl. Phys. Express* 2, 092301 (2009).
75. V. Ryzhii, M. Ryzhii, A. Satou, T. Otsuji, A.A. Dubinov, and V.Y. Aleshkin, Feasibility of terahertz lasing in optically pumped epitaxial multiple graphene layer structures, *J. Appl. Phys.* 106, 084507 (2009).
76. T. Li, L. Luo, M. Hupalo, J. Zhang, M.C. Tringides, J. Schmalian, and J. Wang, Femtosecond population inversion and stimulated emission of dense Dirac Fermions in graphene, *Phys. Rev. Lett.* 108, 167401 (2012).
77. V. Ryzhii, M. Ryzhii, V. Mitin, A. Satou, and T. Otsuji, Effect of heating and cooling of photogenerated electron–hole plasma in optically pumped graphene on population inversion, *Jpn. J. Appl. Phys.* 50, 094001 (2011).
78. E.H. Hwang and S. Das Sarma, Dielectric function, screening, and plasmons in two-dimensional graphene, *Phys. Rev. B* 75, 205418 (2007).
79. M. Jablan, H. Buljan, and M. Soljačić, Plasmonics in graphene at infrared frequencies, *Phys. Rev. B* 80, 245435 (2009).
80. F.H.L. Koppens, D.E. Chang, and F.J. García de Abajo, Graphene plasmonics: a platform for strong light–matter interaction, *Nano Lett.* 11, 3370 (2011).
81. F. Rana, J.H. Strait, H. Wang, and C. Manolatou, Ultrafast carrier recombination and generation rates for plasmon emission and absorption in graphene, *Phys. Rev. B* 84, 045437 (2011).
82. A.Yu. Nikitin, F. Guinea, F.J. Garcia-Vidal, and L. Martin-Moreno, Surface plasmon enhanced absorption and suppressed transmission in periodic arrays of graphene ribbons, *Phys. Rev. B* 85, 081405(R) (2012).
83. B. Wunsch, T. Stauber, F. Sols, and F. Guinea, Dynamical polarization of graphene at finite doping, *New J. Phys.* 8, 318 (2006).
84. L. Brey and H.A. Fertig, Elementary electronic excitations in graphene nanoribbons, *Phys. Rev. B* 75, 125434 (2007).
85. V.V. Popov, T.Yu. Bagaeva, T. Otsuji, and V. Ryzhii, Oblique terahertz plasmons in graphene nanoribbon arrays, *Phys. Rev. B* 81, 073404 (2010).
86. B. Sensale-Rodriguez, R. Yan, M. Zhu, D. Jena, L. Liu, and H.G. Xing, Efficient terahertz electro-absorption modulation employing graphene plasmonic structures, *Appl. Phys. Lett.* 101, 261115 (2012).
87. X.-Y. He, Q.-J. Wang, and S.-F. Yu, Numerical study of gain-assisted terahertz hybrid plasmonic waveguide, *Plasmonics* 7, 571–577 (2012).
88. F. Xia, T. Mueller, Y.-M. Lin, A. Valdes-Garcia, and P. Vouris, Ultrafast graphene photodetector, *Nat. Nanotechnol.* 4, 839–843 (2009).
89. T. Mueller, F. Xia, and P. Avouris, Graphene photodetectors for high-speed optical communications, *Nat. Photon.* 4, 297–301 (2010).
90. M. Dyakonov and M. Shur, Detection, mixing, and frequency multiplication of terahertz radiation by two-dimensional electronic fluid, *IEEE Trans. Electron. Dev.* 43, 1640–1646 (1996).
91. W. Knap, M. Dyakonov, D. Coquillat, F. Teppe, N. Dyakonova, J. Łusakowski, K. Karpierz et al., Field effect transistors for terahertz detection: Physics and first imaging applications, *J. Infrared Millimeter THz Waves* 30, 1319–1337 (2009).

92. W. Knap, S. Nadar, H. Videlier, S. Boubanga-Tombet, D. Coquillat, N. Dyakonova, F. Teppe et al., Field effect transistors for terahertz detection and emission, *J. Infrared Millimeter THz Waves* 32, 618–628 (2011).

93. M. Sakowicz, M.B. Lifshits, O.A. Klimenko, F. Schuster, D. Coquillat, F. Teppe, and W. Knap, Terahertz responsivity of field effect transistors versus their static channel conductivity and loading effects, *J. Appl. Phys.* 110, 054512 (2011).

94. T. Otsuji, T. Watanabe, S. Boubanga Tombet, A. Satou, W. Knap, V. Popov, M. Ryzhii, and V. Ryzhii, Emission and detection of terahertz radiation using two-dimensional electrons in III–V semiconductors and graphene, *IEEE Trans. Terahertz Sci. Technol.* 3, 63–72 (2013).

95. T. Otsuji, T. Watanabe, S.A. Boubanga Tombet, A. Satou, V. Ryzhii, V. Popov, and W. Knap, Emission and detection of terahertz radiation using two dimensional plasmons in semiconductor nano-heterostructures for nondestructive evaluations, *Opt. Eng.* 53, 031206 (2014).

96. S.A. Boubanga Tombet, D. Yadav, T. Watanabe, V. Ryzhii, and T. Otsuji, Terahertz emission in a double-graphene-layer heterostructure, *International Conference on Optical Terahertz Science and Technology Digest (OTST)*, TuS3-3, San Diego, CA (2015).

97. V. Ryzhii, T. Otsuji, M. Ryzhii, and M.S. Shur, *J. Phys. D: Appl. Phys.* 45, 302001 (2012).

98. S.M. Duffy, S. Verghese, K.A. McIntosh, A. Jackson, A.C. Gossard, and S. Matsuura, Accurate modeling of dual dipole and slot elements used with photomixers for coherent terahertz output power, *IEEE Trans. Microw. Theory Tech.* 49, 1032–1038 (2001).

99. A. Stohr, A. Malcoci, A. Sauerwald, I.C. Mayorga, R. Gusten, and D. Jager, Ultrawideband traveling-wave photodetectors for photonic local oscillators, *IEEE J. Lightwave Technol.* 21, 3062–3070 (2003).

100. H. Ito, S. Kodama, Y. Muramoto, T. Furuta, T. Nagatsuma, and T. Ishibashi, High-speed and high-output InP–InGaAs unitraveling-carrier photodiodes, *IEEE J. Sel. Top. Quant. Electron.* 10, 709–727 (2004).

101. V. Ryzhii, I. Khmyrova, M. Ryzhii, A. Satou, T. Otsuji, V. Mitin, and M.S. Shur, Plasma waves in two-dimensional electron systems and their applications, *Int. J. High Speed Electron. Syst.* 17, 521–538 (2007).

2 Microscopic Many-Body Theory for Terahertz Quantum Cascade Lasers

Qi Jie Wang and Tao Liu

CONTENTS

2.1 Introduction 37
2.2 Theoretical Model 39
    2.2.1 Intersubband Semiconductor Bloch Equations 39
    2.2.2 Coupling between Bloch Equation and Maxwell's Equations 45
2.3 Microscopic Analysis on Optical Gain 47
    2.3.1 Effects of Many-Body Interaction and Nonparabolicity on
         Optical Gain 47
    2.3.2 Optical Gain Spectrum as a Function of Bias 49
    2.3.3 Optical Gain Spectrum as a Function of Injection and
         Extraction Coupling Strength 50
    2.3.4 Optical Gain Spectrum as a Function of Doping Density 50
2.4 Microscopic Analysis on Linewidth Enhancement Factor 53
2.5 Summary 56
References 57

2.1 INTRODUCTION

Quantum cascade laser (QCL) was first demonstrated in 1994 [1]. Since the invention, QCLs have become important coherent mid-infrared (mid-IR) [1] and terahertz (THz) [2] radiation sources. Room-temperature continuous-wave operation of mid-IR QCLs has been achieved in the ~3–14 μm wavelength range. However, due to the challenges on building up enough optical gain in long-wavelength emission devices above cryogenic temperatures [3], THz QCLs with wavelengths covering from 60 to 300 μm are still operated below room temperatures. The best temperature performance has been obtained at ~200 K using a resonant-phonon (RP) design [4]. Further improvement on the existing designs requires a better understanding of effects of the electron transport on optical properties, for example, optical gain of THz QCLs.

Up to now, several useful theoretical models, for example, Monte Carlo [5–9], non-equilibrium Green's function [10–14], and simplified density-matrix [15–17] models, have been developed to predict the optical properties and electron transports of QCLs. Although Monte Carlo and nonequilibrium Green's functions analyses show good agreement between theories and experiments in some aspects, implementations of these two models are difficult, requiring intensive numerical computations. Alternatively, the simplified density-matrix model is simple in the analysis and requires much less computation load, while still capturing the essence of coherent effects such as the electron resonant-tunneling (RT) transport. It has been shown as one of the most promising candidates for the study of QCLs. This model is in essence a set of rate equations but includes electron distributions and coherent dynamics in different subbands. Electrons in each subband are assumed to behave the same, regardless of their kinetic energies. Therefore, this model describes the optical properties and electron transport from a *macroscopic* point of view, while the *microscopic* phenomena, for example, the electron dynamics in the in-plane k-space, are neglected. Moreover, the present experiments and theoretical predictions have shown the limitations of this macroscopic model. The gain peak frequency calculated by the macroscopic model is overestimated compared to the experimentally measured lasing frequency [4]. Although coherence effects of RT transport can be described in the macroscopic simplified density-matrix model [16], another aspect that needs to be considered for a more accurate calculation of optical properties and electron transport is the electron–electron Coulomb interaction.

Direct numerical treatment of many-body Coulomb interaction is complex and hence is often handled at the level of the Hartree–Fock approximation [18]. In this case, the set of motion equations, in terms of the diagonal and off-diagonal elements of the reduced single-particle density matrix, are well known as the Hartree–Fock semiconductor Bloch equations that treat Coulomb effects via bandstructure and Rabi frequency renormalizations [19]. For mid-IR QCLs, the effects of many-body Coulomb interactions on the optical properties and electron transports were considered, but the role of coherence of RT transport was neglected [20]. For THz QCLs, many-body effects on population dynamics were investigated [21], but only the injection coupling was considered while the extraction coupling was neglected. Dupont et al. [17] have demonstrated the importance of extraction coupling on optical properties according to the macroscopic simplified density-matrix model but without considering the microscopic properties such as the many-body Coulomb interactions. Since many-body Coulomb interactions result in bandstructure and Rabi frequency renormalizations, and Coulomb-induced subband coupling, they are expected to play an important role in THz QCLs. However, the effects of electron–electron Coulomb interactions on optical properties, for example, optical gain and linewidth enhancement factor (LEF) of THz QCLs, have not yet been reported. In addition, the subband dispersion with different subband effective masses (commonly referred to as nonparabolicity) is known to be important for modifications to gain spectrum. Thus, the effect of the nonparabolicity-induced modifications of subband electrons also needs to be considered. The purpose of this chapter is to study the dependence of the intersubband gain spectrum and LEF on the electron–electron Coulomb interactions, nonparabolicity, RT, and laser device parameters. The effects of device parameters including external electrical bias, injection and extraction coupling strength, doping density, and temperature are taken into account.

In this chapter, we extend the simplified density-matrix model to include the many-body interactions derived from electron Hamiltonian in the second quantization. It not only takes into account the coherent effects in the electron transport through injector and extractor barriers by RT, but also distributions of kinetic energy of electrons and many-body effects based on intersubband semiconductor-Bloch equations. The nonparabolicity effect is also approximately considered. The results show that the gain peak frequency calculated by the many-body model is more close to the experimentally measured lasing frequency compared with existing macroscopic model. Furthermore, this microscopic model can also partially explain the nonzero LEF of QCLs at the gain peak, which existed in the field for a while but cannot be explicitly explained. The proposed model provides a comprehensive picture of optical properties of QCLs, not only enhancing our in-depth understanding of optical gain but also enabling an accurate prediction of the device performances, for example, LEF. More importantly, this model has a low computational load that can greatly simplify the optimization process of active region designs of QCLs compared to other full quantum mechanical models.

2.2 THEORETICAL MODEL

2.2.1 Intersubband Semiconductor Bloch Equations

Currently, the highest temperature operation of THz QCL is achieved by using RT injection design with three quantum wells in each period [4]. We consider the same design in this section. Figure 2.1 shows the conduction band diagram and magnitude squared envelope wave functions of this design in "tight-binding" scheme, where the injector barrier and extractor barrier are made sufficiently thick to allow one period

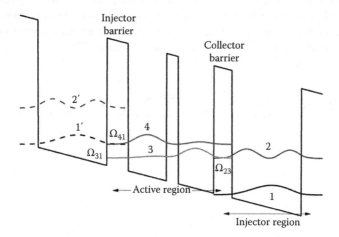

FIGURE 2.1   Conduction band diagram and magnitude squared envelope wave functions of a four level resonant-phonon THz QCL with a diagonal design in the "tight-binding" scheme. Ω_{41}, Ω_{23}, and Ω_{31} are the injection, extraction, and parasitic coupling strength (unit: Hz), respectively. The external electric field is 12.3 kV/cm. The radiative transition is from $4 \rightarrow 3$, and depopulation of the lower laser level is via $3 \rightarrow 2$ (RT) and $(2 \rightarrow 1)$ longitudinal optical phonon scattering. The thickness in angstrom of each layer is given as 49/88/27/82/42/160 starting from the injector barrier. The barriers are indicated in bold fonts. The widest well is doped at 3×10^{10} cm^{-2}.

to be separated into two regions [16,17,22], that is, the active region and the injector region (see Figure 2.1). The energy states within either the active region or the injector region are coupled by scattering processes, but energy states from different regions are coupled by tunneling. This localized wave function analysis defined by each region allows us to investigate the effects of RT on the electron transport and gain spectrum.

In order to conveniently treat the many-body problem, we derive the dynamical equations of motion in the second quantized representation. The Hamiltonian of the system can be divided into four parts [23]:

$$H = H_{el} + H_{rt} + H_0 + H_{Coul} \tag{2.1}$$

The first two terms

$$H_{el} + H_{rt} = -\sum_k \left(\mu E b_{4,k}^\dagger b_{3,k} + c.c \right) - \sum_k \left[\left(\frac{\Delta_{41'}}{2} \right) b_{1',k}^\dagger b_{4,k} + c.c \right]$$

$$- \sum_k \left[\left(\frac{\Delta_{23}}{2} \right) b_{3,k}^\dagger b_{2,k} + c.c \right] - \sum_k \left[\left(\frac{\Delta_{31'}}{2} \right) b_{1',k}^\dagger b_{3,k} + c.c \right] \tag{2.2}$$

are the Hamiltonian for electron–light coupling and the tunneling effects, respectively. The RT terms are written by an analogy with the electron–light term, which is similar to the density matrix model firstly proposed by Kazarinov and Suris [24]. $b_{j,k}^\dagger (b_{j,k})$ is the creation (annihilation) operator of the electron state in subband j with wave vector $\mathbf{k}$. μ is the electron charge times the dipole matrix element of laser transition. $\Delta_{41'}$, Δ_{23}, and $\Delta_{31'}$ are the injection, extraction, and parasitic coupling strengths, which can be derived by a simple tight-binding approach [25].

The last two terms are [26]

$$H_0 + H_{Coul} = \sum_{j,k} \varepsilon_{j,k} b_{j,k}^\dagger b_{j,k} + \frac{1}{2} \sum_{uvv'u'}^{1,2,3,4} \sum_{kk'q} V_q^{uvv'u'} b_{u,k+q}^\dagger b_{v,k'-q}^\dagger b_{v',k'} b_{u',k} \tag{2.3}$$

which describe free electrons and electron–electron Coulomb interactions, respectively. $\varepsilon_{j,k}$ is the jth subband energy, $\mathbf{k}$ is the in plane wave vector, the $V_q^{uvv'u'}$ is the screening Coulomb matrix element that reads (in MKS units)

$$V_q^{uvv'u'} = \frac{e^2}{2A\varepsilon_0\varepsilon_r\varepsilon(q)q} \int dz \int \varphi_u(z)\varphi_{u'}(z)e^{-q|z-z'|}\varphi_v(z')\varphi_{v'}(z')dz' \tag{2.4}$$

where
  A is the quantum well area
  ε_r is background dielectric constant
  $q = |\mathbf{k}-\mathbf{k'}|$

The near-resonant screening is taken into account approximately via the dielectric function $\varepsilon(q)$ calculated by the single subband screening model [18,27]. The Coulomb matrix elements are difficult to evaluate numerically, but they can be simplified according to the approaches proposed in References 28,29 without the loss of high accuracy.

The equations of motion for the polarizations $p_{ij,k} = \langle b_{i,k}^{\dagger} b_{j,k} \rangle$ and electron occupation $n_{i,k} = \langle b_{i,k}^{\dagger} b_{i,k} \rangle$, where the bracket $\langle ... \rangle$ indicates an expectation value, can be derived by using the following Heisenberg equation and the anticommutations relations of fermionic operator [19]:

$$\frac{dO}{dt} = \frac{i}{\hbar}[H,O] \tag{2.5}$$

$$[b_{i,k}, b_{j,k'}^{\dagger}]_{+} = \delta_{ij,kk'}, \quad [b_{i,k}, b_{j,k'}]_{+} = [b_{i,k}^{\dagger}, b_{j,k'}^{\dagger}]_{+} = 0 \tag{2.6}$$

where
O is the operator
H is the Hamiltonian

However, due to the Coulomb interaction terms in Equation 2.3, the result is an infinite hierarchy of coupled differential equations. The hierarchy describes the correlation effect in the Coulomb potential. The first-order correlation is induced by the Hartree–Fock contributions, which results in bandstructure and Rabi frequency renormalizations. Scattering and dephasing contributions cause the second-order correlation in the Coulomb potential, and so on [18,20,30]. In this section, we only include the Hartree–Fock contributions and dephasing and scattering contributions at the level of a relaxation-rate approximation. After a lengthy algebra, we obtain the following Bloch equations of density matrix elements for the polarizations $p_{ij,k}$ and electron occupation $n_{i,k}$ [31]:

$$\frac{dp_{34,k}}{dt} = -\gamma_{34p}p_{34,k} - i\frac{\tilde{\varepsilon}_{43,k}}{\hbar}p_{34,k} - i\tilde{\Omega}_0(n_{4,k} - n_{3,k}) + i\tilde{\Omega}_{41'}^{\dagger}p_{31',k}$$
$$- i\tilde{\Omega}_{23}^{\dagger}p_{24,k} - i\tilde{\Omega}_{31'}p_{41',k}^{\dagger} \tag{2.7}$$

$$\frac{dp_{41',k}}{dt} = -\gamma_{41'p}p_{41',k} - i\frac{\tilde{\varepsilon}_{1'4,k}}{\hbar}p_{41',k} - i\tilde{\Omega}_{41'}(n_{1,k} - n_{4,k}) - i\tilde{\Omega}_0^{\dagger}p_{31',k} + i\tilde{\Omega}_{31'}p_{34,k}^{\dagger} \tag{2.8}$$

$$\frac{dp_{23,k}}{dt} = -\gamma_{23p}p_{23,k} - i\frac{\tilde{\varepsilon}_{32,k}}{\hbar}p_{23,k} - i\tilde{\Omega}_{23}(n_{3,k} - n_{2,k}) + i\tilde{\Omega}_0^{\dagger}p_{24,k} + i\tilde{\Omega}_{31'}^{\dagger}p_{21',k} \tag{2.9}$$

$$\frac{dp_{31',\mathbf{k}}}{dt} = -\gamma_{31'p}p_{31',\mathbf{k}} - i\frac{\tilde{\varepsilon}_{1'3,\mathbf{k}}}{\hbar}p_{31',\mathbf{k}} - i\tilde{\Omega}_0^\dagger p_{41',\mathbf{k}} + i\tilde{\Omega}_{41'}^\dagger p_{34,\mathbf{k}} - i\tilde{\Omega}_{23}^\dagger p_{21',\mathbf{k}}$$

$$- i\tilde{\Omega}_{31'}(n_{1',\mathbf{k}} - n_{3,\mathbf{k}}) \tag{2.10}$$

$$\frac{dp_{24,\mathbf{k}}}{dt} = -\gamma_{24p}p_{24,\mathbf{k}} - i\frac{\tilde{\varepsilon}_{42,\mathbf{k}}}{\hbar}p_{24,\mathbf{k}} + i\tilde{\Omega}_0^\dagger p_{23,\mathbf{k}} + i\tilde{\Omega}_{41'}^\dagger p_{21',\mathbf{k}} - i\tilde{\Omega}_{23}^\dagger p_{34,\mathbf{k}} \tag{2.11}$$

$$\frac{dp_{21',\mathbf{k}}}{dt} = -\gamma_{21'p}p_{21',\mathbf{k}} - i\frac{\tilde{\varepsilon}_{1'2,\mathbf{k}}}{\hbar}p_{21',\mathbf{k}} + i\tilde{\Omega}_{41'}^\dagger p_{24,\mathbf{k}} - i\tilde{\Omega}_{23}^\dagger p_{31',\mathbf{k}} + i\tilde{\Omega}_{31'}^\dagger p_{23,\mathbf{k}} \tag{2.12}$$

$$\frac{dn_{4,\mathbf{k}}}{dt} = -i\left(\tilde{\Omega}_0^\dagger p_{34,\mathbf{k}} - \tilde{\Omega}_0 p_{34,\mathbf{k}}^\dagger\right) - i\left(\tilde{\Omega}_{41'}p_{41',\mathbf{k}}^\dagger - \tilde{\Omega}_{41'}^\dagger p_{41',\mathbf{k}}\right)$$

$$- \gamma_4\left[n_{4,\mathbf{k}} - f_{4,\mathbf{k}}\left(\mu_{4,e}, T_{4,e}\right)\right] - \gamma_{43}\left[n_{4,\mathbf{k}} - f_{4,\mathbf{k}}\left(\mu_{43}, T_l\right)\right] - \gamma_{sp}n_{4,\mathbf{k}} \tag{2.13}$$

$$\frac{dn_{3,\mathbf{k}}}{dt} = -i\left(\tilde{\Omega}_0 p_{34,\mathbf{k}}^\dagger - \tilde{\Omega}_0^\dagger p_{34,\mathbf{k}}\right) - i\left(\tilde{\Omega}_{23}^\dagger p_{23,\mathbf{k}} - \tilde{\Omega}_{23}p_{23,\mathbf{k}}^\dagger\right) - \gamma_3\left[n_{3,\mathbf{k}} - f_{3,\mathbf{k}}\left(\mu_{3,e}, T_{3,e}\right)\right]$$

$$- \gamma_{43}\left[n_{3,\mathbf{k}} - f_{3,\mathbf{k}}\left(\mu_{43}, T_l\right)\right] + \gamma_{sp}n_{4,\mathbf{k}} - i(\tilde{\Omega}_{31'}p_{31',\mathbf{k}}^\dagger - \tilde{\Omega}_{31'}^\dagger p_{31',\mathbf{k}}) \tag{2.14}$$

$$\frac{dn_{2,\mathbf{k}}}{dt} = -i(\tilde{\Omega}_{23}p_{23,\mathbf{k}}^\dagger - \tilde{\Omega}_{23}^\dagger p_{23,\mathbf{k}}) - \gamma_2\left[n_{2,\mathbf{k}} - f_{2,\mathbf{k}}\left(\mu_{2,e}, T_{2,e}\right)\right]$$

$$- \gamma_{21'}\left[n_{2,\mathbf{k}} - f_{2,\mathbf{k}}\left(\mu_{21'}, T_l\right)\right] \tag{2.15}$$

$$\frac{dn_{1',\mathbf{k}}}{dt} = -i\left(\tilde{\Omega}_{41'}^\dagger p_{41',\mathbf{k}} - \tilde{\Omega}_{41'}p_{41',\mathbf{k}}^\dagger\right) - \gamma_{1'}\left[n_{1',\mathbf{k}} - f_{1',\mathbf{k}}\left(\mu_{1',e}, T_{1,e}\right)\right]$$

$$- \gamma_{21'}\left[n_{1',\mathbf{k}} - f_{1',\mathbf{k}}\left(\mu_{21'}, T_l\right)\right] - i\left(\tilde{\Omega}_{31'}^\dagger p_{31',\mathbf{k}} - \tilde{\Omega}_{31'}p_{31',\mathbf{k}}^\dagger\right) \tag{2.16}$$

with

$$\tilde{\varepsilon}_{uv,\mathbf{k}} = \varepsilon_{u,\mathbf{k}} - \varepsilon_{v,\mathbf{k}} - \sum_{\mathbf{k'}\neq\mathbf{k}}\left(V_{\mathbf{k}-\mathbf{k'}}^{uuuu}n_{u,\mathbf{k'}} - V_{\mathbf{k}-\mathbf{k'}}^{vvvv}n_{v,\mathbf{k'}}\right) + \sum_{\mathbf{k'}\neq\mathbf{k}}(n_{u,\mathbf{k'}} - n_{v,\mathbf{k'}})V_{\mathbf{k}-\mathbf{k'}}^{uvuv} \tag{2.17}$$

$$\tilde{\Omega}_0 = \frac{\mu\xi}{2\hbar} + \frac{1}{\hbar}\sum_{k'\neq k}V_{k-k'}^{4334}p_{43,k'} - \frac{2}{\hbar}V_0^{4343}\sum_{k'}p_{43,k'} \tag{2.18}$$

$$\tilde{\Omega}_{uv} = \frac{\Delta_{uv}}{2\hbar} + \frac{1}{\hbar}\sum_{k'\neq k}V_{k-k'}^{uvvu}p_{uv,k'} - \frac{2}{\hbar}V_0^{uvuv}\sum_{k'}p_{uv,k'} \tag{2.19}$$

where

γ_{ijp} is the dephasing rate associated with energy levels i and j
γ_j is the intrasubband electron–electron scattering rate at level j
γ_{ij} is the combined electron–electron and electron–phonon scattering rate between
   levels i and j

The lifetime of the upper laser level as a function of temperature is approximated in
the form of $\gamma_{43} = \gamma_{430}\exp[\hbar(\omega_{LO}-\omega_\lambda)/k_bT_{4e}]$ where γ_{430} is the raw LO phonon scattering rate when the upper-state electrons can sufficiently emit LO phonons and
ω_{LO} is the energy of LO phonon. γ_{sp} is the spontaneous emission rate. $T_{j,e}$ is the
electron temperature at level j; T_l is the lattice temperature. $f_{j,k}$ is the Fermi–Dirac
distribution with chemical potential $\mu_{j,e}$ at level j. The chemical potentials and temperatures are determined by electron number conservation and energy conservation
[19]. The terms with coefficients $V_{k-k'}^{uuuu}$ refer to the exchange self-energy, $V_{k-k'}^{uvvu}$ to the
excitonic enhancement, and V_0^{uvuv} to the depolarization. Processes corresponding to
these contributions are shown in Figure 2.2. The influence of the subband dispersion,
namely, the nonparabolicity, is represented by using effective mass of electrons m^*.
For subband j, we have $\varepsilon_{j,k} = \varepsilon_j + \hbar^2 k^2/2m_j^*$. For our structure, the calculation shows
that $m_1^* \approx 0.0670m_0, m_2^* \approx m_3^* \approx 0.0723m_0,$ and $m_4^* \approx 0.0743m_0$ (m_0 is the free electron
mass) according to Ekenberg's model [32].

Since levels (1′, 4) and (2, 3) are coherently coupled by the tunneling, the coherences
corresponding to the levels (1′, 3), (2, 4), and (1′, 2) have a time-harmonic character

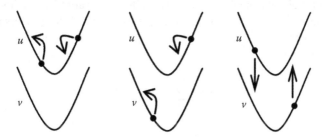

FIGURE 2.2 Graphical representations of the Coulomb interactions between the subbands
u and v (from the left to the right): exchange self-energy, excitonic enhancement, and
depolarization. The exchange shift induces a renormalization of the free-carrier contributions; the excitonic contribution causes a renormalization of the electron–light interaction; the
depolarization contribution is another renormalization of the electron–light interaction with a
very different origin.

due to the time-harmonic (3, 4) coherent coupling. Therefore, we try to look for solutions in the form of

$$p_{41',\mathbf{k}} = p_{41',\mathbf{k}}^{(0)}, \quad p_{34,\mathbf{k}} = p_{34,\mathbf{k}}^{(0)} + \tilde{p}_{34,\mathbf{k}}e^{-i\omega_\lambda t}, \quad p_{31',\mathbf{k}} = p_{31',\mathbf{k}}^{(0)} + \tilde{p}_{31',\mathbf{k}}e^{-i\omega_\lambda t}$$

$$p_{23,\mathbf{k}} = p_{23,\mathbf{k}}^{(0)}, \quad p_{24,\mathbf{k}} = p_{24,\mathbf{k}}^{(0)} + \tilde{p}_{24,\mathbf{k}}e^{-i\omega_\lambda t}, \quad p_{21',\mathbf{k}} = p_{21',\mathbf{k}}^{(0)} + \tilde{p}_{21',\mathbf{k}}e^{-i\omega_\lambda t}$$

(2.20)

where

$p_{ij,\mathbf{k}}^{(0)}$ is the static tunneling induced coherence

$\tilde{p}_{ij,\mathbf{k}}$ is the laser-induced coherence [17]

In the above semiconductor Bloch equations, the relaxation-rate approximation is employed to calculate dephasing and scattering contributions [33], because a full kinetic treatment with dephasing and scattering terms based on the Boltzmann equation requires length computational time. In this relaxation-rate approximation, the dephasing contribution to polarizations is treated as

$$\frac{dp_{ij,\mathbf{k}}}{dt} = -\gamma_{ijp}p_{ij,\mathbf{k}}$$

(2.21)

where γ_{ijp} denotes an effective dephasing rate.

As for scattering contributions, the influence of scattering on the electron distributions is treated by the relaxation of a given population distribution $n_{u,\mathbf{k}}$ to a quasi-equilibrium Fermi–Dirac distribution $f_{u,\mathbf{k}}(T_i,\mu_i)$ with temperature T_i and chemical potential μ_i as $\gamma_u(n_{u,\mathbf{k}}-f_{u,\mathbf{k}})$ at energy level i. The two main scattering contributions are considered in THz QCLs, that is, electron–electron scattering and electron–phonon scattering.

For intrasubband scattering, we only consider the electron–electron scattering due to the much smaller intrasubband electron–phonon scattering rate. The electron distribution of each subband (e.g., subband u) is driven toward Fermi–Dirac distribution by this intrasubband scattering at the corresponding electron temperature $T_{u,e}$ and chemical potential $\mu_{u,e}$. The actual value of electron temperature and chemical potential can be determined by the conditions of particle and energy conservation

$$\sum_{\mathbf{k}} n_{u,\mathbf{k}} = \sum_{\mathbf{k}} f_{u,\mathbf{k}}\left(\mu_{u,e},T_{u,e}\right)$$

(2.22)

$$\sum_{\mathbf{k}} \varepsilon_{u,\mathbf{k}} n_{u,\mathbf{k}} = \sum_{\mathbf{k}} \varepsilon_{u,\mathbf{k}} f_{u,\mathbf{k}}\left(\mu_{u,e},T_{u,e}\right)$$

(2.23)

As to the intersubband scattering, two contributions are considered. First, for the intersubband electron–phonon scattering between levels u and v, energy is dissipated from the electrons to the lattice. In this case, electrons of these two energy levels relax to share a Fermi–Dirac distribution with lattice temperature T_l and chemical potential $\mu_{uv,l}$. The chemical potential can be determined by the particle conservation

$$\sum_k \sum_{i=u,v} n_{i,k} = \sum_k \sum_{i=u,v} f_{i,k}\left(\mu_{uv,l}, T_l\right) \tag{2.24}$$

In case of the intersubband electron–electron scattering between levels u and v, electrons of these two energy levels are driven to share a Fermi–Dirac distribution with electron temperature $T_{uv,e}$ and chemical potential $\mu_{uv,e}$. According to the particle and energy conservation, it has

$$\sum_k \sum_{i=u,v} n_{i,k} = \sum_k \sum_{i=u,v} f_{i,k}\left(\mu_{uv,e}, T_{uv,e}\right) \tag{2.25}$$

$$\sum_k \sum_{i=u,v} \varepsilon_{i,k} n_{i,k} = \sum_k \sum_{i=u,v} \varepsilon_{i,k} f_{i,k}\left(\mu_{uv,e}, T_{uv,e}\right) \tag{2.26}$$

If the electron–electron and electron–phonon scattering between levels u and v occurs on similar timescales, electrons of these two energy levels driven by electron–electron scattering still relax to share a Fermi–Dirac distribution with lattice temperature T_l and chemical potential $\mu_{uv,l}$ [20]. Because the electron–electron and electron–phonon scattering between the upper laser level and lower laser level occur on similar timescales for THz QCLs due to thermally activated phonon scattering in our chosen temperatures, electrons of these two energy levels are assumed to share a Fermi–Dirac distribution at lattice temperature, as shown in Equations 2.13 and 2.14. In addition, the electron–phonon interaction dominates the scattering processes between level 2 and 1 in THz QCLs, so we neglect the electron–electron scattering between levels 2 and 1, as shown in Equations 2.15 and 2.16.

2.2.2 Coupling between Bloch Equation and Maxwell's Equations

A laser field $E(z,t)$ will induce a microscopic electric dipole moment. The summation of these dipoles yields a macroscopic polarization P_{ij}. This polarization then drives the laser field $E'(z,t)$ following the Maxwell's equations. Self-consistency requires $E'(z,t) = E(z,t)$. Therefore, we should know how the polarization affects the slowly varying electric field amplitude in order to obtain the optical gain.

According to the Maxwell's equations, after some algebra, we have the following wave equation [19]:

$$-\nabla^2 \mathbf{E} + \left(\frac{n}{c}\right)^2 \frac{\partial^2 \mathbf{E}}{\partial t^2} = -\mu_0 \frac{\partial^2 \mathbf{P}}{\partial t^2} \tag{2.27}$$

where

n is the refractive index
c is the light speed in vacuum
μ_0 is the permeability of vacuum

Assuming single mode operation in THz QCL, the laser field can be written as

$$\mathbf{E}(z,t) = \frac{1}{2}\hat{i}\xi(z,t)e^{i[kz-\omega_\lambda t-\phi(z)]} + \text{c.c.} \tag{2.28}$$

where
  $\hat{i}$ is the unit vector
  ξ is the slowly varying complex electric field amplitude
  ω_λ is the laser frequency
  k is wave vector
  ϕ is the phase

The induced polarization is

$$\mathbf{P}(z,t) = \frac{1}{2}\hat{i}P(z,t)e^{i[kz-\omega_\lambda t-\phi(z)]} + \text{c.c.} \tag{2.29}$$

The slowly varying polarization amplitude is coupled with complex electric field amplitude as

$$P(z,t) = \varepsilon_0 n^2 \chi(z)\xi(z,t) \tag{2.30}$$

where
  ε_0 is the vacuum permittivity
  $\chi(z)$ is the complex susceptibility of the gain medium.

By inserting Equations 2.28 through 2.30 into Equation 2.27 and neglecting the terms with double derivatives and multiplication of two derivatives (slowly varying envelope approximation), we can get the following equations

$$\frac{d\xi(z,t)}{dz} = -\frac{\omega_\lambda n}{2c}\chi''(z)\xi(z,t) \tag{2.31}$$

$$\frac{d\phi(z,t)}{dz} = -\frac{\omega_\lambda n}{2c}\chi'(z) \tag{2.32}$$

where $\chi(z) = \chi'(z) + i\chi''(z)$. With the definition of the amplitude gain g,

$$\frac{d\xi(z,t)}{dz} = g\xi(z,t) \tag{2.33}$$

we can find the intensity gain $G = 2g$ as

$$G = \frac{\omega_\lambda n}{c}\chi''(z) \tag{2.34}$$

Since the macroscopic polarization can be linked with microscopic through

$$P(z,t) = \frac{1}{V_m} \sum_k \mu_{34} p_{34,k} \qquad (2.35)$$

where V_m is the volume of one period of active region, we can derive the intensity gain G as follows:

$$G = -\frac{2\omega_\lambda}{\varepsilon_0 n c V_m \xi} \, \mathrm{Im} \left(\sum_k \mu_{34} \tilde{p}_{34,k} \right) \qquad (2.36)$$

According to Equation 2.32, we can also obtain the refractive index change δn as

$$\delta n = \frac{\mu}{\varepsilon_0 n V_m \xi} \, \mathrm{Re} \left(\sum_k \mu_{34} \tilde{p}_{34,k} \right) \qquad (2.37)$$

The LEF α can be obtained by the ratio of the change in the real part of the refractive index change δn to the change in the gain G with respect to the carrier density N_0 [19]:

$$\alpha = \frac{-2(\omega_\lambda / c)(d(\delta n)/dN_0)}{dG/dN_0} \qquad (2.38)$$

2.3 MICROSCOPIC ANALYSIS ON OPTICAL GAIN

2.3.1 EFFECTS OF MANY-BODY INTERACTION AND NONPARABOLICITY ON OPTICAL GAIN

The results presented here are obtained by numerically solving the equations of motion (Equations 2.7 through 2.19) for a small laser field (linear absorption). In our calculation, 250 k-points within the each subband are taken. Figure 2.3 shows the computed gain spectra at resonance based on the design shown in Figure 2.1. In order to evaluate the importance of Coulomb interaction and nonparabolicity on optical properties for THz QCLs, we first compare the gain spectra calculated from the microscopic model with "many-body + nonparabolicity" (considering both many-body and nonparabolicity effects), microscopic model with "many-body + parabolicity" (considering both many-body and parabolicity effects), microscopic model with "free-carrier" (considering free carriers and nonparabolicity but neglecting the bandstructure and Rabi frequency renormalization), and the macroscopic density-matrix model (Reference 34), as shown in Figure 2.3a. Comparing the two gain spectra calculated by the microscopic models with "many-body + nonparabolicity" and "many-body + parabolicity," we found that subband dispersion with different subband effective masses (nonparabolicity) causes a slight shift of peak position of gain spectrum to the lower frequency

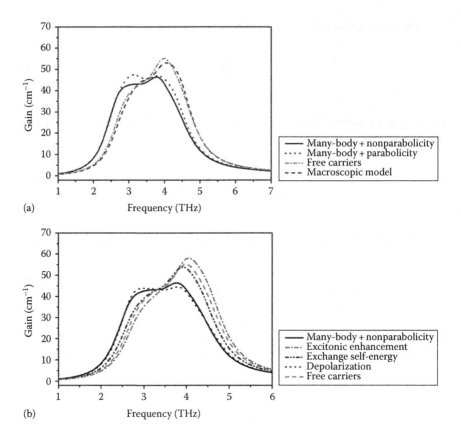

(a)

(b)

FIGURE 2.3 Simulation results for gain spectra at resonance and 100 K. (a) The gain spectra calculated from microscopic model "many-body + nonparabolicity" (solid line), microscopic model "many-body + parabolicity" (dotted lines), microscopic model "free-carriers" (dot-dashed lines), and macroscopic matrix density model (dashed lines). (b) The interplay of different Coulomb interactions. Solid lines, full many-body effects with nonparabolicity; dashed lines, free carriers; dotted lines, depolarization; dot-dashed lines, excitonic enhancement; dot-dot-dashed lines, exchange self-energy.

side and reduces the peak gain due to the redistribution of subband electrons in k space caused by the nonparabolicity effect. Not only the nonparabolicity, but also the many-body Coulomb interactions make the red-shift of gain spectrum and cause the decrease of peak value of optical gain of THz QCLs by comparing microscopic models with "many-body + nonparabolicity" and "free-carrier." The redshift of the gain spectrum is mainly caused by the depolarization terms with the consideration of the interplay of various many-body interactions, as shown in Figure 2.3b. It is shown that, in our population inverted laser system, the depolarization terms cause a red-shift of gain spectrum relative to the "free-carrier" model [35], but it induces the blue-shift in the usual noninverted absorption system. In addition, exchange self-energy terms renormalize the subband energy level and induce nonparabolicity to slightly redshift the gain spectrum, and excitonic enhancement terms give a peak near the higher frequency edge of the spectrum and cause the slight blue-shift of the gain spectrum in THz QCLs. Therefore, the interplay of various many-body effects leads to the red-shift of gain spectrum. In addition, since the macroscopic model does not consider the many-body interactions and nonparabolicity, the obtained peak value and peak position of the optical gain are overestimated when comparing the results obtained from the microscopic "many-body + nonparabolicity" model and the macroscopic model. Moreover, the spectral lineshape, which is important for calculating, for example, the LEF, is not accurately predicted by the macroscopic model. Overall, the different spectral features, for example, different peak values, peak frequency positions, and spectrum lineshapes, from the two models demonstrate that the microscopic density-matrix model can enrich an in-depth understanding of optical properties of THz QCLs and enables a more accurate prediction of the gain spectrum.

2.3.2   Optical Gain Spectrum as a Function of Bias

Figure 2.4 shows the gain spectra as increasing biases calculated from the microscopic model (in the following discussions, unless otherwise specified, "microscopic model" refers to the microscopic model "many-body + nonparabolicity") and macroscopic one. Due to the coherence of RT transport across the injector and extractor barriers, the effects of resonant tunneling in THz QCLs on the broadening mechanism of gain spectrum are complicated. As shown in Figure 2.4, the gain spectrum is broadened with the increase of the applied bias. This broadening results from the contributions of the polarization $p_{31,k}$ and $p_{24,k}$ to polarization $p_{34,k}$ due to the indirect $1 \rightarrow 3$ and $4 \rightarrow 2$ radiative transition formed by the coherent coupling $1 \leftrightarrow 4$, $3 \leftrightarrow 4$ and $3 \leftrightarrow 2$. As the bias increases, levels 1–4 and 3–2 are in resonance, then levels 1–3 and 4–2 become more coherent; hence, $1 \rightarrow 3$ and $4 \rightarrow 2$ radiative transition becomes stronger and contributes the spectrum broadening as the bias increases.

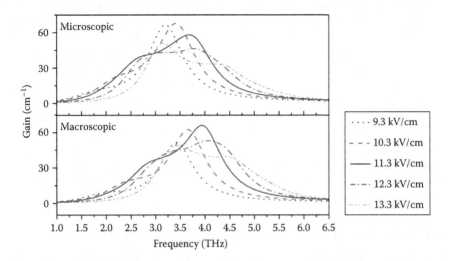

FIGURE 2.4   Effects of bias on gain spectra at 100 K. The designed bias is 12.3 kV/cm.

It is noted that the gain peak frequency calculated by the microscopic model is lower (~0.3 THz) than that by the macroscopic model, as shown in Figure 2.4. Because, as demonstrated in Reference 4, the macroscopic density matrix model overestimates (~0.6 THz) the gain peak frequency compared to the experimentally measured lasing frequency; therefore, the microscopic many-body model predicts a better result closer to the experimental lasing frequency. However, the gain peak frequency calculated by the microscopic model is still ~0.3 THz higher than the experimental value. This discrepancy is probably caused by the neglected intermodule electron–light scattering in this particular density matrix model [4]. Due to the choice of basis states from the two isolated modules (the active and the injector modules) in the "tight-binding" scheme (see Figure 2.1), this model considers only intramodule scatterings and hence only one intramodule dipole moment z_{43} but neglects other direct dipole moment contributions, for example, z_{42}. This limit is not inherent to the density matrix model, but is due to the choice of the basis states. Further work could be carried out to include this intermodule electron–light scattering.

2.3.3 OPTICAL GAIN SPECTRUM AS A FUNCTION OF INJECTION AND EXTRACTION COUPLING STRENGTH

In order to illustrate the effects of coherence of RT, we simulate the gain spectra at different injection, extraction, and parasitic coupling strengths. Figure 2.5 shows the effects of injection coupling strength on optical gain when the extraction coupling strength is set as 4.94 and 2 meV, respectively. Both Figure 2.5a and b show that, as the injection coupling strength gradually increases toward the value of extraction coupling strength, the gain spectrum tends to be enhanced, peak frequency follows the variation of coupling strength, and spectrum width is slightly broadened. When the injection coupling strength is small enough (smaller than extraction coupling strength), double-peaked gain spectrum is generated and obviously shown. Moreover, as the comparison of microscopic model and free-carrier one shows, the many-body interaction tends to suppress the high-frequency side of gain spectrum and enhance the low-frequency side, as shown in Figure 2.5a.

In contrast, since the extraction coupling strength is larger than the injection coupling strength, as the extraction coupling strength increases, as shown in Figure 2.6, the extraction electron transport tends to be more coherent and the gain spectrum is additionally broadened, and the peak value is reduced owing to the interplay of Coulomb interaction and indirectly coherent interaction p_{24}. When the extraction coupling strength is large enough (larger than injection coupling strength), double-peaked gain spectrum is generated. Similarly, owing to modifications to gain spectrum by many-body interactions, the peak at high-frequency side is suppressed and low-frequency side is enhanced according to the results of microscopic model, as compared with free-carrier one.

2.3.4 OPTICAL GAIN SPECTRUM AS A FUNCTION OF DOPING DENSITY

Since many-body Coulomb interaction strongly depends on the doping density, we anticipate that doping will strongly affect the gain spectra, as shown in Figure 2.7.

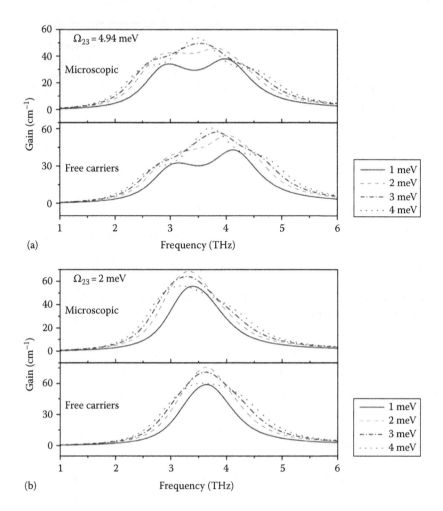

FIGURE 2.5 Effects of injection coupling strength with and without many-body interactions on gain spectra at resonance and 100 K. (a) Gain spectra at different injection coupling strength with the extraction coupling strength of 4.94 meV. (b) Gain spectra at different injection coupling strength with the extraction coupling strength of 2 meV.

As the doping density increases, the many-body interactions become stronger, and hence the spectra are further redshifted. Moreover, not only the peak value with doping density is enhanced by many-body interactions, the lineshape of spectrum is modified, as shown in Figure 2.7b. Figure 2.7c shows the interplay of various Coulomb interactions at doping density 3×10^{10} and 5.4×10^{10} cm^{-2}, respectively. As shown by this figure, the redshift and spectrum lineshape modification are mainly attributed to depolarization, as demonstrated in Reference 36. In contrast, the spectrum lineshape calculated from free-carrier model is not changed with the rise in doping density.

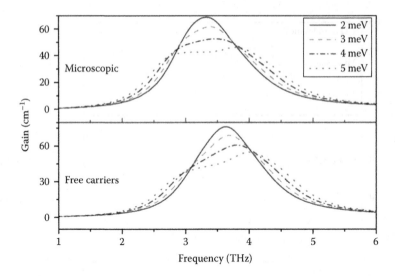

FIGURE 2.6 Gain spectra with and without many-body interactions at different extraction coupling strength at resonance and 100 K.

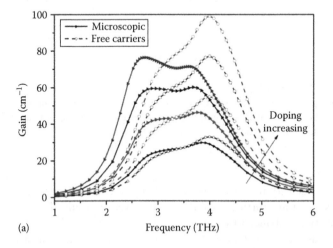

(a)

FIGURE 2.7 (a) Effects of doping density on gain spectra at resonance and 100 K. The solid lines with right triangle and dashed lines with open circles show the results of many-body and free-carrier models, respectively. From the bottom to the top for each kind of color line, the doping density is 1.8×10^{10}, 3×10^{10}, 4.2×10^{10}, and 5.4×10^{10} cm^{-2}. The effects of doping density on lifetimes of energy levels are neglected.   (*Continued*)

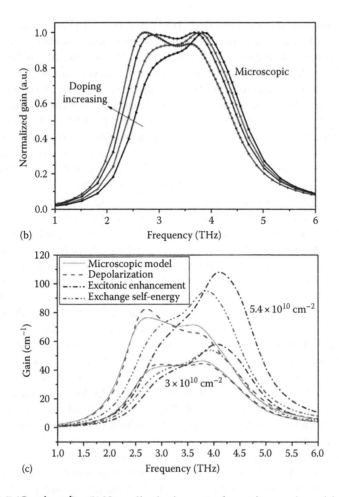

FIGURE 2.7 (*Continued*) (b) Normalized gain spectra from microscopic model at different doping density. (c) The interplay of various Coulomb interactions at doping density 3×10^{10} and 5.4×10^{10} cm^{-2}, respectively. Solid lines, full many-body effects with nonparabolicity; dashed lines, depolarization; dot-dashed lines, excitonic enhancement; dot-dot-dashed lines, exchange self-energy.

2.4 MICROSCOPIC ANALYSIS ON LINEWIDTH ENHANCEMENT FACTOR

LEF, α, plays an important role in determining the optical emission linewidth of the laser systems [37] and the frequency responses [38]. It is defined as the ratio of the carrier-induced variation of the real and imaginary parts of the susceptibility, and describes the amplitude-phase coupling [37]. Owing to this coupling, a variation of the intensity of optical field will induce an excess perturbation of the phase of a laser mode, which influences the performance of lasers, for example, causing the laser linewidth increase well beyond the Schawlow–Townes limit. Typical value of

LEF in interband semiconductor diode lasers is about 2–7 [39], resulting from the asymmetric differential gain spectrum caused by two bands associated with the laser transition with opposite curvature in the k-space. In contrast, due to the intersubband transition characteristics, QCLs are expected to have a narrow and symmetric gain spectrum, hence, according to the Kramers–Kronig relation, resulting in a zero LEF at the peak gain wavelength predicted by Faist et al. [1]. However, the currently reported experiments [40–45] have demonstrated that the LEFs in both mid-IR and THz QCLs are not zero, although they are much smaller than those of diode lasers. For example, an LEF of up to −0.5 and 0.5 at the gain peak was reported at a lasing wavelength of 8.22 μm [40] and ~116 μm (~2.55 THz) [41], respectively. In addition, a strong dependence of the LEF on frequency detuning is observed, and a large value of up to −2 is reported for distributed feedback mid-IR QCLs [44].

Although several experimental works have been carried out in studying the LEF of QCLs, few theoretical investigations on LEF have been reported [45] and the reasons behind the observed nonzero LEF at the gain peak have not been completely understood for both mid-IR and THz QCLs. In the previous work, the nonzero LEF was investigated in mid-IR QCLs by considering the refractive index change due to the device self-heating and the transitions not involved in the laser action based on the experiments and the macroscopic theoretical analysis [45]. The results show that device self-heating is the dominate factor compared with other factors. In addition to the device self-heating effect, other mechanisms such as many-body Coulomb interactions, coherence of RT, and nonparabolicity can cause the nonzero LEF of QCLs at the gain peak, which cannot be considered in the *macroscopic* picture. Furthermore, a theoretical analysis on the LEF of THz QCLs lacks. Since the active region structures of THz QCLs are different from those of mid-IR QCLs, they show different characteristic of nonzero LEF. To disclose more physical underlining mechanisms of nonzero LEF value in THz QCLs, a microscopic model is required. In the earlier section, we developed a microscopic density matrix model to examine the optical gain of THz QCLs demonstrating that the many-body effects, which lead to renormalization of band structure and Rabi frequency, have significant modifications to the gain profile. In this section, we extend this microscopic model to further investigate the role of many-body Coulomb interactions, coherence of RT, and nonparabolicity on the LEF of THz QCLs.

Figure 2.8 shows the α-spectrum at different biases calculated from the microscopic model including the many-body Coulomb interactions, coherence of RT transport and nonparabolicity. The points indicate the positions of the gain peak. In contrast to the macroscopic calculations of zero LEFs, the value of α-spectrum at gain peak cannot be neglected due to the asymmetry of gain spectrum caused by the interplay of many-body interactions, nonparabolicity, and tunneling effects. The obtained LEF is around −0.7 when electric pumping does not exceed the designed bias. The sign of α-value at gain peak reflects the asymmetry of gain spectrum relative to the lasing central frequency ω_λ. It is noted that, when the bias goes above the designed value and becomes 13.3 kV/cm, the LEF becomes −0.3. This is mainly attributed to the symmetric changes of gain spectrum with the extraction detuning due to tunneling effects, that is, the energy splitting due to the coupling between the lower laser level and the extraction level. It has shown that the peak position and

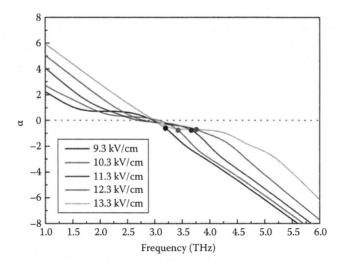

FIGURE 2.8   Linewidth enhancement factor (LEF) including many-body Coulomb interactions, coherence of resonant-tunneling transport, and nonparabolicity for different biases at 100 K. The points indicate the values of LEF at the gain peak.

the lineshape of gain spectrum strongly follow the variations of extraction detuning. When the extraction detuning changes to a positive value from a negative one, the peak frequency is redshifted. Therefore, the absolute value of LEF is reduced. In addition, although the LEF at gain peak is smaller than 1, the LEF shows a large value as the lasing frequency is slightly away from the central frequency, which means that LEF of THz QCLs strongly depends on the frequency detuning.

Figure 2.9 shows the details of α-parameter at frequencies of gain peak, 0.2 THz redshift and 0.2 THz blueshift relative to the peak position under different biases, computed from the microscopic model with "many body + nonparabolicity," microscopic free-carrier model, the macroscopic models with and without RT, respectively. According to the macroscopic models with and without RT, the coherence of RT contributes to a significant increase of the LEF at gain peak. Since tunneling exhibits an increasing broadening and modification for gain spectrum when injection level (level 1) and upper laser level (level 4), and lower laser (level 3) and extraction level (level 2) are in resonance simultaneously, the absolute α value rises with the increasing bias in the regime of negative injection and extraction detunings according to macroscopic model with RT. Once the operation bias exceeds the designed bias, the sign of LEF is changed due to the variation of the symmetry of gain spectrum relative to the lasing central frequency ω_λ. Furthermore, the nonparabolicity can only induce a slight influence on the LEF according to the comparisons between free-carrier model and macroscopic one with RT, but the many body Coulomb interaction causes a large variation of LEF at gain peak with the comparison of the model "many-body + nonparabolicity" and free-carrier one due to its strong modifications to gain spectrum. Overall, the nonzero LEF of THz QCLs is mainly due to the combined impacts from the Coulomb interaction and coherence of RT effects.

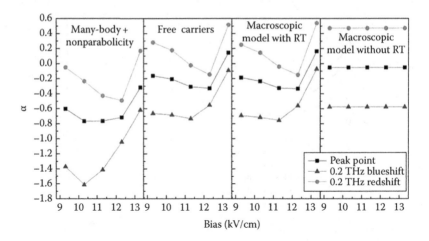

FIGURE 2.9 Linewidth enhancement factor as a function of the applied bias at the gain peak, 0.2 THz redshift, and 0.2 THz blueshift points, respectively, at 100 K. From the left to the right, the curves correspond to microscopic model "many body + nonparabolicity," microscopic free-carrier model, and macroscopic model with and without resonant tunneling, respectively.

2.5 SUMMARY

We have established the Hartree–Fock semiconductor Bloch equations with dephasing and scattering contributions treated at the level of the relaxation-rate approximation, which describes the electron–electron Coulomb interaction, nonparabolicity, and coherence of RT transport. We use the developed model to investigate the optical gain and LEF of THz QCLs. The simulation results calculated from the microscopic model with "many-body + nonparabolicity," the microscopic model with "many-body + parabolicity," the microscopic model with "free-carrier," and the macroscopic model are compared to demonstrate the importance of those parameters in the simulation of optical properties of THz QCLs. The effects of the external bias, injection and extraction coupling strength, and doping density on optical gain are also systematically investigated. The results show that the gain peak frequency calculated by the many-body model is closer to the experimentally measured lasing frequency, compared with the macroscopic model. The interplay of the many-body interactions, coherence of RT, and nonparabolicity plays an important role in the nonzero LEF of THz at gain peak. It can partially explain the nonzero LEF at the gain peak, which existed in the field for a while but cannot be explicitly explained.

The many-body model discussed in this chapter provides a relatively comprehensive picture of optical properties of THz QCLs and enables a more accurate and faster prediction and calculation of the device performance, for example, gain spectra and LEF. It also provides an essential design tool for the optimization of the quantum wells/barriers thicknesses to improve the performance of THz QCLs, such as the maximum operation temperature. There are still certain aspects in which the model can be further improved. First, this model takes into account dephasing mechanisms in laser transition and transport coherence, but only considers them under the relaxation-rate approximation using phenomenological values, where the nondiagonal scattering contributions are neglected. Nondiagonal scattering contribution is expected to yield not only a smaller linewidth but also a reshaping of the gain spectrum. A more stringent treatment of the dephasing parameters is required for future theoretical studies. Another limitation is that the present implementation of many-body model considers only intramodule electron–light scatterings but neglects intermodule dipole moment contributions due to the choice of the basis states from the isolated modules in the "tight-binding" scheme. Further study is required to include intermodule electron–light scattering.

REFERENCES

1. J. Faist, F. Capasso, D. L. Sivco, C. Sirtori, A. L. Hutchinson, and A. Y. Cho, Quantum cascade laser, *Science* **264**, 553–556 (1994).
2. R. Kohler, A. Tredicucci, F. Beltram, H. E. Beere, E. H. Linfield, A. G. Davies, D. A. Ritchie, R. C. Iotti, and F. Rossi, Terahertz semiconductor-heterostructure laser, *Nature* **417**, 156–159 (2002).
3. B. S. Williams, Terahertz quantum-cascade lasers, *Nat. Photon.* **1**, 517–525 (2007).
4. S. Fathololoumi, E. Dupont, C. W. I. Chan, Z. R. Wasilewski, S. R. Laframboise, D. Ban, A. Mátyás, C. Jirauschek, Q. Hu, and H. C. Liu, Terahertz quantum cascade lasers operating up to ~200 K with optimized oscillator strength and improved injection tunneling, *Opt. Express* **20**, 3866–3876 (2012).
5. H. Callebaut, S. Kumar, B. S. Williams, Q. Hu, and J. L. Reno, Analysis of transport properties of tetrahertz quantum cascade lasers, *Appl. Phys. Lett.* **83**, 207–209 (2003).
6. C. Jirauschek and P. Lugli, Monte-Carlo-based spectral gain analysis for terahertz quantum cascade lasers, *J. Appl. Phys.* **105**, 123102 (2009).
7. A. Matyas, M. A. Belkin, P. Lugli, and C. Jirauschek, Temperature performance analysis of terahertz quantum cascade lasers: Vertical versus diagonal designs, *Appl. Phys. Lett.* **96**, 201110 (2010).
8. Y. J. Han, W. Feng, and J. C. Cao, Optimization of radiative recombination in terahertz quantum cascade lasers for high temperature operation, *J. Appl. Phys.* **111**, 113111 (2012).
9. P. Borowik, J. L. Thobel, and L. Adamowicz, Monte Carlo based microscopic description of electron transport in GaAs/Al$_{0.45}$Ga$_{0.55}$As quantum-cascade laser structure, *J. Appl. Phys.* **108**, 073106 (2010).
10. T. Kubis, C. Yeh, P. Vogl, A. Benz, G. Fasching, and C. Deutsch, Theory of nonequilibrium quantum transport and energy dissipation in terahertz quantum cascade lasers, *Phys. Rev. B* **79**, 195323 (2009).
11. S. C. Lee and A. Wacker, Nonequilibrium Green's function theory for transport and gain properties of quantum cascade structures, *Phys. Rev. B* **66**, 245314 (2002).
12. T. Liu, T. Kubis, Q. J. Wang, and G. Klimeck, Design of three-well indirect pumping terahertz quantum cascade lasers for high optical gain based on nonequilibrium Green's function analysis, *Appl. Phys. Lett.* **100**, 122110 (2012).
13. T. Kubis, S. R. Mehrotra, and G. Klimeck, Design concepts of terahertz quantum cascade lasers: Proposal for terahertz laser efficiency improvements, *Appl. Phys. Lett.* **97**, 261106 (2010).
14. T. Schmielau and M. F. Pereira, Nonequilibrium many body theory for quantum transport in terahertz quantum cascade lasers, *Appl. Phys. Lett.* **95**, 231111 (2009).
15. R. Terazzi and J. Faist, A density matrix model of transport and radiation in quantum cascade lasers, *New J. Phys.* **12**, 033045 (2010).

16. S. Kumar and Q. Hu, Coherence of resonant-tunneling transport in terahertz quantum-cascade lasers, *Phys. Rev. B* **80**, 245316 (2009).
17. E. Dupont, S. Fathololoumi, and H. C. Liu, Simplified density-matrix model applied to three-well terahertz quantum cascade lasers, *Phys. Rev. B* **81**, 205311 (2010).
18. H. Haug and S. W. Koch, *Quantum Theory of the Optical Electronic Properties of Semiconductors* (World Scientific Publishing, Singapore, 2009).
19. W. W. Chow, S. W. Koch, and M. Sargent, III, *Semiconductor-Laser Physics* (Springer-Verlag, Berlin, Germany, 1994).
20. I. Waldmueller, W. W. Chow, E. W. Young, and M. C. Wanke, Nonequilibrium many-body theory of intersubband lasers, *IEEE J. Quant. Electron.* **42**, 292–301 (2006).
21. F. Wang, X. G. Guo, and J. C. Cao, Many-body interaction in resonant tunneling of terahertz quantum cascade lasers, *J. Appl. Phys.* **108**, 083714 (2010).
22. G. Beji, Z. Ikonic, C. A. Evans, D. Indjin, and P. Harrison, Coherent transport description of the dual-wavelength ambipolar terahertz quantum cascade laser, *J. Appl. Phys.* **109**, 013111 (2011).
23. T. Liu, K. E. Lee, and Q. J. Wang, Microscopic density matrix model for optical gain of terahertz quantum cascade lasers: Many-body, nonparabolicity, and resonant tunneling effects, *Phys. Rev. B* **86**, 235306 (2012).
24. R. F. Kazarinov and R. A. Suris, Electric and electromagnetic properties of semiconductors with a superlattice, *Sov. Phys. Semicond.* **6**, 120 (1972).
25. G. Bastard, *Wave Mechanics Applied to Semiconductor Heterostructures* (Les Editions de Physique, Paris, France, 1988).
26. D. E. Nikonov, A. Imamoglu, L. V. Butov, and H. Schmidt, Collective intersubband excitations in quantum wells: Coulomb interaction versus subband dispersion, *Phys. Rev. Lett.* **79**, 4633–4636 (1997).
27. J. T. Lu and J. C. Cao, Coulomb scattering in the Monte Carlo simulation of terahertz quantum-cascade lasers, *Appl. Phys. Lett.* **89**, 211115 (2006).
28. M. F. Pereira, S. C. Lee, and A. Wacker, Controlling many-body effects in the midinfrared gain and terahertz absorption of quantum cascade laser structures, *Phys. Rev. B* **69**, 205310 (2004).
29. O. Bonno, J. L. Thobel, and F. Dessenne, Modeling of electron-electron scattering in Monte Carlo simulation of quantum cascade lasers, *J. Appl. Phys.* **97**, 043702 (2005).
30. I. Waldmueller, J. Forstner, and A. Knorr, *Nonequilibrium Physics at Short Time Scales* (Springer, Berlin, Germany, 2004).
31. T. Liu, K. E. Lee, and Q. J. Wang, Importance of the microscopic effects on the linewidth enhancement factor of quantum cascade lasers, *Opt. Express* **21**, 27804 (2013).
32. U. Ekenberg, Nonparabolicity effects in a quantum well: Sublevel shift, parallel mass, and Landau levels, *Phys. Rev. B* **40**, 7714–7726 (1989).
33. W. W. Chow, H. C. Schneider, S. W. Koch, C. H. Chang, L. Chrostowski, and C. J. Chang-Hasnain, Nonequilibrium model for semiconductor laser modulation response, *IEEE J. Quant. Electron.* **38**, 402–409 (2002).
34. E. Dupont, S. Fathololoumi, and H. C. Liu, Simplified density-matrix model applied to three-well terahertz quantum cascade lasers, *Phys. Rev. B* **81**, 205311 (2010).
35. H. C. Liu and A. J. SpringThorpe, Optically pumped intersubband laser: Resonance positions and many-body effects, *Phys. Rev. B* **61**, 15629–15632 (2000).
36. M. V. Kisin, M. A. Stroscio, G. Belenky, and S. Luryi, Electron-plasmon relaxation in quantum wells with inverted subband occupation, *Appl. Phys. Lett.* **73**, 2075–2077 (1998).
37. C. H. Henry, Theory of the linewidth of semiconductor-lasers, *IEEE J. Quant. Electron.* **18**, 259–264 (1982).
38. T. Chattopadhyay and P. Bhattacharyya, Role of linewidth enhancement factor on the frequency response of the synchronized quantum cascade laser, *Opt. Commun.* **309**, 349–354 (2013).

39. M. Osinski and J. Buus, Linewidth broadening factor in semiconductor-lasers—An overview, *IEEE J. Quant. Electron.* **23**, 9–29 (1987).
40. M. Lerttamrab, S. L. Chuang, C. Gmachl, D. L. Sivco, F. Capasso, and A. Y. Cho, Linewidth enhancement factor of a type-I quantum-cascade laser, *J. Appl. Phys.* **94**, 5426–5428 (2003).
41. R. P. Green, J. H. Xu, L. Mahler, A. Tredicucci, F. Beltram, G. Giuliani, H. E. Beere, and D. A. Ritchie, Linewidth enhancement factor of terahertz quantum cascade lasers, *Appl. Phys. Lett.* **92**, 071106 (2008).
42. T. Aellen, R. Maulini, R. Terazzi, N. Hoyler, M. Giovannini, J. Faist, S. Blaser, and L. Hvozdara, Direct measurement of the linewidth enhancement factor by optical hetero-dyning of an amplitude-modulated quantum cascade laser, *Appl. Phys. Lett.* **89**, 091121 (2006).
43. J. von Staden, T. Gensty, W. Elsäßer, G. Giuliani, and C. Mann, Measurements of the α factor of a distributed-feedback quantum cascade laser by an optical feedback self-mixing technique, *Opt. Lett.* **31**, 2574–2576 (2006).
44. N. Kumazaki, Y. Takagi, M. Ishihara, K. Kasahara, A. Sugiyama, N. Akikusa, and T. Edamura, Detuning characteristics of the linewidth enhancement factor of a midin-frared quantum cascade laser, *Appl. Phys. Lett.* **92**, 121104 (2008).
45. J. Kim, M. Lerttamrab, S. L. Chuang, C. Gmachl, D. L. Sivco, F. Capasso, and A. Y. Cho, Theoretical and experimental study of optical gain and linewidth enhancement factor of type-I quantum-cascade lasers, *IEEE J. Quant. Electron.* **40**, 1663–1674 (2004).

3 Photonic Engineering of Terahertz Quantum Cascade Lasers

Qi Jie Wang and Guozhen Liang

CONTENTS

3.1 Introduction 61
3.2 Concentric-Circular-Grating THz QCLs 62
    3.2.1 Grating Design 63
    3.2.2 Fabrication, Experimental Results, and Analysis 68
3.3 Plasmonic Waveguide Structures for THz QCLs 71
    3.3.1 In-Plane Integration of the SSP Structure with THz QCL 73
    3.3.2 Fabrication, Experimental Results, and Analysis 75
References 81

3.1 INTRODUCTION

The terahertz (THz) frequency range, lying between the microwave and the mid-infrared frequencies (~0.3–10 THz), remains one of the least-developed regions in the electromagnetic spectrum. THz radiation has, however, proven potentially widespread and important applications in, for example, spectroscopy, heterodyne detection, imaging, and communications [1]. In many of these applications, suitable high-power coherent THz sources with low beam divergence and single-mode operation are highly desirable. Therefore, much emphasis has been placed on the development of new and appropriate THz sources, and in particular with a focus on the development of electrically pumped, semiconductor-based light sources that can be mass-produced. A breakthrough occurred in 2002, when the first THz quantum cascade lasers (QCLs) were demonstrated [2].

QCLs are semiconductor lasers comprising multiple strongly coupled quantum wells, exploiting electron transitions between subbands of the conduction band. By adjusting the widths of the quantum wells/barriers, the emission frequency and performance of the laser can be tailored, adding additional flexibility in the laser design. Without the assistance of an external magnetic field, THz QCLs have covered the frequency range from 1.2 to 5 THz [3]. With respect to the optical power, THz QCLs can provide over 240 mW peak power in pulse mode, and 130 mW in continuous-wave, at 10 K heatsink temperature, in devices using a single-plasmon waveguide [4].

Although this waveguide design gives high output power, the developed metal–metal waveguide, where the THz QCL active region is sandwiched between two metal plates, enables a better temperature performance to be achieved (~200 K) [5]. However, owing to the subwavelength optical mode confinement of the metal–metal waveguide, conventional edge-emitting ridge THz QCLs are characterized by an extremely wide beam divergence (~180°), and a low output power, which is a result of a large impedance mismatch between the modes inside the cavity and in free space. Moreover, without a mechanism for mode-selection, ridge-waveguide lasers suffer from multiple-mode operation. These problems impede the practical applications of THz QCLs.

Various approaches have been developed to address these challenges, involving both edge-emitting and surface-emitting devices. For the former, early works employed additional optical components, such as silicon lenses [6] and horn antennas [7,8] to collimate the divergent output beam. These solutions are not easily integrated with the laser and require careful alignment, but Yu et al. subsequently developed an integrated spoof surface plasmon (SSPs) collimator for THz QCLs, and successfully reduced the beam divergence to ~12° [9,10]. None of these techniques provide optical mode selection, only optical collimation and beam engineering. However, in 2009, Amanti et al. showed that a third-order distributed feedback (DFB) grating could simultaneously enable single-mode operation and collimate the edge-emitting beam tightly (~10°) [11]. For surface-emitting devices, linear second-order DFB grating THz QCLs have been demonstrated to couple the optical power vertically out of a laser cavity [12,13]. However, although the beam divergence is narrow along the direction parallel to the ridge axis, it is still highly divergent perpendicular to the ridge axis. Therefore, to achieve a narrow and symmetric far-field profile in two dimensions, some researchers tried bending the laser ridges into rings, resulting in second-order DFB ring-grating THz QCLs [14,15]. Nevertheless, the reported far-field patterns of these structures remained distorted. As an alternative approach, 2D photonic crystals were investigated for controlling the surface emission from THz QCLs [16–18]. This allows the far-field profile to be engineered and provides controllable emission properties, and a large emission area. Indeed, single-mode devices with a range of far-field patterns have been reported, and beam divergences as low as ~12° × 8° have been obtained [18]. However, it is not simple to achieve single-mode operation with these photonic crystal cavities because the design usually relies on the band structure of a unit cell of the photonic crystal with periodic boundary conditions, without considering the actual finite scale of the device. Moreover, it is not easy to predict and control the radiation loss (power) of the device without 3D full-wave simulations of the whole structure, which are often computationally intensive.

3.2 CONCENTRIC-CIRCULAR-GRATING THz QCLs

Compared with photonic crystal structures, concentric-circular-grating (CCG) is simpler both in design and in analysis, thanks to its axisymmetric geometry, whereas it possesses all the advantages of the photonic crystals. As it will be shown in the following text, the design can be undertaken through 2D numerical simulations, taking into account the boundary condition of the actual finite structure. Moreover, the emission loss (power) is predictable.

3.2.1 GRATING DESIGN

The scheme for implementing the CCG QCLs is shown in Figure 3.1a, where the 10 μm thick active region is sandwiched between the metallic CCGs on top and a metal plate on the bottom. To achieve surface emitting, the gratings are designed to be second order. Inside the active region, the electric field profile can be approximately expressed as a superposition of Bessel functions of the first and second kind, J_m and Y_m, respectively [19]:

$$E_z(r) = [A(r)J_m(kr) + B(r)Y_m(kr)]e^{im\varphi}, \tag{3.1}$$

where
- $A(r)$ and $B(r)$ are the slowly varying mode amplitudes
- k is the wave vector in the active region
- φ is the azimuthal coordinate
- $m = 0, 1, 2, \ldots$, denotes the order of the Bessel functions and also the order of the azimuthal modes

The design of the structure was undertaken through numerical simulations using a commercial finite element method solver: *Comsol Multiphysics*. This solver is very effective in finding the eigenfrequencies and mode distributions of a resonator, although 3D full-wave simulations of the proposed structure are computationally intensive, and thus, it is impractical to perform 3D simulations throughout the design. However, we can take advantage of the mode's known azimuthal dependence ($e^{im\varphi}$) and simplify the problem to a 2D cross-sectional simulation, as shown in Figure 3.1b; the magnified views show the widths of the metal in each period and the boundary region without metal coverage, denoted as R_1, R_2, $\ldots$,

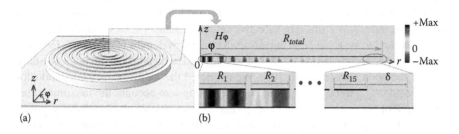

(a)   (b)

FIGURE 3.1 (a) Three dimensional schematic representation of the THz concentric-circular-grating quantum cascade laser design. The semitransparent rectangle represents the 2D simulation plane. (b) Two dimensional simulated magnetic field distribution making use of the 360° rotational symmetry of the designed structure. The magnified views show the widths of the metal in each period and the boundary region without metal coverage, indicated as R_1, R_2, $\ldots$, R_{15}, and δ, respectively. The black lines represent metal.

and δ, respectively. Using the 2D "partial differential equation (PDE) mode" of the *Comsol Multiphysics*, which allow the simulation based on user's own equations, one is able to calculate the optical modes with arbitrary azimuthal order (*m*) by solving the following equations:

$$\nabla \times \left(\epsilon^{-1} \nabla \times \overrightarrow{H} \right) - \omega^2 c^{-2} \overrightarrow{H} = 0 \qquad (3.2)$$

$$\overrightarrow{H}(r,z,\varphi) = \overrightarrow{H}(r,z)e^{-im\varphi} \qquad (3.3)$$

where
  $\overrightarrow{H}$ is the magnetic field
  ϵ is the relative permittivity
  c is the speed of light in vacuum
  ω is the angular frequency of the light

Without loss of generality, we restrict our design to $m = 0$ (the fundamental azimuthal mode) because the surface-emitting beams of these modes give radial polarization, which is highly desirable when coupling light to a THz metal-wire waveguide or the metallic tip of a near-field imaging system.

The structure was designed to operate at ~3.75 THz. At this frequency, the real part of the refractive index of the active region was calculated to be ~3.6, from the emission spectra of a ridge laser with the same active region, using $n_{active} = c/(2L\Delta f)$, where L is the length of the laser ridge and Δf is the Fabry–Pérot mode spacing in the spectra. To calculate the emission loss only, the imaginary part of the refractive index of the active region was set to zero and the metal was considered to be perfect. The 2D structure was then surrounded by an absorbing boundary that can isotropically absorb the emitted radiation without reflection.

The design began by determining the grating period and the widths of the open slits. For large area electric pumping via one or a few connected wires, it is important to guarantee uniform current injection over the pumping area, otherwise the mode behavior of the QCL may be disturbed. Therefore, the width of each open slits on the grating was predetermined as 2 μm. The effective refractive index of the CCG structure was then estimated to be $n_{eff} = 3.57$, and so for a target wavelength of 80 μm (3.75 THz), a second-order grating period of 22.7 μm is required, which makes the mark-to-space ratio of the open slits less than 10%. We restricted the radius of the semiconductor active region R_{total} to be ~16 grating periods, taking into account the maximum current output of commercial power supplies and the heat extraction capacity of a typical cryostat. To suppress unwanted whispering-gallery-like modes, an annular boundary region of width δ = 22.7 μm was left uncovered by metal [20].

The grating was initially designed as a standard second-order DFB grating. The whole grating structure starting from the center to the boundary is as follows: 20.7/**2**/20.7/**2**/20.7/**2**/20.7/**2**/20.7/**2**/20.7/**2**/20.7/**2**/20.7/**2**/20.7/**2**/20.7/**2**/20.7/**2**/20.7/**2**/20.7/**2**/20.7/**2**/22.7/**22.7** μm, where the bold number indicates the open region without metal coverage. The corresponding electromagnetic field distributions and spectra

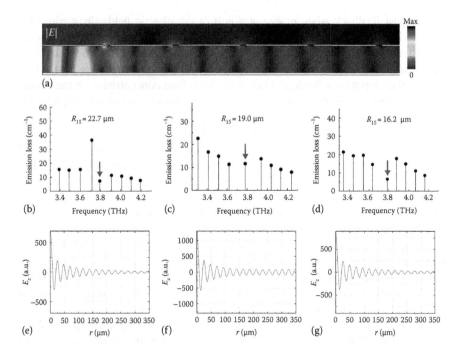

FIGURE 3.2 (a) Electric field distribution of the standard second-order DFB CCG THz QCL in the central region. (b–d) mode spectra of the structures for different R_{15} with other parameters unchanged. (e–g) The corresponding electric field (E_z) profiles of the expected modes, highlighted by the red arrows in (b–d), respectively, obtained at the middle height of the active region.

are shown in Figure 3.2a and b, respectively. A red arrow highlights the mode with the lowest loss, which is expected to be the lasing mode. However, since neighboring modes have similar predicted losses, this structure may not achieve single-mode operation. In linear DFB gratings, the width of the grating boundary plays a critical role in determining the mode behavior [12], and this also holds for our CCG design here. As shown in Figure 3.2c and d, by varying R_{15} while keeping all other parameters unchanged, the corresponding mode spectrum changes significantly. The best result was achieved when $R_{15} = 16.2$ μm and correspondingly, $R_{total} = 356.7$ μm, as the expected lasing mode was the most distinct from neighboring ones. In the modifications that will be discussed in the following texts, δ, R_{15}, and R_{total} are left unchanged.

To exploit the gain of the active region fully, one should consider the electric field profile inside the cavity, with an evenly distributed electric field profile being desirable. A uniform electric field distribution also helps improve the beam divergence. We thus investigated the E_z (z-polarized electric field) profile inside the cavity, as the QCL active region can only provide gain for the TM mode (Figure 3.2e through g). For our initial designs, it can be seen that E_z is concentrated in the central region (also shown in Figure 3.2a). This is because the first few slits in the central region provide a very strong feedback. In fact, the strength of feedback and the emission property of a slit depend critically on the location of the slit relative to the electric field: the slit provides the strongest feedback when placed at a null of the electric field, albeit with high emission loss from this slit. In contrast, a slit placed at an extremum of the electric field provides the weakest feedback and the lowest emitting loss. From Figure 3.2a, one can see that the first few slits are located near the nulls of the electric field, which explains their strong feedback and hence the high field concentration in the central region. To mitigate this, the positions of the first few slits were systematically adjusted to reduce their feedback.

Figure 3.3 shows the simulated results following adjustment of slit locations. The slits are numbered outward from the center as slits 1, 2, 3.... First, slits 1 and 2 were moved outward by 3.4 μm; this aligned them with the maximum of the electric field (Figure 3.3a(i)). E_z thus becomes more evenly distributed along the radius (Figure 3.3a(iii)), compared with Figure 3.2g, and the expected lasing mode is also more distinct from others (Figure 3.3a(ii)) because of the reduction in emission loss after this adjustment, as a result of the suppressed light emission from slits 1 and 2. To optimize the structure further, slit 3 was moved outward by 1.9 μm (Figure 3.3b), and then slits 4 and 5 were moved outward by 0.7 μm (Figure 3.3c). Figure 3.3c shows the final design. It has the greatest mode distinction, and most evenly distributed E_z field. Furthermore, the calculated emission loss is 3.9 cm^{-1}, comparable to that of a typical ridge THz QCL. The grating parameters starting from the center are 24.1/**2**/20.7/**2**/19.2/**2**/19.5/**2**/20.7/**2**/20/**2**/20.7/**2**/20.7/**2**/20.7/**2**/20.7/**2**/20.7/**2**/20.7/**2**/20.7/**2**/16.2/**22.7** μm, where the bold number indicates the open regions (slit or boundary region).

Since the CCG is designed for an electrically pumped QCL, a three-spoke bridge structure was employed to connect the concentric rings, as illustrated in Figure 3.4a, and to enable electric pumping of the whole grating.

The design discussed earlier considers only the fundamental azimuthal modes. However, higher azimuthal modes ($m \geq 1$) are also eigen-solutions of the circular structure, similar to the higher lateral modes in a ridge laser. To take into account these higher azimuthal modes and, more importantly, the influence of the three-spoke bridge, we performed a 3D full-wave simulation of the final structure. The computation took over 35 h and 180 GB RAM, running parallelly on a computing cluster node with 16 processes. Figure 3.4b presents the mode spectra of the first four azimuthal modes; modes $m > 4$ are not shown because they have higher loss. It can be seen that there usually exists a second-order azimuthal mode accompanying each fundamental mode with a similar loss. As the electric field in the active region can be described as a superposition of the Bessel functions of the first and second kind (Equation 3.1), we compare the fundamental ($J_0(r)$, $Y_0(r)$) and second-order ($J_2(r)$, $Y_2(r)$) Bessel functions in Figure 3.4c and d. From Figure 3.4c, we can see that the $-J_2(r)$ almost coincides with $J_0(r)$ except at the first two or three peaks, and the same is true for $-Y_2(r)$ and $Y_0(r)$ (Figure 3.4d). Therefore, the CCG provides similar feedback and emission loss for the fundamental and second-order azimuthal modes, and hence, the second-order azimuthal modes accompany the fundamental azimuthal modes, with similar emission losses. Nevertheless, the CCG is specifically optimized for the expected fundamental mode, and it consistently has a lower loss than the corresponding second-order mode. It should also

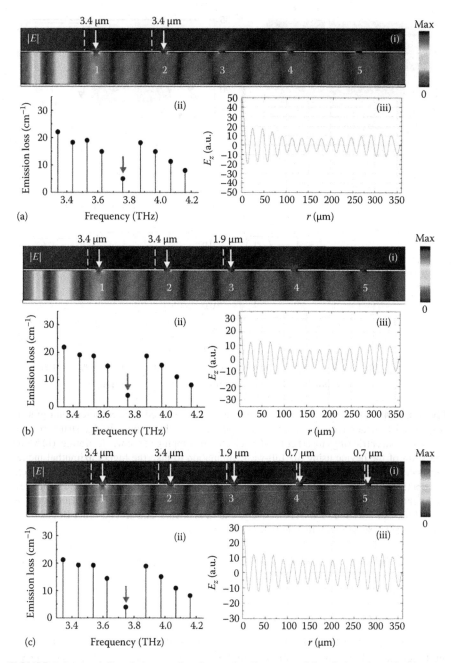

FIGURE 3.3 (a–c) Simulation results after each adjustment of the slit locations. (i) Electric field (|E|) distribution of the expected mode at the central part of the grating. The first five slits are numbered 1, 2, ..., 5, and their locations are indicated by the white arrows. White dashed lines near the arrows indicate the original locations of the slits. The displacement distance of each slit is given above the arrow. (ii) Mode spectrum of the adjusted structure, a arrow indicates the expected mode. (iii) The electric field (E_z) profile.

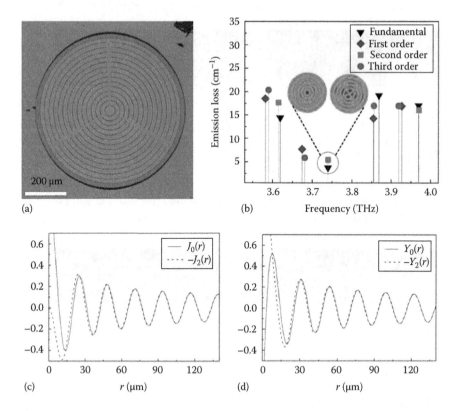

FIGURE 3.4 (a) Optical microscope image of a fabricated concentric-circular-grating distributed feedback terahertz quantum cascade laser. A three-spoke bridge structure connects the concentric rings together to allow electrical pumping of the whole grating. (b) Mode spectrum of the device using 3D full-wave simulation to take the higher azimuthal modes into account. Only the first four lowest-order azimuthal modes are plotted. The electric field distributions of the modes with lowest losses (the fundamental and second azimuthal modes, highlighted by a circle) are shown in the inset. (c, d) Comparison of the Bessel function $J_0(r)$ and $-J_2(r)$, and $Y_0(r)$ and $-Y_2(r)$, respectively. The fundamental and second-order Bessel functions coincide except in the first few peaks.

be noted, from the inset of Figure 3.4b, that the perturbation of the three-spoke structure on the optical modes in the active region is minor.

3.2.2 FABRICATION, EXPERIMENTAL RESULTS, AND ANALYSIS

The active region of the THz QCLs used for this work is a regrowth of that reported in Reference 21 but with a slightly higher doping, whose heterogeneous GaAs/$Al_{0.15}Ga_{0.85}$, as structure was grown on a semi-insulating GaAs substrate by molecular beam epitaxy. A Ti/Au (15/550 nm) layer was deposited on both the active wafer and a receptor n+ GaAs wafer by electron-beam evaporation. The active wafer was inverted onto the receptor wafer, carefully aligning the crystal orientation of the wafer pair to facilitate cleaving. The wafer pairs were bonded together under application of pressure (~5.4 MPa) at 300°C for 1 hour in a vacuum environment. The original substrate of the active wafer was then removed by lapping and selective chemical etching; this was followed by the removal of the highly absorptive contact layer of the active region using a H_2SO_4:H_2O_2:H_2O (1:7:80) solution, to prevent significant loss for the THz radiation coupled out through the grating slits. Top metal gratings (Ti/Au 15/200 nm) were defined by standard optical lithography and lift-off, after which the active region was wet-etched to form a disk structure using a H_3PO_4:H_2O_2:H_2O (1:1:10) solution with photoresist (AZ5214) used as the mask. The backside substrate was thinned down to ~150 μm to improve heat dissipation, and a Ti/Au (20/300 nm) layer was employed for the bottom contact. The sample was then cleaved, mounted onto Cu submounts, wire bonded, and finally put on the cold finger of a cryostat for measurement. Figure 3.4a shows the optical microscope image of a fabricated device.

Light–current–voltage characterization was carried out at different heatsink temperatures under pulsed mode operation (200 ns pulses at 10 kHz repetition), and results are shown in Figure 3.5a. The devices operate up to 110 K. For comparison, the maximum operation temperature, under the same operation conditions, of double-metal waveguide, ridge lasers of similar size fabricated from the same epitaxial wafer is around 130 K. Figure 3.5b shows the emission spectra of the device at a 9 K heatsink temperature for various injection currents, ranging from threshold to the rollover. The detected side-mode suppression ratio is around 30 dB, limited by the noise floor of the room-temperature pyroelectric detector, as shown in the inset of Figure 3.5b. This, together with the robust single-mode operation at 3.73 THz under all injected currents, reflects the strong single-mode selectivity of the structure. Moreover, the collected power of the device is ~5 times higher than that of its ridge laser counterpart, indicating efficient light coupling out of the cavity by the grating.

To investigate the electric field distribution in the grating cavity, the 2D far-field emission pattern was measured. The radiation emitted by the device was sampled through a small aperture in front of a parabolic mirror, which was scanned over part of a spherical surface centered on the device. The parabolic mirror reflected the radiation into a liquid helium–cooled bolometer for detection. The small aperture was used to increase the measurement angular resolution. The measured far-field pattern (Figure 3.6a) exhibits double lobes, with full-width-at-half-maximum values of 12.5° × 6.5° and 13.5° × 7°. Similar results were obtained in several devices. This implies that the mode in the CCG resonator is not the expected fundamental mode, which should have a ring-shaped far-field pattern (Figure 3.6b). Based on the method described in Reference 22, the calculated far-field profiles of the fundamental, first-order, and second-order azimuthal modes are shown in Figure 3.6b through d, respectively; none has a double-lobe far-field profile.

It was found that the deviation of the far-field pattern from the expected ring shape is due to an anisotropic sidewall profile caused by wet chemical etching. As shown in Figure 3.7a, the wet-etched sidewall is more vertical for the upper and lower boundaries than the left and right boundaries, which makes the shape of the disk slightly elliptical (the conjugate radius along the y-axis is 356.7 μm, and the transverse radius along the x-axis is 358.7 μm). It should be noted that this deformation actually stems from the anisotropic crystal structure of the

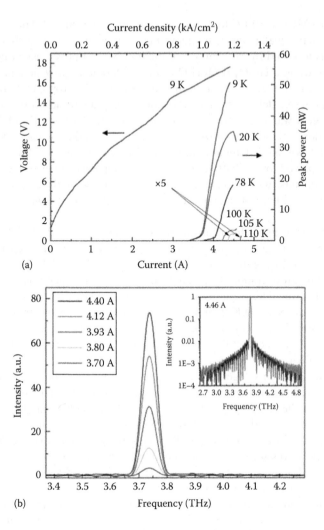

FIGURE 3.5 (a) Light–current–voltage characterization of the concentric-circular-grating device at different heatsink temperatures under pulsed mode operation. (b) Emission spectra of the device at 9 K heatsink temperature as a function of injected current, from threshold to the rollover. The inset shows a logarithmic scale plot of the spectrum, demonstrating a side-mode suppression ratio of around 30 dB.

active region semiconductor, which makes the wet-etched sidewall profiles and the undercuts different along the different crystallographic directions. Therefore, all the devices have similar, if not the same, boundary deformation. Although slight, this deformation breaks the circular symmetry of the fundamental mode (Figure 3.7b), and consequently, the simulated far-field pattern is as shown in Figure 3.7b, which agrees well with the measured result. It is worth mentioning that although the deformation of the boundary has such an effect on the funda-mental azimuthal modes, the simulations show that there is negligible effect on

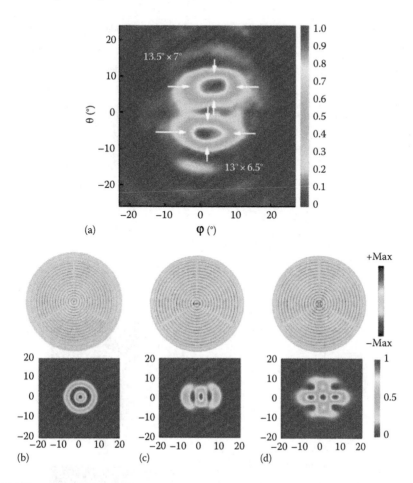

FIGURE 3.6 (a) Measured far-field pattern of a typical terahertz concentric-circular-grating quantum cascade laser. (b–d) Electric field distributions and the corresponding far-field patterns for the fundamental, first-order, and second-order azimuthal modes, respectively.

the higher azimuthal modes. This is because higher-order azimuthal modes have a lower degree of circular symmetry and are less sensitive to this deformation.

3.3 PLASMONIC WAVEGUIDE STRUCTURES FOR THz QCLs

Surface plasmons (SPs) are bound surface electromagnetic waves existing at the interface between metallic and dielectric materials. With the flexibility of confining and manipulating the optical wave at a subwavelength level by structuring the metallic surface, SPs have boosted numerous studies in compact photonic circuits [23], enhanced light-matter interactions [24], near-field imaging system [25,26], and beam shaping [10,27,28], etc. However, although great success have been achieved in the optical region ranging from visible to mid-infrared, the concept of SPs fails

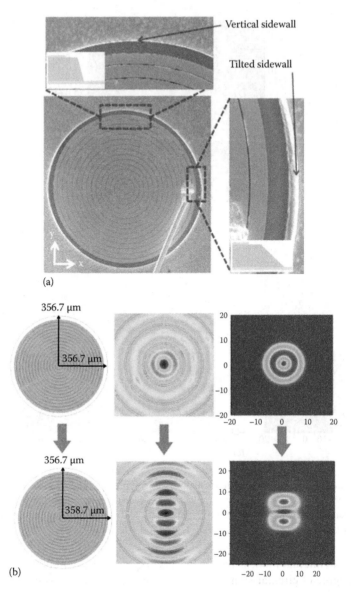

FIGURE 3.7 (a) Scanning electron-beam microscope images of the anisotropic sidewall profiles caused by wet chemical etching. The sidewalls of the upper and lower boundaries are more vertical than those of the left and right boundaries. (b) The change of electric field distribution and far-field pattern caused by a slight (2 μm) deformation of the circular active region disk.

in the THz region, where the metal behaves more like a perfect conductor and, thus, the penetration depth of the THz field into the metal is negligible (two or three orders of magnitude shorter than the wavelength), which renders the electromagnetic field loosely bound onto the flat surface. Fortunately, it has been recently found that metal surface corrugated at subwavelength scales can support tightly confined THz surface waves, just like the SPs behavior at shorter wavelength [29–37]. Such surface waves are hence called "spoof" surface plasmons (SSPs). As mentioned previously, SPP structure has been fabricated on the facet of THz QCLs for collimation purpose, achieving a beam divergence of ~11.7° × 16° [9]. The drawbacks of this scheme are the limited area of the laser facet and the incompatibility of the fabrication with conventional optical lithography, all of which limit the further application of the SSP structure. The more promising method is to integrate SSP structures planarly with THz QCLs. This is possible as QCLs are intrinsically transverse magnetic (TM)-polarized, which matches with the polarization of the SSPs. Therefore, one can directly and efficiently couple the THz wave out of the laser cavity as SSP waves using a planar integrated SSP waveguide, which can then be manipulated by employing the well established SP techniques. More importantly, since all the structures are planar, this scheme provides a platform where an integrated THz photonic circuit may be built by integrated other optoelectronic devices like Schottky diode THz mixer [38,39], graphene modulator, and/or graphene detector.

The work that will be introduced in the following text reports the planar integration of tapered THz QCLs with SSP waveguides, which are further processed into unidirectional SSP out-couplers by introducing periodically arranged scatterers. The resulting surface-emitting THz beam is highly collimated with a beam divergence as narrow as 3.6° × 9.7°. As the beam divergence is proportional to the light emission area, this low divergence indicates the good waveguiding property of the SSP waveguide, while the low optical background of the beam implies a high coupling efficiency of the light from the laser cavity to the SSPs.

3.3.1 In-Plane Integration of the SSP Structure with THz QCL

Figure 3.8a shows a 3D schematic cross-section view of our device along the white dashed line in Figure 3.8b, which is a scanning electron microscope image of a fabricated device. The 10 μm thick active region, whose band structure design is similar with that in Reference 21 but with a gain peaking at ~3 THz (100 μm wavelength), is sandwiched between two gold plates. THz radiation will be emitted from the laser facet facing the SSP waveguide. The emission from the other facet is blocked by a DBR mirror formed by metal patterning on top of the active region. To facilitate the fabrication, the SSP waveguide and scatterers are realized by gold coating the whole grooved active region with a gold thickness much larger than the skin depth (Figure 3.8a).

Bearing in mind that the ideal input for this 1D SSP waveguide is a parallel light, the THz QCL is designed with a tapered structure and a curved front facet to collimate the emitting light in the lateral direction. Figure 3.9a shows the electric field (E_z) distribution of the tapered THz QCL. The tapered structure has a tapering angle of 36.5° and a length of 800 μm, which quickly extends the width of the output facet from 50 to 500 μm. To make the output beam collimated, the shape of the taper facet is chosen as an arc centered 20 μm behind the ridge-taper interface. It is worth mentioning that the usually adopted flat taper facet [40] is not applicable for such a large tapering angle because it will distort the optical due to a too large mismatch between the shape of the facet and the wave front of the light

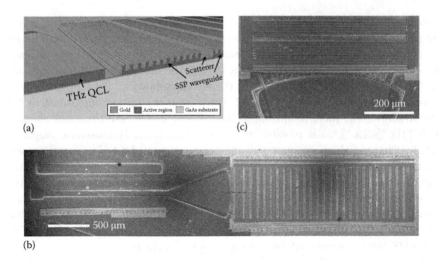

(a)

(c)

(b)

FIGURE 3.8 (a) Three dimensional schematic cross-sectional view of our device along the middle dashed line in (b). (b) Scanning electron microscope image of a fabricated device. Also shown is a ridge laser fabricated by the side of the device. (c) Enlarged view of the central region of the device.

coming from the laser ridge. In consequence, this will lead to uncollimated beams (Figure 3.9b). Finally, to suppress the higher-order lateral modes, side-absorbers (black region) were employed by leaving the active region uncovered by the metal. The exposed topmost n^+-GaAs contact layer of the active region acts as an efficient absorber [16].

Figure 3.10a shows the electric field distribution of the whole device, obtained by a 2D simulation performed in the cross-section plane along the symmetry line of the structure. The detailed geometrical parameters of the SSP collimator are presented in Figure 3.10b. As shown in the figure, the excitation and scattering of the SSPs can be clearly observed. The wavelength of the SSP varies from 90.4 to 92.7 μm as the wave propagates away from the laser fact, whereas it is 96 μm in free space. The $1/e$ evanescent tail of the SSP is around 50 μm. This matches well with the value calculated by $1/\sqrt{k_{SSP}^2 - k_0}$, where k_{SSP} is the SSP wave vector and k_0 is the free-space wave vector. The increased k vector and the resulting confined surface mode, together with the minor effect of the scatterers in this case, improve the laser output by a factor of 1.4, as they reduce the k and spatial mismatch between the modes inside and outside the laser cavity. As for the coupling efficiency, approximately 41% of the laser output is coupled into the SSPs, with the remaining 59% being radiated directly into free space. Most of the SSP light are then scattered out by 25 scatterers, which are periodically grooved on the waveguide with a period of 78 μm so that the scattered SSP light and the uncoupled light interfere constructively, resulting in a narrow (2.7°) single lobe in the far field (Figure 3.10c). The geometric details of the SSP structure can be found in

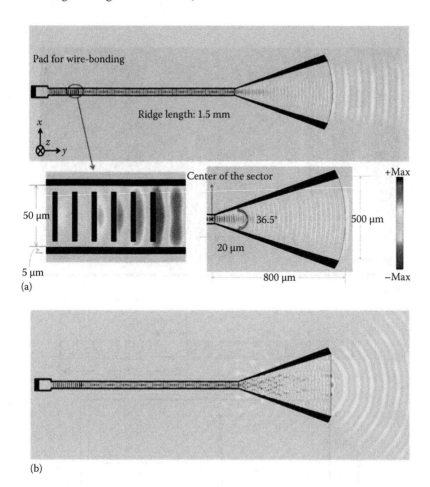

FIGURE 3.9   (a) Electric field (E_z) distribution of the tapered terahertz quantum cascade laser with a curved front facet. The zoom-in views show the details of the DBR region and the taper, respectively. The black regions represent the active region without metal coverage, which are highly absorptive. (b) E_z distribution of a tapered structure with a flat facet. The electric field in the laser cavity is distorted, resulting in uncollimated beams.

Figure 3.10b. As a comparison, the laser without the SSP structure emits rather uniformly in all directions (Figure 3.10d).

3.3.2   FABRICATION, EXPERIMENTAL RESULTS, AND ANALYSIS

The fabrication began with an Au–Au thermocompression bonding of the active region to an n⁺ GaAs receptor wafer. The original active region substrate was removed by a combination of lapping and selective chemical etching. Top contact metal

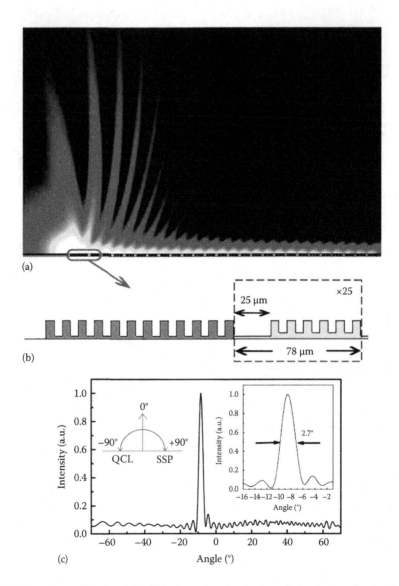

FIGURE 3.10 (a) Simulated 2D light intensity distribution of the device, the simulation was performed in the plane of symmetry of the device. (b) Cross section of the spoof surface plasmon (SSP) collimator design. All the narrow grooves have a width of 5 μm and a period of 10 μm. The narrow grooves are 9 μm deep in the orange region and 7 μm deep in the yellow region. The 25 μm groove, which acts as a scatterer to scatter out the SSPs, is 9 μm in depth. The yellow region represents a periodic unit, which was repeated 25 times. The narrow grooves in the orange region are deeper to enhance the SSP coupling. (c) Calculated far-field intensity profile along the SSP waveguide direction. The enlarged view of the central lobe in the inset exhibits that the beam divergence is as narrow as 2.7°.   (*Continued*)

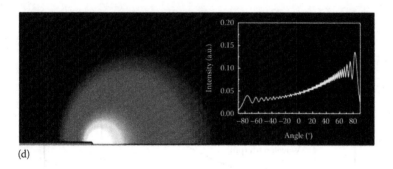

(d)

FIGURE 3.10 (Continued) (d) Simulated 2D light intensity of a laser without the SSP structure. The inset shows that the light emission is rather uniform in all directions.

(Ti/Au 15/350 nm) was then defined by standard optical lithography and lift-off process. The laser cavity, shallow (SSP waveguide) and deep (scatterer) grooves, were etched by reactive ion etching (RIE) using a SiO_2 film as the masks. After that, without removing the SiO_2 mask, a multiple-angle electron-beam evaporation procedure was used to ensure full coverage of the whole structure by a Ti/Au layer. The Ti/Au layer on the laser top and sidewalls was then removed by a gold etchant with a thick AZ4620 photoresist covering the SSP region. The SiO_2 mask on top of the laser was etched away by plasmons and a dilute HF solution. The substrate was thinned to 120 μm and deposited a 20/300 nm Ti/Au layer for the bottom contact. The samples were cleaved, mounted on Cu submounts, wire bonded, and finally put on the cold finger of a cryostat for measurement.

Devices with two different groove widths were fabricated, labeled as A and B (Figure 3.12a). The groove widths of devices A are 600 μm, while devices B have a wider groove width of 1000 μm. Both kinds of devices show similar LIV characteristics. Figure 3.4 shows the LIV curves of a typical device under different temperatures, the insets show the laser spectra of devices A and B at 4.3 A and 9 K heatsink temperature, where their far fields will be measured. Operating under pulse mode with 500 ns pulses at 10 kHz repetition, the device's maximum operating temperature is around 118 K. This is comparable with that of the ridge laser fabricated by the side of the device, which operated up to 136 K.

Two dimensional far-field patterns of the devices have been measured by scanning a pyroelectric detector on a spherical surface centered at the laser facet. Three dimensional simulations were also performed using a finite-difference time-domain solver. Figure 3.12c presents the measured far-field pattern of device A. Figure 3.12b, which shows the horizontal and vertical line scans of Figure 3.12c across the peak value, reveals that the beam divergence in the vertical direction is as low as 3.6°, while it is 9.7° in the horizontal direction. Moreover, the intensities of the background are less than 10% of the peak value, indicating a high coupling efficiency of the light into SSP. Note that the measured beam divergence in the vertical direction is larger than the simulated results at 94 μm wavelength

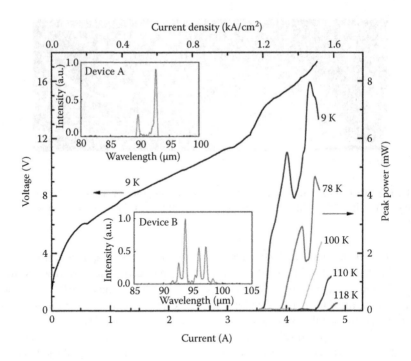

FIGURE 3.11 Pulsed light–current–voltage characteristics of the laser under different heat-sink temperatures. Inset: laser spectra of devices A and B at 4.3 A and 9 K, where their far-field patterns were measured.

(Figure 3.12c and e). This is probably due to the multimode emission of our devices (inset of Figure 3.11), as the position of the far-field lobe shifts in the vertical direction as the laser wavelength changes. In simulation, a change of 1 μm in wavelength leads to ~1° shift of the far field. Therefore, the measured far-field pattern is actually the superimposition of several slightly shifted far-field patterns if the device is not single-mode. This effect is more apparent for a device B that has a larger laser bandwidth (inset of Figure 3.11), whose measured and simulated far-field patterns are shown in Figure 3.12d and f, respectively. Here, the simulation considered the emission bandwidth from 92 to 98 μm. Additionally, in contrast to the common understanding that a broader laser emission area gives a narrower beam divergence, device B has a larger beam divergence in the horizontal direction than device A.

To clarify the question, we investigated the near-field distributions of the devices, as presented in Figure 3.13a and b. As it is shown, the SSP can spread in the lateral direction so that device B does provide a broader laser emission area compared with device A. However, the spread of the SSP in the lateral direction induces phase delay of the electric field. As we can see from Figure 3.13c and d, which plot the amplitude and phase distribution of the radiative electric field (E_y) along the white dashed

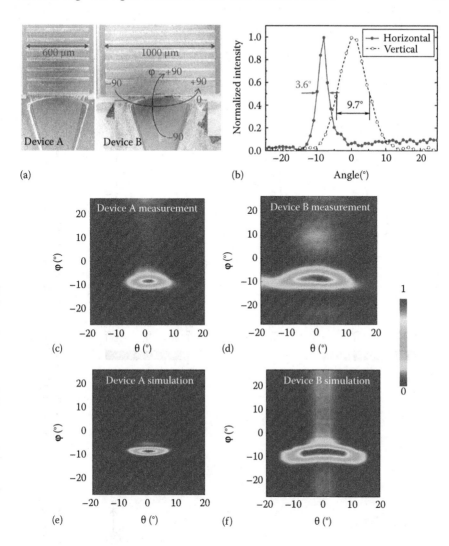

FIGURE 3.12 (a) Scanning electron microscope images of the devices A and B showing that the groove width of device A is 600 μm and that of device B is 1000 μm. (b) Horizontal and vertical line scans of the far-field pattern in (c) across the peak value. (c, d) Measured far-field patterns of devices A and B, respectively. (e) Simulated far-field pattern of device A at 94 μm wavelength. (f) Simulated far-field patterns of device B considering the multimode emission from 92 to 98 μm.

lines in Figure 3.13c and d, the phase of E_y of device B varies largely in the lateral direction, whereas it is uniform for device A. The large phase difference (>180°) of the electric field between the central region and the lateral boundary region of device B leads to destructive interference in the far field, thereby giving a larger beam divergence.

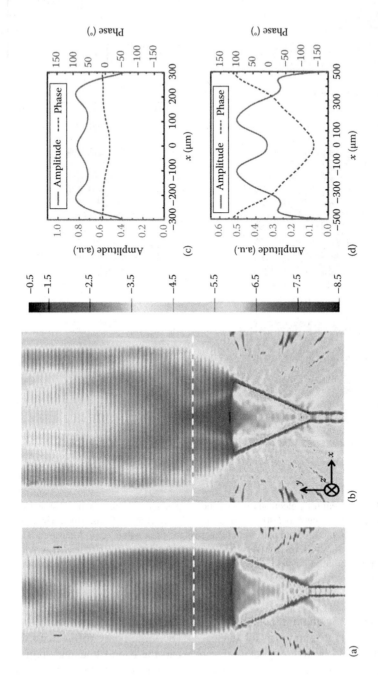

FIGURE 3.13 Near-field light intensity distributions of (a) device A and (b) device B in logarithmic scale, taken on a plane 2 μm above the device surface. (c, d) Amplitude and phase distributions of the radiative electric field (E_y) along the white dashed line in (a) and (b), respectively.

REFERENCES

1. M. Tonouchi, Cutting-edge terahertz technology, *Nat. Photon.* **1**, 97–105 (2007).
2. R. Köhler, A. Tredicucci, F. Beltram, H. E. Beere, E. H. Linfield, A. G. Davies, D. A. Ritchie, R. C. Iotti, and F. Rossi, Terahertz semiconductor-heterostructure laser, *Nature* **417**, 156–159 (2002).
3. S. Kumar, Recent progress in terahertz quantum cascade lasers, *IEEE J. Sel. Top. Quant. Electron.* **17**, 38–47 (2011).
4. B. Williams, S. Kumar, Q. Hu, and J. Reno, High-power terahertz quantum-cascade lasers, *Electron. Lett.* **42**, 18–19 (2006).
5. S. Fathololoumi, E. Dupont, C. W. I. Chan, Z. R. Wasilewski, C. Jirauschek, Q. Hu, and D. Ban, Terahertz quantum cascade lasers operating up to ~200 K with optimized oscillator strength and improved injection tunneling, *Opt. Exp.* **20**, 3331–3339 (2012).
6. A. Wei Min Lee, Q. Qin, S. Kumar, B. S. Williams, Q. Hu, and J. L. Reno, High-power and high-temperature THz quantum-cascade lasers based on lens-coupled metal-metal waveguides, *Opt. Lett.* **32**, 2840–2842 (2007).
7. M. Amanti, M. Fischer, C. Walther, G. Scalari, and J. Faist, Horn antennas for terahertz quantum cascade lasers, *Electron. Lett.* **43**, 125–126 (2007).
8. W. Maineult, P. Gellie, A. Andronico, P. Filloux, G. Leo, C. Sirtori, S. Barbieri et al., Metal–metal terahertz quantum cascade laser with micro-transverse-electromagnetic-horn antenna, *Appl. Phys. Lett.* **93**, 183508 (2008).
9. N. Yu, Q. J. Wang, M. A. Kats, J. A. Fan, S. P. Khanna, L. Li, A. G. Davies, E. H. Linfield, and F. Capasso, Designer spoof surface plasmon structures collimate terahertz laser beams, *Nat. Mater.* **9**, 730–735 (2010).
10. N. Yu, Q. Wang, and F. Capasso, Beam engineering of quantum cascade lasers, *Laser Photon. Rev.* **6**, 24–46 (2012).
11. M. I. Amanti, M. Fischer, G. Scalari, M. Beck, and J. Faist, Low-divergence single-mode terahertz quantum cascade laser, *Nat. Photon.* **3**, 586–590 (2009).
12. S. Kumar, B. S. Williams, Q. Qin, A. W. Lee, Q. Hu, and J. L. Reno, Surface-emitting distributed feedback terahertz quantum-cascade lasers in metal–metal waveguides, *Opt. Exp.* **15**, 113–128 (2007).
13. J. A. Fan, M. A. Belkin, F. Capasso, S. Khanna, M. Lachab, A. G. Davies, and E. H. Linfield, Surface emitting terahertz quantum cascade laser with a double-metal waveguide, *Opt. Exp.* **14**, 11672–11680 (2006).
14. L. Mahler, M. I. Amanti, C. Walther, A. Tredicucci, F. Beltram, J. Faist, H. E. Beere, and D. A. Ritchie, Distributed feedback ring resonators for vertically emitting terahertz quantum cascade lasers, *Opt. Exp.* **17**, 13031–13039 (2009).
15. E. Mujagić, C. Deutsch, H. Detz, P. Klang, M. Nobile, A. M. Andrews, W. Schrenk, K. Unterrainer, and G. Strasser, Vertically emitting terahertz quantum cascade ring lasers, *Appl. Phys. Lett.* **95**, 011120 (2009).
16. Y. Chassagneux, R. Colombelli, W. Maineult, S. Barbieri, H. E. Beere, D. A. Ritchie, S. P. Khanna, E. H. Linfield, and A. G. Davies, Electrically pumped photonic-crystal terahertz lasers controlled by boundary conditions, *Nature* **457**, 174–178 (2009).
17. Y. Chassagneux, R. Colombelli, W. Maineults, S. Barbieri, S. P. Khanna, E. H. Linfield, and A. G. Davies, Predictable surface emission patterns in terahertz photonic-crystal quantum cascade lasers, *Opt. Exp.* **17**, 9491–9502 (2009).
18. Y. Chassagneux, R. Colombelli, W. Maineult, S. Barbieri, S. P. Khanna, E. H. Linfield, and A. G. Davies, Graded photonic crystal terahertz quantum cascade lasers, *Appl. Phys. Lett.* **96**, 031104 (2010).
19. J. Scheuer and A. Yariv, Annular Bragg defect mode resonators, *J. Opt. Soc. Am. B* **20**, 2285–2291 (2003).

20. G. Liang, H. Liang, Y. Zhang, S. P. Khanna, L. Li, A. Giles Davies, E. Linfield et al., Single-mode surface-emitting concentric-circular-grating terahertz quantum cascade lasers, *Appl. Phys. Lett.* **102**, 031119 (2013).

21. S. Kumar, Q. Hu, and J. L. Reno, 186 K operation of terahertz quantum-cascade lasers based on a diagonal design, *Appl. Phys. Lett.* **94**, 131105 (2009).

22. S.-H. Kim, S.-K. Kim, and Y.-H. Lee, Vertical beaming of wavelength-scale photonic crystal resonators, *Phys. Rev. B* **73**, 235117 (2006).

23. T. W. Ebbesen, C. Genet, and S. I. Bozhevolnyi, Surface-plasmon circuitry, *Phys. Today* **61**, 44 (2008).

24. B. Ng, J. Wu, S. M. Hanham, A. I. Fernández-Domínguez, N. Klein, Y. F. Liew, M. B. H. Breese, M. Hong, and S. A. Maier, Spoof plasmon surfaces: A novel platform for THz sensing, *Adv. Opt. Mater.* **1**, 543–548 (2013).

25. A. J. Huber, F. Keilmann, J. Wittborn, J. Aizpurua, and R. Hillenbrand, Terahertz near-field nanoscopy of mobile carriers in single semiconductor nanodevices, *Nano Lett.* **8**, 3766–3770 (2008).

26. X. Luo and T. Ishihara, Surface plasmon resonant interference nanolithography technique, *Appl. Phys. Lett.* **84**, 4780 (2004).

27. L. Li, T. Li, S. Wang, C. Zhang, and S. Zhu, Plasmonic airy beam generated by in-plane diffraction, *Phys. Rev. Lett.* **107**, 1–4 (2011).

28. H. J. Lezec, A. Degiron, E. Devaux, R. A. Linke, L. Martin-Moreno, F. J. Garcia-Vidal, and T. W. Ebbesen, Beaming light from a subwavelength aperture, *Science* **297**, 820–822 (2002).

29. J. B. Pendry, Mimicking surface plasmons with structured surfaces, *Science* **305**, 847–848 (2004).

30. F. García de Abajo and J. Sáenz, Electromagnetic surface modes in structured perfect-conductor surfaces, *Phys. Rev. Lett.* **95**, 233901 (2005).

31. F. J. García de Abajo, Colloquium: Light scattering by particle and hole arrays, *Rev. Mod. Phys.* **79**, 1267–1290 (2007).

32. M. Navarro-Cía, M. Beruete, S. Agrafiotis, F. Falcone, M. Sorolla, and S. A. Maier, Broadband spoof plasmons and subwavelength electromagnetic energy confinement on ultrathin metafilms, *Opt. Exp.* **17**, 18184–18195 (2009).

33. X. Shen, T. Jun, D. Martin-Cano, and F. J. Garcia-Vidal, Conformal surface plasmons propagating on ultrathin and flexible films, *Proc. Natl. Acad. Sci.* 110, 40–45 (2013).

34. D. Martin-Cano, M. L. Nesterov, A. I. Fernandez-Dominguez, F. J. Garcia-Vidal, L. Martin-Moreno, and E. Moreno, Domino plasmons for subwavelength terahertz circuitry, *Opt. Exp.* **18**, 754–764 (2010).

35. C. R. Williams, S. R. Andrews, S. A. Maier, A. I. Fernández-Domínguez, L. Martín-Moreno, and F. J. García-Vidal, Highly confined guiding of terahertz surface plasmon polaritons on structured metal surfaces, *Nat. Photon.* **2**, 175–179 (2008).

36. S. Maier, S. Andrews, L. Martín-Moreno, and F. García-Vidal, Terahertz surface plasmon-polariton propagation and focusing on periodically corrugated metal wires, *Phys. Rev. Lett.* **97**, 176805 (2006).

37. D. Martin-Cano, O. Quevedo-Teruel, E. Moreno, L. Martin-Moreno, and F. J. Garcia-Vidal, Waveguided spoof surface plasmons with deep-subwavelength lateral confinement, *Opt. Lett.* **36**, 4635–4637 (2011).

38. M. Wanke, E. Young, and C. Nordquist, Monolithically integrated solid-state terahertz transceivers, *Nature* **4**, 565–569 (2010).

39. H.-W. Hübers, Terahertz technology: Towards THz integrated photonics, *Nat. Photon.* **4**, 503–504 (2010).

40. Y. Li, J. Wang, N. Yang, J. Liu, T. Wang, F. Liu, Z. Wang, W. Chu, and S. Duan, The output power and beam divergence behaviors of tapered terahertz quantum cas-cade lasers, **21**, 15998–16006 (2013).

Section II

Imaging Sensors and Systems

4 Intense Few-Cycle Terahertz Pulses and Nonlinear Terahertz Spectroscopy

Ibraheem Al-Naib and Tsuneyuki Ozaki

CONTENTS

4.1 Introduction 85
4.2 Generation Techniques 86
    4.2.1 Optical Rectification Technique 86
    4.2.2 Gas Plasma Technique.. 88
4.3 Detection Techniques 91
    4.3.1 Electro-Optic Sampling... 91
    4.3.2 Spectral Domain Interferometry 93
    4.3.3 Air-Breakdown Coherent Detection 94
4.4 Nonlinear Terahertz Spectroscopy 95
    4.4.1 Nonlinear THz Response of Monolayer Graphene 95
    4.4.2 Nonlinear THz Response of Silicon Waveguide 96
    4.4.3 Biomedical-Induced Nonlinearity 97
    4.4.4 Real-Time Near-Field THz Imaging... 98
4.5 Summary and Outlook 100
References.. 100

4.1 INTRODUCTION

Since the early days of physics, physicists and engineers have been endeavoring to examine the interaction of electromagnetic waves with matter. Developing new sources has been one of the most important factors in exploring novel phenomena in physics and chemistry, which has enabled innovations in linear and nonlinear spectroscopy techniques. While such development at microwave and optical frequencies has been rapidly advanced in the last century, terahertz (THz) technology has only been around for about three decades [1,2]. Nevertheless, as illustrated in the previous chapters of this book, the wide range of possible applications of today's THz technology is stimulating enormous attention from specialists in various fields, such as imaging, sensing, quality control, wireless communication, and basic science [3–9]. The development of this technology depends on realizing efficient and robust sources and detectors. For example, the development of terahertz time-domain spectroscopy (THz-TDS) [10] more than two decades ago opened a new chapter in THz science, initiating great efforts to develop THz applications to exploit the unique opportunities that THz waves offer. For instance, observation of intermolecular vibrations in some chemicals and organic molecules due to the low THz photon energy (~4 meV at 1 THz) is just one example of a broad range of applications [8,9].

Nevertheless, there is an urgent need to develop nonlinear THz spectroscopy techniques and intense THz sources that have the potential to reveal a new category of nonlinear phenomena and explore nonlinear effects in various materials. For example, intense THz pulses can induce an ultrafast electric- or magnetic-field switching at timescales of tens of femtoseconds to picoseconds, which is much faster than what can be achieved through conventional electronics. Such nonlinear phenomena have been enabled through the development of ultrafast nonlinear THz spectrometers and have been a subject of several excellent reviews [11–14].

In this chapter, we briefly discuss the techniques to generate and detect intense THz pulses. We will review some of the most recent developments and breakthroughs, starting with the generation methods of intense THz field including optical rectification (OR) and gas plasma techniques. Next, detection techniques such as electro-optic (EO) sampling, spectral domain interferometry (SDI), and air-breakdown coherent detection (ABCD) are elucidated. Finally, various high-field-strength THz experiments will be illustrated. It is worth mentioning that photoconductive antennas can also be used to generate intense THz pulses, but we will not discuss this topic here and refer the readers to Chapter x.

4.2 GENERATION TECHNIQUES

We discuss in this section the two main categories of techniques to generate THz pulses, namely, OR [1] and gas plasma [15,16]. It is safe to say that none of these methods would have been feasible without the invention of ultrafast lasers, as high light intensity is necessary to drive nonlinear optical processes. More specifically, the recent development of Ti–sapphire lasers has enabled scientists to build new THz sources and achieve levels of THz energies that were not attainable before. After explaining the basic idea of each technique, we will overview the most recent developments.

4.2.1 OPTICAL RECTIFICATION TECHNIQUE

One of the most known techniques to generate intense THz waves is OR, which is based on the excitation of nonlinear effects in crystals. Mixing different frequency components within the spectrum of the laser pulse leads to the generation of THz radiation. Consequently, this process leads to a distribution that appears in the time domain as an electric field transient with a shape similar to the envelope of the laser pulse, as shown in Figure 4.1.

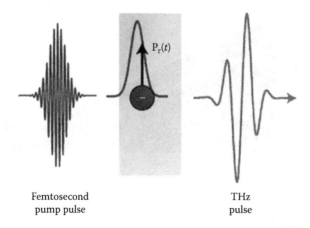

Femtosecond
pump pulse

THz
pulse

FIGURE 4.1 Optical rectification scheme, where a femtosecond pulse (left pulse) propagates through the nonlinear medium, and the induced polarization inside the medium gives rise to the terahertz electric field (right pulse). (From Kampfrath, T. et al., *Nat. Photonics*, 7, 680, 2013.)

The bandwidth of the excitation pulse and the phase matching between the near-infrared pump beam and the generated THz field represent the two main factors that influence the bandwidth and the temporal shape of the generated THz pulse. Several nonlinear crystals have been used for the generation and detection of THz radiation, such as GaAs, GaP, InP, GaSe, $LiNbO_3$, and $LiTaO_3$ [17]. However, the most commonly used crystal is ZnTe [18], since there is very good phase-matching conditions of the generated THz radiation with the Ti–sapphire laser pulses near 800 nm at normal incidence. A bandwidth of 0.1–3.5 THz is typically achieved, with a dynamic range of 60 dB or more. However, THz generation using ZnTe crystals suffers from the three main drawbacks. First, it has a relatively low bandgap of 2.3 eV at 300 K [19], and thus, two-photon absorption of the Ti–sapphire pump takes place at low fluences compared to the damage threshold of the ZnTe crystals [20]. Free carriers generated through two-photon absorption would then result in THz absorption, thus lowering the THz yield. Second, the lowest optical phonon mode of ZnTe is at 5.3 THz [21,22], which limits the usable bandwidth. This can be extended using GaP as its lowest phonon mode occurs at 11 THz [23,24], but at the cost of poorer performance at low frequencies and lower field strength. Finally, the relevant nonlinear coefficient of ZnTe of 68.5 pm/V is less than half that of some other crystals such as lithium niobate (LN), which does not have any of the aforementioned pitfalls. However, LN has poor phase-matching conditions at Ti–sapphire wavelengths. Therefore, special pumping configurations are required for intense THz generation from LN, where the wave front of the optical pump pulse should be tilted relative to its propagation direction [25,26]. THz field strengths of greater than 1.2 MV/cm [27] and THz energies at the sub-mJ level (up to 125 and 400 μJ) have been achieved [28,29], and it has been estimated to increase by a factor of three to five by cryogenically cooling the crystal [30]. Figure 4.2 depicts an informative presentation of how the tilted-pulse-front pumping (red/solid line inside the LN prism) coincides with the

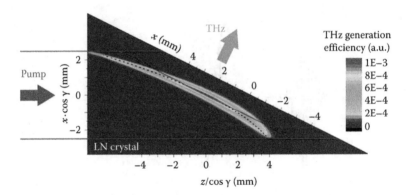

FIGURE 4.2 Illustration of the tilted-pulse-front technique. The color plot inside the crystal prism shows the spatial distribution of terahertz generation efficiency. (From Fülöp, J.A., et al., *Opt. Express*, 18, 12311, 2010.)

image of the grating (black/dashed line) at the pump beam center [31]. The color plot inside the crystal prism shows the spatial distribution of the THz generation efficiency [31].

A summary of achievable THz energy is shown in Figure 4.3a for a variety of sources based on ZnTe (collinear) and LN (noncollinear) crystals [12]. While quadratic scaling can be noticed in either of the crystals at low optical pulse energies, saturation and linear scaling behavior is observed in ZnTe when the optical pulse energy exceeds 50 μJ as a result of free-carrier absorption due to two-photon absorption of the pump light. In LN crystal, however, the scaling behavior does not exhibit saturation until very high pump energies are reached [12]. The highest THz energy achieved to date at room temperature is 436 μJ using 58 mJ pump energy [29], which corresponds to a conversion efficiency of 0.77%. Moreover, it is expected that the conversion efficiency in LN crystals can exceed 10% if it is cooled down to 10 K if 400 fs pump pulse duration is used as shown in Figure 4.3b [28,32]. Such pulse duration is much longer than what is typically used (about 100 fs). However, it was shown later on that a factor of fivefold reduction in the maximum conversion efficiency is expected as a result of spectral reshaping of the optical pump pulse caused by THz generation (cascading effects) [33]. Nevertheless, a more recent investigation showed that cooling down the LN crystal to a cryogenic temperature of 23 K enhanced the generated THz energy by a factor of 2.7, from 68 to 186 μJ [29]. Finally, it is worth mentioning that 8.3 GV/m THz level is predicted and measured using a large-size, partitioned nonlinear DSTMS organic crystal assembly in a collinear setup [34,35].

4.2.2 GAS PLASMA TECHNIQUE

THz generation using ionized gas has been studied more than 20 years ago, using 50 mJ driving lasers at near-infrared wavelengths [36], but the conversion efficiency was less than 10^{-6}. One to two orders of magnitude improvement in efficiency was later achieved using the fundamental and its second harmonic (SH) [16,37,38]. These two methods are conventionally referred to as the "one-color" and "two-color"

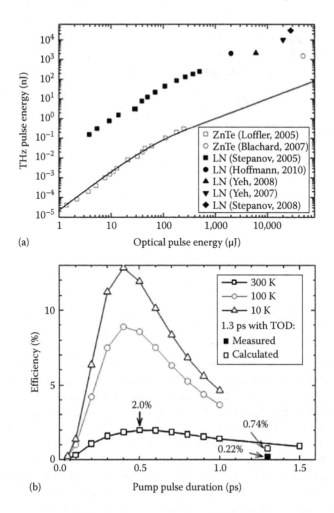

FIGURE 4.3 (a) Terahertz (THz) pulse energy output versus the pump pulse energy; (b) calculated THz-generation efficiency versus pump pulse duration for various crystal temperatures. (a: From Hoffmann, M.C. and Fülöp, J.A., *J. Phys. D: Appl. Phys.*, 44, 083001, 2011; b: From Fülöp, J.A. et al., *Opt. Lett.*, 37, 557, 2012.)

schemes, respectively. Unlike using a specific nonlinear medium where its inherent properties would influence THz generation, the principle here is to "engineer" a medium in air or other gases in order to generate extremely broadband single-cycle THz pulses. It is based on nonlinear processes in gas plasmas ionized by femtosecond lasers, and hence, there is no damage threshold. The bandwidth of the THz pulses is limited by the duration of the pump laser pulse only [12]. Figure 4.4 shows a schematic for a typical reflective THz air-breakdown coherent detection (R-THz-ABCD) experimental setup. The BBO crystal (typically type I) is usually used for SH generation as it features a broad phase-matching range from 409.6 to 3500 nm and high damage threshold [39]. When the polarizations of the fundamental and the

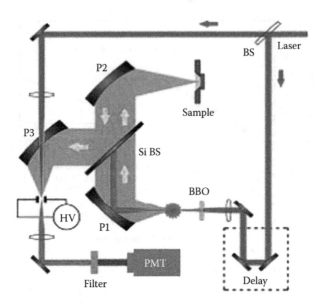

FIGURE 4.4   Schematic of the gas plasma setup. (From Ho, I.-C. et al., *Opt. Express*, 18, 2872, 2010.)

SH are parallel, THz generation is maximized [37,40]. Ambient air or pure nitrogen is usually used as the gas target for THz generation. Using the latter increases the generation efficiency and avoids absorption of the THz radiation [18].

Unlike THz sources based on OR, gas plasma provides a spectral range covering from 0.2 to over 30 THz and beyond as shown in Figure 4.5, which is only limited by the optical pulse duration, free from problems such as phonon absorption and significant dispersion that exist in semiconductors and nonlinear crystals [42]. The bandwidth can be extended up to 200 THz [43], which provides an attractive characteristic for broadband spectroscopic studies [44]. Recently, high field strengths were demonstrated when driving lasers with wavelengths longer than 800 nm are used. The maximum THz yield was found for driving laser wavelengths at 1.8 μm, with the THz field estimated at 4.4 MV/cm with only 400 μJ of the total pump energy [45]. Interestingly, the decrease in the conversion efficiency for wavelengths longer than 1.8 μm was predicted and attributed to a decrease in the peak intensity, while the medium is completely ionized for shorter wavelengths [45]. Table 4.1 compares plasma source to its counterpart based on ZnTe crystal and Fourier transform infrared spectroscopy (FTIR) with emphasis on dynamic range, bandwidth, resolution, peak power, and data acquisition time [41].

Systems based on this technique require laser pulses that are generally quite bulky, although they can still be tabletop systems and thus much smaller than free-electron lasers. More recent efforts have shown that one could replace the elongated plasmas (which usually have lengths ranging from few millimeters to several centimeters) with microplasmas as THz emitters of only a few tens of microns [46]. Although it is limited to one-color approach, the authors could measure the THz signal with pump energy of only 660 nJ, which is more than one order of magnitude

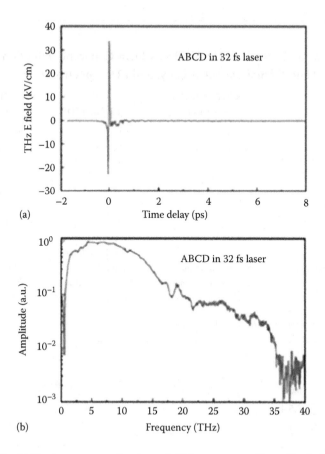

FIGURE 4.5 (a) Terahertz temporal pulse and (b) its spectrum. (From Ho, I.-C. et al., *Opt. Express*, 18, 2872, 2010.)

lower than previously reported values [46]. Moreover, the THz waves are radiated almost orthogonally to the laser propagation direction, and the generation volume is subwavelength compared to the emission [37] (Figure 4.6).

Spectroscopy is one of the main applications of the plasma systems to study carrier dynamics of semiconductors, vibration relaxation in large molecules, and biological complexes [41]. Moreover, subwavelength resolution THz imaging techniques have been recently demonstrated [47]. In this case, the femtosecond laser filament forms a waveguide for the THz wave, which varies between 20 and 50 μm that is significantly smaller than the wavelength of the THz wave [47].

4.3 DETECTION TECHNIQUES

4.3.1 Electro-Optic Sampling

With EO detection, the THz pulse is measured by employing the Pockels effect. A small portion of the laser beam is directed toward an EO crystal, which is followed by

TABLE 4.1

Comparisons of Reflective THz Air-Breakdown Coherent Detection, Traditional Time-Domain Spectroscopy, and FTIR Spectroscopy

	R-THz-ABCD (85 fs Laser)	Traditional TDS	FTIR (Bruker IFS 66v/S)
Source	Dry nitrogen	ZnTe	Mercury lamp
Detector	Dry nitrogen	ZnTe	DTGS
DR of power (<3 THz)	>10^6	>10^8	~300
Bandwidth (10% or greater of peak amplitude)	0.5–10 THz	0.1–3 THz	Far to midinfrared
Resolution	~0.1 cm^{-1}	~0.1 cm^{-1}	~0.1 cm^{-1}
Peak power	6×10^4 W	2×10^3 W (amplified laser)	1×10^{-7} W
Data acquisition time	Seconds or minutes	Minutes	Minutes
Uniqueness	Time resolved and broad bandwidth	Time resolved and high SNR	Broad bandwidth

Source: Ho, I.-C. et al., *Opt. Express*, 18, 2872, 2010.

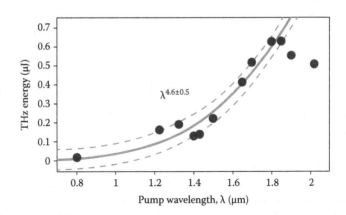

FIGURE 4.6   Recorded terahertz energy for 12 different pump wavelengths between 0.8 and 2.2 μm. (From Clerici, M. et al., *Phys. Rev. Lett.*, 110, 253901, 2013.)

a quarter wave plate and a Wollaston prism, as shown in Figure 4.7. When there is no THz signal, the voltage across both detectors is balanced and the difference between them is set to zero. When the probe beam overlaps with the THz signal, there will be polarization rotation provoked in the nonlinear crystal due to the changes in the electric field of the THz pulse, which is sensed by the difference of the reading of the photodiode detectors. This allows for the THz electric field to be directly sensed as a function of time. The absorption in the EO crystals and group velocity mismatch between the THz and probe waves are the main drawbacks of this technique. ZnTe crystals, for example, display good phase matching up to a few THz due to different

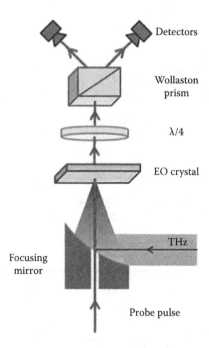

FIGURE 4.7   Balanced detection scheme for electro-optic sampling. (From Stojanovic, N. and Drescher, M., *J. Phys. B: At. Mol. Opt. Phys.*, 46, 192001, 2013.)

frequency modes and the strong absorption and dispersion of the phonon at 5.3 THz [21,22,48]. GaP crystal has a broader bandwidth of up to 7 THz [49]. Nevertheless, EO sampling is widely used and especially in conjunction with the OR technique.

4.3.2 Spectral Domain Interferometry

In order to obtain a high spectral resolution, a long scanning time is required. This is typically achieved by employing thick detection crystals. However, when the THz field strength is high enough, this leads to overrotation and induces a phase difference of more than 90°. Recently, a novel technique based on SDI has been proposed [51,52]. In this technique, instead of using two crossed polarizers, the change in the phase difference is measured using SDI, as shown in Figure 4.8a. The complete details of the mechanism are given in Reference 51. In short, the beam splitter (BS2) splits the probe beam into two equal parts. The reflected signal of the beam is sent to a glass plate. Next, the reflected signal from the glass plate consists of "front" and "back" pulses from the front and back surfaces of the glass plate. The two probe pulses are focused onto the ZnTe crystal at the same position and then sent to a spectrometer, and the interference between the front and back pulses is observed. Once the THz pulse is temporally matched with the optical back probe pulse, the refractive index of the ZnTe crystal is modulated by the THz electric field. As a result, there will be a phase difference between the front and back optical probe pulses as the front pulse is not affected. This phase difference can then be measured

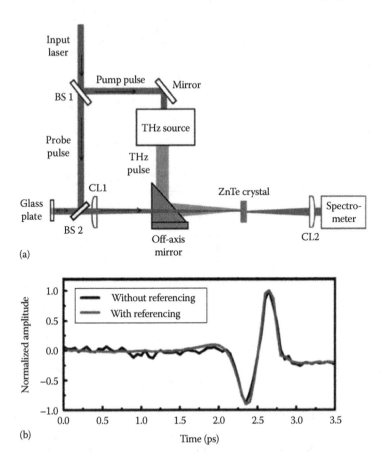

FIGURE 4.8 (a) Experimental setup for the spectral domain interferometry (SDI) technique, (b) measured terahertztemporal shape using SDI, without (black line) and with (gray line) referencing. (From Sharma, G. et al., *Opt. Lett.*, 37, 4338, 2012; Sharma, G. et al., *Opt. Lett.*, 38, 2705, 2013.)

using a spectrometer, which is proportional to the THz electric field. The maximum phase that can be measured is limited by the depth range of the SDI. In experiments, a phase difference of approximately 8898π can be measured, which is 18,000 times larger than the phase difference that could be measured using EO sampling method. The reconstructed THz pulse is shown in Figure 4.8b.

4.3.3 AIR-BREAKDOWN COHERENT DETECTION

Gasses are known to be free from phonon resonances and echoes due to the multiple THz. In turn, THz spectroscopy across very wide frequency band can be achieved [53] as mentioned in Section 2.2. The same concept can be utilized for detection as shown in Figure 4.9 [39,54]. Those unique features make the THz wave ABCD system an ideal system for ultrawideband spectroscopy. This technique has been recently employed in THz remote sensing via coherent

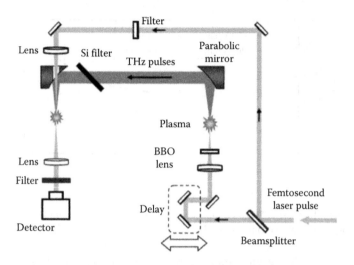

FIGURE 4.9 Schematic diagram of the air-plasma experimental setup. (From Dai, J. et al., *Phys. Rev. Lett.*, 97, 8, 2006.)

manipulation of the UV fluorescence emission from an asymmetrically ionized plasma from a distance of 10 m [55].

4.4 NONLINEAR TERAHERTZ SPECTROSCOPY

The capability of generating intense THz pulses with a field amplitude of 100s of kV/cm and MV/cm has been long awaited and has enabled a wide range of nonlinear THz spectroscopic techniques that were not possible before. Technically, these techniques can be classified as (1) intensity dependent, (2) THz pump/THz probe, and (3) 2D technique. The intensity of the incident THz pulse in the first technique is simply controlled by two wire-grid polarizers where the first one is rotated and the second one is kept fixed to maintain polarization [56] or through z-scan by scanning the sample through the focus [57]. THz pump/THz probe is employed by utilizing a time-delayed second THz pulse in order to probe not only the nonlinear THz response but also the dynamics of the induced response [58]. Moreover, using the 2D technique, it is possible to map the nonlinear optical response as a function of two frequency coordinates, the excitation frequency and the detection frequency [59–61]. The Fourier transform along the delay time between the two incident pulses generates the excitation frequency, while the detection frequency is derived from the real time [61]. In the following, we discuss some of the recent investigations in this regard.

4.4.1 NONLINEAR THz RESPONSE OF MONOLAYER GRAPHENE

The unique properties of graphene, such as zero-gap band structure, have attracted great attention [62,63]. It has been proposed as a promising material for potential applications in THz photodetection with an operating frequency higher than

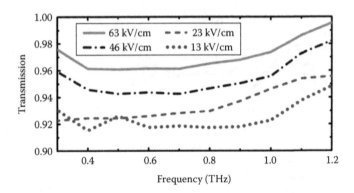

FIGURE 4.10   Terahertz (THz) transmission through the graphene sample at four different THz electric field levels. (From Hafez, H.A. et al., *AIP Adv.*, 4, 117118, 2014.)

100 GHz [64,65]. Thus, understanding these dynamics is crucial to the development of this technology [66,67]. Figure 4.10 shows a typical frequency domain response for nonlinear THz transmission of epitaxial graphene sample compared to the response of a bare SiC substrate [67]. One can notice a clear increase in the transmission through graphene when the THz field is increased. It is worth mentioning that different experiments showed different levels of transmission enhancement depending on the substrate that the graphene is fabricated on and the corresponding Fermi level [66–68]. It is mainly interpreted due to an increase in the scattering rates as a result of transient heating of the electrons and reduction in the THz conductivity.

Interestingly, a much higher field amplitude (peak field of 300 kV/cm) is found to increase the sample transmittance as well, but at a wavelength of 800 nm by as much as 17% on a timescale of 1 ps. However, impact ionization is found to play an important role in the carrier-generation process and THz-induced hot-carrier dynamics [69]. High-harmonic generation in graphene is another field of theoretical [70–75] and experimental studies [76] that have shown a great potential for graphene, where further experiments are required to improve the conversion efficiency.

4.4.2   NONLINEAR THz RESPONSE OF SILICON WAVEGUIDE

Various nonlinear phenomena in the THz frequency range have been recently demonstrated from different semiconductors such as GaAs, Si, InAs, superconductors, and VO_2 [77–81]. More recently, enhanced nonlinear effects at THz frequencies have been demonstrated by concentrating a THz pulse into a 2 cm long silicon waveguide [82]. A cross-sectional micrograph of fabricated silicon ridge waveguide and calculated TE eigenmode at 0.5 THz are shown in Figure 4.11. The measured and calculated transmission through this waveguide versus the input pulse energy is shown in Figure 4.11c, where more than a twofold increase in transmission is demonstrated as the pulse energy is increased. The effect is basically absent in the high-resistivity waveguide, which confirms the role of hot carriers in the nonlinear response [82].

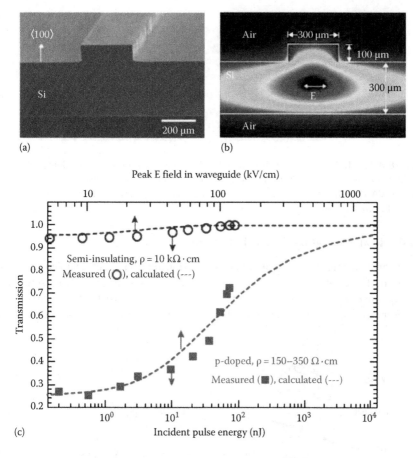

FIGURE 4.11 (a) Cross-sectional micrograph of fabricated silicon ridge waveguide, (b) calculated TE eigenmode at 0.5 THz, and (c) normalized power transmission for semi-insulating (circle) and doped (square) waveguides and corresponding calculated (dashed lines) pulse energy transmission. (From Li, S. et al., *Optica*, 2, 553, 2015.)

4.4.3   BIOMEDICAL-INDUCED NONLINEARITY

The absorption of THz radiation by water in biological tissues can lead to thermal effects. Hence, exposure risks are considered to be thermal in nature [83]. Hence, it has been generally accepted that THz waves are safe. Nevertheless, recent experiments have revealed that the exposure of lab-grown tissue to intense THz pulses may cause significant induction of phosphorylation of histone (H2AX) [83], which indicates that a DNA damage has occurred, as shown in Figure 4.12. Furthermore, it is found that a cellular response to THz waves is considerably different from that stimulated by exposure to 400 nm UVA. This study suggests that intense THz pulses should be probably studied for future therapeutic applications [84].

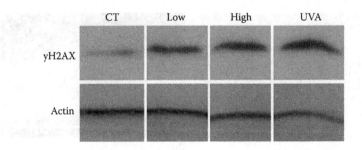

FIGURE 4.12 Induction of γH2AX in artificial human skin tissues equivalents following 10 minute exposure to either high (1.0 μJ) or low (0.1 μJ) energy THz pulses or 2 minute exposure to UVA (400 nm) pulses (0.080 μJ), as compared to control (CT) samples. (From Titova, L.V. et al., *Biomed. Opt. Express*, 4, 559, 2013.)

4.4.4 REAL-TIME NEAR-FIELD THZ IMAGING

THz near-field microscope is another application that has been enabled after the invention of intense THz sources. It is based on an LN tilted-pulse-front setup that was explained in Section 4.2. The rest of the microscope consists of two parts (sample observation and analyzer part) as shown in Figure 4.13 [85]. For detection of the

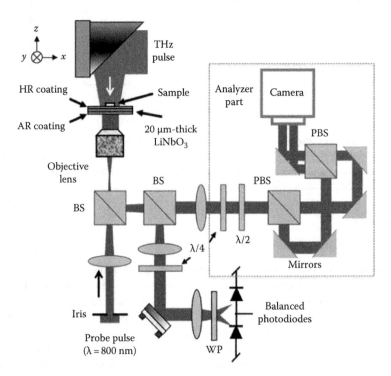

FIGURE 4.13 Schematic of terahertz near-field microscope experimental setup: BS, nonpolarized beam splitter; PBS, polarized beam splitter; WP, Wollaston prism. (From Blanchard, F. et al., *Opt. Express*, 19, 8277, 2011.)

THz wave, EO sampling is carried out in an x-cut LN crystal with a thickness of 20 μm. The latter was mounted on a 0.5 mm thick glass [85].

To achieve high linear spatial resolution by 2D EO THz imaging, two conditions must be satisfied [86]. The performances of the probe pulse imaging unit must agree with the desired spatial resolution, and the THz pulses must be captured in the near-field region before diffraction. To demonstrate the microscope's performance, the THz electric field in the near-field region of a metallic dipole antenna after the

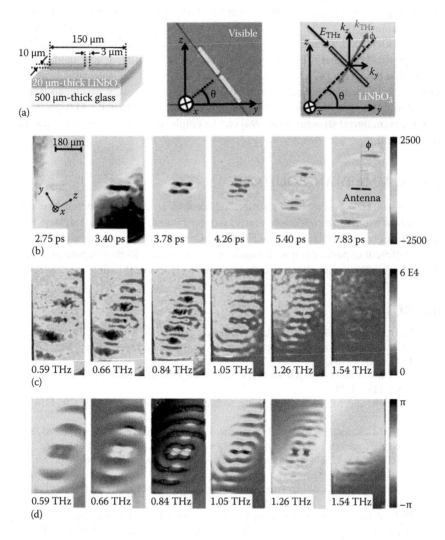

FIGURE 4.14 (a) Schematic and visible representations of dipole antenna together with the emitted terahertz (THz) field relative to the dipole antenna and lithium niobate crystal orientations, (b) temporal snapshots of THz electric field, (c) amplitude, and (d) phase components of Fourier transforms obtained from time-dependent field distribution for each pixel. (From Blanchard, F. et al., *Opt. Express*, 19, 8277, 2011.)

irradiation of a uniform THz pulse is shown in Figure 4.14. In part (a), a schematic description and a visible image of the sample are shown.

Figure 4.14b shows six temporal snapshots of the electric field passing through and reemitted by a dipole antenna. The calculated spectrally resolved amplitude and phase are shown in Figure 4.14c and d, respectively. They are calculated by taking the Fourier transforms at each pixel position along their time-dependent field distribution. The achieved resolution was shown to be $\lambda/30$, which is strongly influenced by the EO crystal thickness [85]. Further studies were carried out to map radiation patterns of a split-ring resonator (SRR), which is the most common unit cell of metamaterials [86,87]. The two lowest resonance modes of an SRR were successfully identified. In addition, by combining the near-field THz maps with TDS, it was possible to clarify the transition from reactive to radiative near-field zones of the SRR at 0.45 THz [87]. Moreover, this study showed that it is possible to achieve significant THz enhancement by utilizing such SRRs. Indeed, it was recently demonstrated to be very effective [14]. Furthermore, nanoslits [88,89] and high-quality factor metamaterial structures [90–94] can be employed to further enhance the THz electric field.

4.5   SUMMARY AND OUTLOOK

As intense THz sources become more accessible, it can be expected that completely new avenues in condensed matter will be opened, and we shall witness many breakthroughs in this field soon. Various nonlinear phenomena can be probed now using different experimental techniques, which provide a higher degree of control. We hope that this book chapter stimulates more efforts in this field.

REFERENCES

1. D. H. Auston, K. P. Cheung, J. A. Valdmanis, and D. A. Kleinman, Cherenkov radiation from femtosecond optical pulses in electro-optic media, *Phys. Rev. Lett.* **53**, 1555–1558 (1984).
2. C. Fattinger and D. Grischkowsky, Point source terahertz optics, *Appl. Phys. Lett.* **53**, 1480–1482 (1988).
3. B. B. Hu and M. C. Nuss, Imaging with terahertz waves, *Opt. Lett.* **20**, 1716–1718 (1995).
4. D. M. Mittleman, M. Gupta, R. Neelamani, R. G. Baraniuk, J. V Rudd, and M. Koch, Recent advances in terahertz imaging, *Appl. Phys. B* **68**, 1085–1094 (1999).
5. Z. Jiang and X. C. Zhang, Single-shot spatiotemporal terahertz field imaging, *Opt. Lett.* **23**, 1114–1116 (1998).
6. J. O'Hara and D. Grischkowsky, Quasi-optic synthetic phased-array terahertz imaging, *J. Opt. Soc. Am. B* **21**, 1178–1191 (2004).
7. J. A. Zeitler and L. F. Gladden, In-vitro tomography and non-destructive imaging at depth of pharmaceutical solid dosage forms, *Eur. J. Pharm. Biopharm.* **71**, 2–22 (2009).
8. M. Tonouchi, Cutting-edge terahertz technology, *Nat. Photonics* **1**, 97–105 (2007).
9. P. U. Jepsen, D. G. Cooke, and M. Koch, Terahertz spectroscopy and imaging—Modern techniques and applications, *Laser Photon. Rev.* **5**, 124–166 (2011).

10. D. Grischkowsky, S. Keiding, M. van Exter, and C. Fattinger, Far-infrared time-domain spectroscopy with terahertz beams of dielectrics and semiconductors, *J. Opt. Soc. Am. B* **7**, 2006–2015 (1990).
11. K. Tanaka, H. Hirori, and M. Nagai, THz nonlinear spectroscopy of solids, *IEEE Trans. Terahertz Sci. Technol.* **1**, 301–312 (2011).
12. M. C. Hoffmann and J. A. Fülöp, Intense ultrashort terahertz pulses: Generation and applications, *J. Phys. D: Appl. Phys.* **44**, 083001 (2011).
13. T. Kampfrath, K. Tanaka, and K. A. Nelson, Resonant and nonresonant control over matter and light by intense terahertz transients, *Nat. Photonics* **7**, 680–690 (2013).
14. H. Y. Hwang, S. Fleischer, N. C. Brandt, B. G. Perkins, M. Liu, K. Fan, A. Sternbach, X. Zhang, R. D. Averitt, and K. A. Nelson, A review of non-linear terahertz spectroscopy with ultrashort tabletop-laser pulses, *J. Mod. Opt.* **62**, 1447–1479 (2014).
15. H. Hamster, A. Sullivan, S. Gordon, W. White, and R. W. Falcone, Subpicosecond, electromagnetic pulses from intense laser-plasma interaction, *Phys. Rev. Lett.* **71**, 2725–2728 (1993).
16. D. J. Cook and R. M. Hochstrasser, Intense terahertz pulses by four-wave rectification in air, *Opt. Lett.* **25**, 1210–1212 (2000).
17. Y.-S. Lee, *Principles of Terahertz Science and Technology* (Springer, New York, 2009).
18. K.-E. Peiponen, J. A. Zeitler, and M. Kuwata-Gonokami, *Terahertz Spectroscopy and Imaging*, Vol. 171 (Springer-Verlag, Berlin, Heidelberg, 2013).
19. M. Schall and P. U. Jepsen, Above-band gap two-photon absorption and its influence on ultrafast carrier dynamics in ZnTe and CdTe, *Appl. Phys. Lett.* **80**, 4771 (2002).
20. G. Ramian, FEL table. http://sbfel3.ucsb.edu/www/fel_table.html. Accessed June 3, 2016.
21. G. Gallot, J. Zhang, R. W. Mcgowan, T. Jeon, and D. Grischkowsky, Measurements of the THz absorption and dispersion of ZnTe and their relevance to the electro-optic detection of THz radiation, *Appl. Phys. Lett.* **74**, 3450 (1999).
22. M. Schall, M. Walther, and P. Uhd Jepsen, Fundamental and second-order phonon processes in CdTe and ZnTe, *Phys. Rev. B* **64**, 094301 (2001).
23. D. Stanze, A. Deninger, A. Roggenbuck, S. Schindler, M. Schlak, and B. Sartorius, Compact cw Terahertz spectrometer pumped at 1.5 μm wavelength. *J. Infrared, Millimeter, Terahertz Waves* **32**, 225–232 (2010).
24. A. C. Tropper, H. D. Foreman, A. Garnache, K. G. Wilcox, and S. H. Hoogland, Vertical-external-cavity semiconductor lasers, *J. Phys. D: Appl. Phys.* **37**, R75–R85 (2004).
25. A. J. Deninger, T. Göbel, D. Schönherr, T. Kinder, A. Roggenbuck, M. Köberle, F. Lison, T. Müller-Wirts, and P. Meissner, Precisely tunable continuous-wave terahertz source with interferometric frequency control, *Rev. Sci. Instrum.* **79**, 044702 (2008).
26. M. Scheller, J. M. Yarborough, J. V Moloney, M. Fallahi, M. Koch, and S. W. Koch, Room temperature continuous wave milliwatt terahertz source, *Opt. Express* **18**, 27112–27117 (2010).
27. A. V. Kavokin, I. A. Shelykh, T. Taylor, and M. M. Glazov, Vertical cavity surface emitting terahertz laser, *Phys. Rev. Lett.* **108**, 197401 (2012).
28. J. A. Fülöp, L. Pálfalvi, S. Klingebiel, G. Almási, F. Krausz, S. Karsch, and J. Hebling, Generation of sub-mJ terahertz pulses by optical rectification, *Opt. Lett.* **37**, 557–559 (2012).
29. J. A. Fülöp, Z. Ollmann, C. Lombosi, C. Skrobol, S. Klingebiel, L. Pálfalvi, F. Krausz, S. Karsch, and J. Hebling, Efficient generation of THz pulses with 0.4 mJ energy, *Opt. Express* **22**, 20155–20163 (2014).
30. M. Scheller, S. W. Koch, and J. V Moloney, Grating-based wavelength control of single- and two-color vertical-external-cavity-surface-emitting lasers, *Opt. Lett.* **37**, 25–27 (2012).

31. J. A. Fülöp, L. Pálfalvi, G. Almási, and J. Hebling, Design of high-energy terahertz sources based on optical rectification, *Opt. Express* **18**, 12311–12327 (2010).

32. J. A. Fülöp, L. Pálfalvi, M. C. Hoffmann, and J. Hebling, Towards generation of mJ-level ultrashort THz pulses by optical rectification, *Opt. Express* **19**, 15090–15097 (2011).

33. K. Ravi, W. R. Huang, S. Carbajo, and X. Wu, Limitations to THz generation by optical rectification using tilted pulse fronts, *Opt. Express* **22**, 20239–20251 (2014).

34. C. Vicario, B. Monoszlai, and C. P. Hauri, GV/m single-cycle terahertz fields from a laser-driven large-size partitioned organic crystal, *Phys. Rev. Lett.* **112**, 213901 (2014).

35. M. Shalaby and C. P. Hauri, Demonstration of a low-frequency three-dimensional terahertz bullet with extreme brightness, *Nat. Commun.* **6**, 5976 (2015).

36. H. Hamster, A. Sullivan, S. Gordon, and R. W. Falcone, Short-pulse terahertz radiation from high-intensity-laser-produced plasmas, *Phys. Rev. E* **49**, 671–678 (1994).

37. M. Kress, T. Löff, S. Eden, M. Thomson, and H. G. Roskos, Terahertz-pulse generation by photoionization of air with laser pulses composed of both fundamental and second-harmonic waves, *Opt. Lett.* **29**, 1120–1122 (2004).

38. T. Bartel, P. Gaal, K. Reimann, M. Woerner, and T. Elsaesser, Generation of single-cycle THz transients with high electric-field amplitudes, *Opt. Lett.* **30**, 2805–2807 (2005).

39. F. Chen, W. W. Wang, and J. Liu, Diode single-end-pumped AO Q-switched Nd:GdVO4 266 nm laser, *Laser Phys.* **20**, 454–457 (2009).

40. X. Xie, J. Dai, and X.-C. Zhang, Coherent control of THz wave generation in ambient air, *Phys. Rev. Lett.* **96**, 1–4 (2006).

41. I.-C. Ho, X. Guo, and X.-C. Zhang, Design and performance of reflective terahertz air-biased-coherent-detection for time-domain spectroscopy, *Opt. Express* **18**, 2872–2883 (2010).

42. J. Dai, J. Liu, and X.-C. Zhang, Terahertz wave air photonics: Terahertz wave generation and detection with laser-induced gas plasma, *IEEE J. Sel. Top. Quantum Electron.* **17**, 183–190 (2011).

43. E. Matsubara, M. Nagai, and M. Ashida, Ultrabroadband coherent electric field from far infrared to 200 THz using air plasma induced by 10 fs pulses, *Appl. Phys. Lett.* **101**, 011105 (2012).

44. T. I. Oh, Y. S. You, N. Jhajj, E. W. Rosenthal, H. M. Milchberg, and K. Y. Kim, Intense terahertz generation in two-color laser filamentation: Energy scaling with terawatt laser systems, *New J. Phys.* **15**, 075002 (2013).

45. M. Clerici, M. Peccianti, B. E. Schmidt, L. Caspani, M. Shalaby, M. Giguère, A. Lotti et al., Wavelength scaling of terahertz generation by gas ionization, *Phys. Rev. Lett.* **110**, 253901 (2013).

46. F. Buccheri and X.-C. Zhang, Terahertz emission from laser-induced microplasma in ambient air, *Optica* **2**, 366–369 (2015).

47. J. Zhao, W. Chu, L. Guo, Z. Wang, J. Yang, W. Liu, Y. Cheng, and Z. Xu, Terahertz imaging with sub-wavelength resolution by femtosecond laser filament in air, *Sci. Rep.* **4**, 3880 (2014).

48. M. Schall and P. U. Jepsen, Freeze-out of difference-phonon modes in ZnTe and its application in detection of THz pulses, *Appl. Phys. Lett.* **77**, 2801 (2000).

49. Q. Wu and X.-C. Zhang, 7 terahertz broadband GaP electro-optic sensor, *Appl. Phys. Lett.* **70**, 1784 (1997).

50. N. Stojanovic and M. Drescher, Accelerator- and laser-based sources of high-field terahertz pulses, *J. Phys. B: At. Mol. Opt. Phys.* **46**, 192001 (2013).

51. G. Sharma, K. Singh, I. Al-Naib, R. Morandotti, and T. Ozaki, Terahertz detection using spectral domain interferometry, *Opt. Lett.* **37**, 4338–4340 (2012).

52. G. Sharma, K. Singh, A. Ibrahim, I. Al-Naib, R. Morandotti, F. Vidal, and T. Ozaki, Self-referenced spectral domain interferometry for improved signal-to-noise measurement of terahertz radiation, *Opt. Lett.* **38**, 2705–2707 (2013).

53. X.-C. Zhang and J. Xu, *Introduction to THz Wave Photonics* (Springer, New York, 2010).

54. J. Dai, X. Xie, and X.-C. Zhang, Detection of broadband terahertz waves with a laser-induced plasma in gases, *Phys. Rev. Lett.* **97**, 8–11 (2006).

55. J. Liu, J. Dai, S. L. Chin, and X. Zhang, Broadband terahertz wave remote sensing using coherent manipulation of fluorescence from asymmetrically ionized gases, *Nat. Photonics* **4**, 627–631 (2010).

56. F. Blanchard, G. Sharma, L. Razzari, X. Ropagnol, H.-C. Bandulet, F. Vidal, R. Morandotti et al., Generation of intense terahertz radiation via optical methods, *IEEE J. Sel. Top. Quantum Electron.* **17**, 5–16 (2011).

57. L. Razzari, F. Su, G. Sharma, F. Blanchard, A. Ayesheshim, H.-C. Bandulet, R. Morandotti et al., Nonlinear ultrafast modulation of the optical absorption of intense few-cycle terahertz pulses in n-doped semiconductors, *Phys. Rev. B* **79**, 193–204 (2009).

58. M. C. Hoffmann, J. Hebling, H. Y. Hwang, K. Yeh, and K. A. Nelson, Impact ionization in InSb probed by terahertz pump—Terahertz probe spectroscopy, *Phys. Rev. B* **79**, 161201(R) (2009).

59. W. Kuehn, K. Reimann, M. Woerner, and T. Elsaesser, Phase-resolved two-dimensional spectroscopy based on collinear n-wave mixing in the ultrafast time domain., *J. Chem. Phys.* **130**, 164503 (2009).

60. M. Woerner, W. Kuehn, P. Bowlan, K. Reimann, and T. Elsaesser, Ultrafast two-dimensional terahertz spectroscopy of elementary excitations in solids, *New J. Phys.* **15**, 025039 (2013).

61. T. Elsaesser, K. Reimann, and M. Woerner, Focus: Phase-resolved nonlinear terahertz spectroscopy—From charge dynamics in solids to molecular excitations in liquids, *J. Chem. Phys.* **142**, 212301 (2015).

62. R. R. Nair, P. Blake, A. N. Grigorenko, K. S. Novoselov, T. J. Booth, T. Stauber, N. M. R. Peres, and A. K. Geim, Fine structure constant defines visual transparency of graphene, *Science* **320**, 1308 (2008).

63. K. S. Novoselov, A. K. Geim, S. V Morozov, D. Jiang, Y. Zhang, S. V Dubonos, I. V Grigorieva, and A. A. Firsov, Electric field effect in atomically thin carbon films, *Science* **306**, 666–669 (2004).

64. F. Xia, T. Mueller, Y.-M. Lin, A. Valdes-Garcia, and P. Avouris, Ultrafast graphene photodetector, *Nat. Nanotechnol.* **4**, 839–843 (2009).

65. L. Vicarelli, M. S. Vitiello, D. Coquillat, A. Lombardo, A. C. Ferrari, W. Knap, and M. Polini, Graphene field-effect transistors as room-temperature terahertz detectors, *Nat. Mater.* **11**, 865–872 (2012).

66. M. J. Paul, Y. C. Chang, Z. J. Thompson, A. Stickel, J. Wardini, H. Choi, E. D. Minot, T. B. Norris, and Y.-S. Lee, High-field terahertz response of graphene, *New J. Phys.* **15**, 085019 (2013).

67. H. A. Hafez, I. Al-Naib, K. Oguri, Y. Sekine, M. M. Dignam, A. Ibrahim, D. G. Cooke, et al., Nonlinear transmission of an intense terahertz field through monolayer graphene, *AIP Adv.* **4**, 117118 (2014).

68. H. Y. Hwang, N. C. Brandt, H. Farhat, A. L. Hsu, J. Kong, and K. A. Nelson, Nonlinear THz conductivity dynamics in CVD-grown graphene, *J. Chem. Phys. B* **117**, 15819–15824 (2011).

69. S. Tani, F. Blanchard, and K. Tanaka, Ultrafast carrier dynamics in graphene under a high electric field, *Phys. Rev. Lett.* **109**, 166603 (2012).

70. S. A. Mikhailov, Non-linear graphene optics for terahertz applications, *Microelectron. J.* **40**, 712–715 (2009).

71. K. L. Ishikawa, Nonlinear optical response of graphene in time domain, *Phys. Rev. B* **82**, 201402 (2010).

72. I. Al-Naib, J. E. Sipe, and M. M. Dignam, High harmonic generation in undoped graphene: The interplay of interband and intraband dynamics, *Phys. Rev. B* **90**, 245423 (2014).

73. H. A. Hafez, I. Al-Naib, M. M. Dignam, Y. Sekine, K. Oguri, F. Blanchard, D. G. Cooke et al., Nonlinear terahertz field-induced carrier dynamics in photoexcited epitaxial monolayer graphene, *Phys. Rev. B* **91**, 035422 (2015).

74. I. Al-Naib, M. Poschmann, and M. M. Dignam, Optimizing third-harmonic generation at terahertz frequencies in graphene, *Phys. Rev. B* **91**, 205407 (2015).

75. I. Al-Naib, J. E. Sipe, and M. M. Dignam, Nonperturbative model of harmonic generation in undoped graphene in the terahertz regime, *New J. Phys.* **17**, 113018 (2015).

76. P. Bowlan, E. Martinez-Moreno, K. Reimann, T. Elsaesser, and M. Woerner, Ultrafast terahertz response of multilayer graphene in the nonperturbative regime, *Phys. Rev. B* **89**, 041408 (2014).

77. F. H. Su, F. Blanchard, G. Sharma, L. Razzari, A. Ayesheshim, T. L. Cocker, L. V Titova et al., Terahertz pulse induced intervalley scattering in photoexcited GaAs, *Opt. Express* **17**, 9620–9629 (2009).

78. I. Al-Naib, G. Sharma, M. M. Dignam, H. Hafez, A. Ibrahim, D. G. Cooke, T. Ozaki, and R. Morandotti, Effect of local field enhancement on the nonlinear terahertz response of a silicon-based metamaterial, *Phys. Rev. B* **88**, 195203 (2013).

79. K. Fan, H. Y. Hwang, M. Liu, A. C. Strikwerda, A. Sternbach, J. Zhang, X. Zhao, X. Zhang, K. A. Nelson, and R. D. Averitt, Nonlinear terahertz metamaterials via field-enhanced carrier dynamics in GaAs, *Phys. Rev. Lett.* **110**, 217404 (2013).

80. M. Liu, H. Y. Hwang, H. Tao, A. C. Strikwerda, K. Fan, G. R. Keiser, A. J. Sternbach et al., Terahertz-field-induced insulator-to-metal transition in vanadium dioxide metamaterial, *Nature* **487**, 345–348 (2012).

81. C. Zhang, B. Jin, J. Han, I. Kawayama, H. Murakami, J. Wu, L. Kang, J. Chen, P. Wu, and M. Tonouchi, Terahertz nonlinear superconducting metamaterials, *Appl. Phys. Lett.* **102**, 081121 (2013).

82. S. Li, G. Kumar, and T. E. Murphy, Terahertz nonlinear conduction and absorption saturation in silicon waveguides, *Optica* **2**, 553–557 (2015).

83. L. V. Titova, A. K. Ayesheshim, A. Golubov, D. Fogen, R. Rodriguez-juarez, F. A. Hegmann, and O. Kovalchuk, Intense THz pulses cause H2AX phosphorylation and activate DNA damage response in human skin tissue, *Biomed. Opt. Express* **4**, 559–568 (2013).

84. H. Ito, Breakthroughs in photonics 2013: Terahertz wave photonics, *IEEE Photonics J.* **6**, 1–5 (2014).

85. F. Blanchard, A. Doi, T. Tanaka, H. Hirori, H. Tanaka, Y. Kadoya, and K. Tanaka, Real-time terahertz near-field microscope, *Opt. Express* **19**, 8277–8284 (2011).

86. F. Blanchard, A. Doi, T. Tanaka, and K. Tanaka, Real-time, subwavelength terahertz imaging, *Annu. Rev. Mater. Res.* **43**, 237–259 (2013).

87. F. Blanchard, K. Ooi, T. Tanaka, A. Doi, and K. Tanaka, Terahertz spectroscopy of the reactive and radiative near-field zones of split ring resonator, *Opt. Express* **20**, 19395–19403 (2012).

88. M. A. Seo, H. R. Park, S. M. Koo, D. J. Park, J. H. Kang, O. K. Suwal, S. S. Choi et al., Terahertz field enhancement by a metallic nano slit operating beyond the skin-depth limit, *Nat. Photonics* **3**, 152–156 (2009).

89. M. Shalaby, H. Merbold, M. Peccianti, L. Razzari, G. Sharma, T. Ozaki, R. Morandotti et al., Concurrent field enhancement and high transmission of THz radiation in nanoslit arrays, *Appl. Phys. Lett.* **99**, 041110 (2011).

90. W. Cao, R. Singh, I. A. I. Al-Naib, M. He, A. J. Taylor, and W. Zhang, Low-loss ultra-high-Q dark mode plasmonic Fano metamaterials, *Opt. Lett.* **37**, 3366–3368 (2012).
91. I. Al-Naib, Y. Yang, M. M. Dignam, W. Zhang, and R. Singh, Ultra-high Q even eigenmode resonance in terahertz metamaterials, *Appl. Phys. Lett.* **106**, 011102 (2015).
92. I. Al-Naib, E. Hebestreit, C. Rockstuhl, F. Lederer, D. Christodoulides, T. Ozaki, and R. Morandotti, Conductive coupling of split ring resonators: A path to THz metamaterials with ultrasharp resonances, *Phys. Rev. Lett.* **112**, 183903 (2014).
93. R. Singh, I. A. I. Al-Naib, Y. Yang, D. Roy Chowdhury, W. Cao, C. Rockstuhl, T. Ozaki, R. Morandotti, and W. Zhang, Observing metamaterial induced transparency in individual Fano resonators with broken symmetry, *Appl. Phys. Lett.* **99**, 201107 (2011).
94. I. Al-Naib, C. Jansen, R. Singh, M. Walther, and M. Koch, Novel THz metamaterial designs: From near- and far-field coupling to high-Q resonances, *IEEE Trans. Terahertz Sci. Technol.* **3**, 772–782 (2013).

5 Applications of Terahertz Technology for Plastic Industry

Daniel M. Hailu and Daryoosh Saeedkia

CONTENTS

5.1 Introduction 107
5.2 Identifying and Sorting Black Plastic Polymers 108
5.3 Measuring Multilayer Plastic Bottles and Preforms 110
5.4 Measuring Wall Thickness of Hose and Tubes 114
5.5 Inspecting IV Bags for Leakage 116
5.6 Conclusion 117
References 117

5.1 INTRODUCTION

Terahertz (THz) technology [1] has come a long way since Fleming [2] in 1974 used the term "terahertz" (THz) to describe spectral line frequency coverage of a Michelson interferometer. THz is a form of light, part of the electromagnetic spectrum that fills the wavelength range from 1 mm to 100 µm (300 GHz–3 THz). Compared to x-ray energy, THz waves provide noninvasive and nonionizing imaging due to their low photon energy. THz technology [1,6–8] has opened a new horizon of remarkable opportunities within ever-growing industrial applications, wireless-data communications [7,9,10], material spectroscopy and sensing [6,16], monitoring and spectroscopy in science and the pharmaceutical industry [16–18], food industry, security, aerospace, oil industry, hidden object detection [21], and medical imaging.

Over the past two decades, intense research and development activities in academia and industry have closed the gap between the microwave and infrared spectra. Compact THz sources and detectors have been developed to generate, detect, and manipulate coherent THz signals, thanks to the growth of telecom semiconductor, and laser markets, as well as recent advances in semiconductor, laser and optical technologies. Photonic-based THz sources and detectors have become candidates for many industrial applications. Terahertz photoconductive antennas (THz-PCAs) are at the heart of THz time-domain systems widely used to generate THz broadband pulses and THz narrowband continuous-wave (CW) signals [3,6]. THz sensing and imaging systems are now commercially available, and THz wireless communication is on the horizon. It is expected that within the next 5–10 years, terabits and terabytes will become an everyday reality to meet consumer demands. Although major gains in performance and functionality are still anticipated, commercially available THz devices and systems have already made the THz spectrum accessible to many scientists and technologists in diverse areas, ranging from biology and medicine to chemical, pharmaceutical, and environmental sciences.

THz-PCAs are widely used to generate THz broadband pulses in terahertz time-domain spectroscopy (THz-TDS) systems [4–6,16]. Broadband THz pulses can be generated by exciting THz-PCAs with a femtosecond short pulse laser. Using a femtosecond laser with ~100 fs optical pulse duration, THz pulses with their frequency content extended up to around 5 THz and an average power of a few μW can be achieved [5,6]. These THz-PCAs are also used in an all-fiber, compact, THz spectrometer [13] used to demonstrate black plastic sorting. Since they have been demonstrated as practical THz sources and detectors, THz-PCAs have been the subject of a vast number of scientific and industrial reports investigating their application as THz wave transmitters and receivers. In CW mode, two CW laser beams, with their frequency difference in the THz range, combined inside an optical fiber, are mixed in a photo-absorbing medium (photomixer) and generate a beat frequency signal [6,14,15]. THz signals with the frequency linewidth as low as a few kHz can be generated by photomixers. The frequency of the THz signal can be tuned by changing the wavelengths of the lasers. The output power in conventional photomixers falls from 2 μW at 1 THz to below 0.1 μW at 3 THz. The use of photomixers sensor system is discussed and demonstrated.

5.2 IDENTIFYING AND SORTING BLACK PLASTIC POLYMERS

Plastics have a wide variety of uses and are a recoverable resource through recycling programs. However, the plastic recycling industry's current sorting technologies are challenged by an inability to sort the entire mixed streams of discarded plastic products into the primary resin type, thus restricting optimal recycling. Black plastics account for around 50% of the plastics that stream out of e-scrap, and the figure is rising with the development of new devices. These realities hold important consequences for recyclers because black plastics are not detectable using conventional separation processes. TeTechS has developed a sensor system that can enhance current optical sorting equipment by identifying the different plastic resins effectively and sort black plastic resins or composites that cannot be differentiated by any existing equipment. TeTechS is commercializing its THz-based plastic sorting system for the recovery of black plastic resins from electronic waste (e-waste), municipal, and industrial recycling streams as well as automotive shredder residue. The system represents a patented technology [11] in which a laser beam is converted into a THz waveform used to sort light and dark plastics. The system offers the capability for industrial users to characterize waste plastics coming from product packaging, electronics, and vehicles. By creating cost-effective recovery of black plastics in sorting processes, the increased ability to recover and reuse these polymers in a form that meets industry standards will increase the sustainability of our global resources.

The problem addressed here is the differentiation and identification of black plastics based on their polymer type. Current optical sorting devices use the near-infrared (NIR) portion of the electromagnetic spectrum to characterize plastic materials in waste streams by reflection, but NIR is only effective for light-colored plastics. If the waste plastics component is black, dark-colored, or painted, the NIR is absorbed by the black plastic and the NIR separation technology does not work. Other separation technology called float sink requires heavy water usage, water treatment, size reduction, final washing to eliminate salt residues, and final drying of the recovered plastic stream. The float sink process is much more energy intensive than optical systems. In e-waste, the typical black plastic portion of the plastic stream has been about 50%, and this black portion is increasing with the continuing development of new devices. Additionally, the majority of plastics used in automotive applications are black in color. Current technology of separation and recovery does not effectively work for the black fractions in the e-waste stream.

The proposed technique uses the properties of THz waves and THz-TDS of black plastic polymers [11,12] to sort them. TeTechS' Rigel system converts a laser beam into a THz waveform and uses THz waves to sort light- and dark-colored plastics. Rigel provides the ability to characterize waste plastics and specifically black plastics coming from e-waste and ASR. TeTechS' THz plastic sorting system provides the technology to global markets for the recovery of black plastic components from municipal and e-waste streams and from ASR streams.

Rigel system shown in Figure 5.1 uses TeTechS' T-Era THz transmitters and receivers and has the capability of obtaining a spectrum as the plastic is passing down a chute, determining the spectrum and differentiating based on a database of spectrum for various polymers. The absorption coefficient in the THz range is employed to sort the plastics where two THz pulses propagating through air and sample, respectively, are recorded and their spectral amplitudes and phases are compared: $\tilde{E}_{sam}(\omega)/\tilde{E}_{ref}(\omega) = T(\omega)e^{j\varphi(\omega)}$. The refractive index and absorption coefficient are found from the transmission coefficient [6,7,16], where d is the thickness of the sample:

$$n(\omega) = 1 + \frac{\varphi(\omega)c}{\omega d} \qquad (5.1)$$

and

$$\alpha(\omega) = -\frac{2}{d}\ln\left(\frac{(n+1)^2}{4n}T(\omega)\right) \qquad (5.2)$$

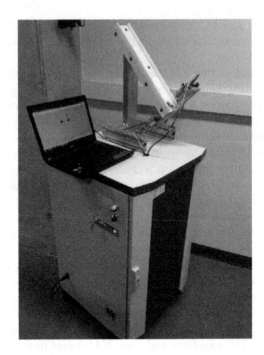

FIGURE 5.1   TeTechS' Rigel™ terahertz spectrometer [13] system is used to identify black plastic polymers.

THz passes through black plastics and the spectral information about the plastic samples gives the user the option to differentiate and sort the black plastics. Figure 5.2 shows the absorption coefficient for common black plastics. The sensors have the capability of scanning the moving plastic flakes with adjustable beam size between 50 mm for large pieces to 2 mm beam size for certain resolutions and sensitivity. Also, an array of sensors can cover a line scanner so that a number of plastics are sorted simultaneously.

5.3   MEASURING MULTILAYER PLASTIC BOTTLES AND PREFORMS

THz technology based on THz-TDS setup in reflection as well as transmission mode can be used for thickness measurement such as tablet coating thickness [22] measurement, paper thickness measurement [23], and bottle wall thickness measurement. We present the use of TeTechS' PlasThick™ THz sensor system for wall thickness measurement of an opaque plastic bottle. The problem addressed here is the wall thickness measurement and multilayer thickness measurement of plastic bottles and preforms, using a noncontact, noninvasive, and nondestructive measurement technique. Currently, to measure the wall thickness of transparent plastic bottles, various methods are being used. One is to use an infrared interferometry that

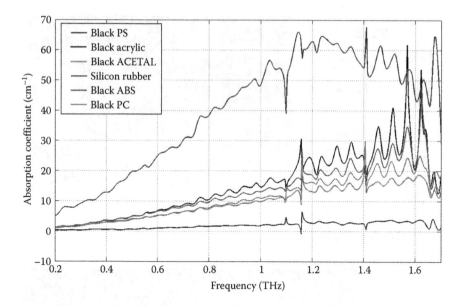

FIGURE 5.2 Terahertz measurement and absorption coefficients calculated from measurement data for plastic samples tested using a THz-based material identification system.

cannot measure opaque materials and low-contrast multilayer structures. An other method is to use Hall effect (Magna-Mike) measurement probes. Measurements are made when a magnetic probe is held or scanned on one side of the test material and a small target ball (or disk or wire) is placed on the opposite side or dropped inside a container. The probe's Hall effect sensor measures the distance between the probe tip and the target ball. This technique is time consuming, only measures overall wall thickness, cannot measure multilayer structures, and, is incapable of integrating to manufacturing lines. Thus, there is a need for a fast, noncontact, nondestructive, and high-precision measurement (~10 μm) technique for thickness measurement of plastic bottles and preforms, which also has the capability of integrating to manufacturing lines for advanced manufacturing.

The proposed reflection-mode time-domain THz pulse measurement setup measures the thickness of the walls of plastic bottles or the thickness of individual layers in multilayer structures. The THz pulse reflected from the walls of the plastic bottles has a specific time delay that allows users to calculate the thickness of each wall of an opaque plastic bottle. Figure 5.3 shows the schematic configuration of a reflection measurement setup for conducting thickness measurement using TeTechS' THz sensor system.

The echoes of the incident THz pulses are reflected from the walls and layers of the preform or bottle under test. THz pulses penetrate into plastics and are reflected at each plastic/air or multilayer boundary. The THz pulses from the transmitter go to the multilayer structure or sample under test and the reflected pulses from the sample are coupled into the THz detector. The reflected THz pulses from the multilayer

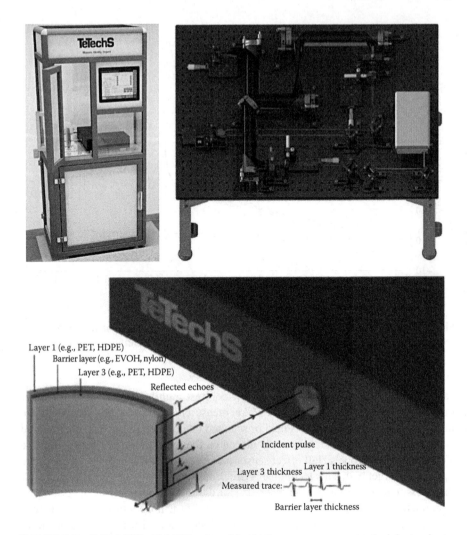

FIGURE 5.3 TeTechS PlasThick™ unit and its thickness measurement principle: terahertz (THz) pulse time-of-flight tomography and a THz time-domain spectroscopy measurement kit for reflection and transmission measurements.

sample have their time delay measured which corresponds to the thickness of the layers of the plastic bottle. The peak amplitudes of the reflected pulses also decrease as they experience absorption loss and Fresnel reflections. Figure 5.4 depicts the THz reflection-mode measurement for a transparent PET preform that shows the reflection from the first interface between air and PET (outside interface), reflections from PET and barrier interface, and the reflection from the last interface between air and PET (inside surface).

This method is independent of the opposite side wall of the bottle or preform and can be used for any multilayer structure. For the opposite wall of the bottle, the bottle can be rotated by 180° and the measurement can be conducted. The THz beam can

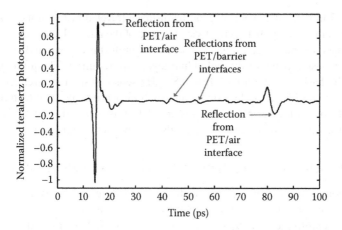

FIGURE 5.4 Terahertz pulse measurement trace for a PET multilayer preform.

be focused on the sample with a spot size of 1 mm to enhance the spatial resolution. For the monolayers seen in Figure 5.5, the overall thickness was measured based on the reflection of the THz pulses and using the formula

$$d = \frac{\Delta t \times c}{2n}$$

wherein the thickness in millimeters is related to the time delay Δt between peaks of the pulse in picoseconds; the refractive index of the material PET, n; the speed of light c,

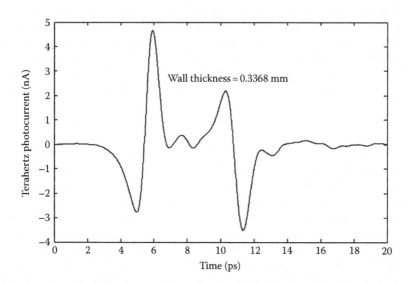

FIGURE 5.5 Terahertz pulse measurement trace for a PET monolayer bottle.

which is 0.3 mm/ps; and factor of 2 for the distance traveled by the probe beam in the THz time-domain setup that occurs twice because of the retroreflector delay line.

Multilayer samples such as plastic bottles and preforms appear challenging when thickness measurements of the individual layers are required. The aforementioned presents the basics for the use of TeTechS' THz sensor system in a reflection-mode measurement configuration for conducting such measurements. The technique uses THz pulses on multilayer plastic bottles and determines the thickness of each wall based on the time delay of the reflected pulses with a very high precision.

5.4   MEASURING WALL THICKNESS OF HOSE AND TUBES

When it comes to measuring the wall thickness of hose and tubes, there is a clear need for a measurement technique that is both noninvasive and nondestructive. Currently, measurement of such materials is done by methods including the widely used ultrasonic bath measurement technique which is a contact-based method that is highly sensitive to temperature change and requires the use of a couplant medium such as water to operate. This technique presents limitations for the applications wherein a couplant material in contact with the product is not desirable, such as in pharmaceutical tube manufacturing. The measurement is also very sensitive to temperature fluctuation. Thus, there is a need for a fast, noncontact, nondestructive, reliable, and high-precision measurement technique for thickness measurement of tubes and hose, which also has the capability of integrating to manufacturing lines for advanced manufacturing.

Figure 5.3 shows the schematic configuration of a reflection measurement setup for conducting thickness measurement using TeTechS' PlasThick THz measurement system. The incident THz pulses are reflected from the walls of the water hose shown in Figure 5.6. The THz pulses penetrate into the water hose and are reflected at each of the rubber/air or rubber/rubber boundaries. The pulses from the transmitter go to the multilayer structure and the reflected pulses from the sample are coupled back into the THz detector. The reflected THz pulses from the multilayer sample have their time delay measured which corresponds to the thickness of the layers. The peak amplitudes of the reflected pulses also decrease as they experience reflections consistent with Fresnel reflections. From the difference between the time delays of the reflected pulses, one can calculate the thickness of the layers.

Figure 5.7 shows the THz reflection-mode measurement for a multilayer water hose with the first reflected pulse from the surface from the air/hose interface and then the second reflection from the surface of the plastic outer layer to the rubber inner layer interface. The third reflection is from the rubber inner layer to the air interface. The refractive index of the plastic was 1.5 in the THz frequency range. The phase of the third pulse is reversed because the pulse is reflected from the dense layer/air interface. The difference between the first and second peaks is 1.55 mm, which corresponds to the actual thickness of the first plastic layer of the hose being 0.88 mm. The difference between the first and the third peak is 3.32 mm, which corresponds to the actual total thickness of the wall of the water hose being 1.89 mm. Table 5.1 summarizes the measurement results.

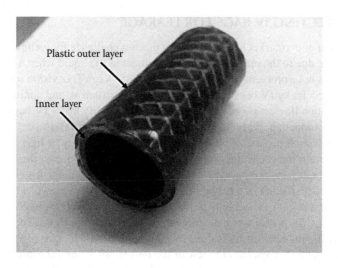

FIGURE 5.6   Multilayer water hose under test.

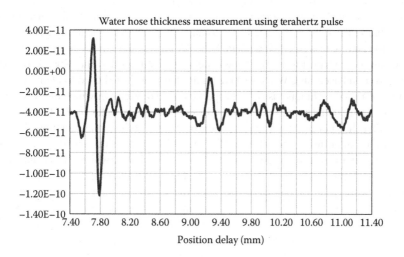

FIGURE 5.7   Terahertz pulse measurement trace for a multilayer water hose.

TABLE 5.1

	Layer Thickness (mm)
Plastic outer layer (green layer)	0.88
Inner layer of hose (foam plus liner)	1.01
Total wall thickness of hose	1.89

5.5 INSPECTING IV BAGS FOR LEAKAGE

The detection of premarket leaks in IV bags poses a challenge for techniques such as x-ray imaging due to the inability to see the contrast of the liquid when a leak is present; but this is not a problem for THz waves. TeTechS' Vega THz vision sensor system is able to detect leaky IV bags in a noncontact, nondestructive, and noninvasive way.

In automation lines, it is necessary to detect and identify leaky IV bags to remove them from production with high accuracy, but in industrial inspection applications, this detection process is currently performed using x-ray or light-based imaging. With certain materials, this method proves challenging due to the low contrast between the liquid and the plastic IV bag although in some cases, this challenge can be overcome with the use of a backlight. However, there are cases when the IV bag is cloudy and as a result the backlight will not suffice instead, THz would be needed. In some instances, the problem of detecting and identifying leaky IV bags on an automation line during manufacturing is done by a human visually identifying the presence of the liquid. The IV bags pass in view of the human for 1 s and the human then has to decide to reject it within that time frame and do it consistently for hours; as a result, this approach has a rejection rate that is prone to human error. There is zero tolerance for error as the presence of a leaky IV bag at even just 5% is unacceptable. Other techniques such as x-ray or other modalities of vision sensor systems cannot see the liquid leaking from an IV bag because of low contrast between the liquid substance and the plastic bags. Thus, there is a need to automate the process by including a sensor system in the automation line that detects the leaky bags with high accuracy using a THz vision sensor system. Figure 5.1 shows the leaky IV bags and the Vega THz vision sensor system based on CW mode THz photomixers [6,14,15] used to identify them.

The proposed technique uses the properties of THz waves in which the waves pass through plastic material and get absorbed by the liquid or water in the IV bags. The use of TeTechS' noncontact Vega THz sensor system verifies and identifies the presence of the liquid inside enclosures or IV bags. Vega is a vision sensor system with a transmitter and a receiver head connected to a control box, which can be integrated into a typical manufacturing automation line. Figure 5.8 shows the THz vision sensor system at work as an IV bag passes between the sensors and the system identifies the leaky IV bags as the THz signal drops and an LED lights up to indicate a leak. The Vega[TM] THz vision sensor system has a single-point, fiber-coupled transmitter sensor that emits a THz beam that after passing through the sample, such as the IV bags, is collected by the receiver sensor. The THz photocurrent signal acquired by the receiver is based on the transmitted THz field through the IV bags under test. The value of the THz photocurrent signal is based on the properties of the material under test such as the clear nonleaky IV bag versus a leaky IV bag. The signal passing through a nonleaky IV bag experiences a small drop in amplitude as a result of its interaction with the plastic bag, while the signal passing through the liquid in the leaky IV bag drops significantly. The level of the drop in signal depends on the amount of the liquid that has leaked from the internal IV bag into the outside bag. The software for the Vega THz vision sensor system does the data acquisition and processes the data in real time. The IV bags are identified while moving at a speed of 1 per second between the sensors. It also has the threshold levels adjusted to be able to detect the amount and the presence of the liquid

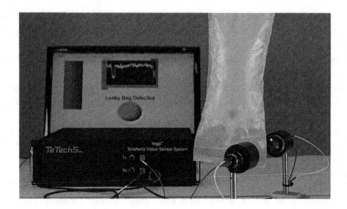

FIGURE 5.8 TeTechS' Vega™ terahertz vision sensor system for leaky IV bag detection. Here, the terahertz transmitter and receiver modules are placed at about 20 cm away from the IV bags. The graph depicted is the amplitude of the detected terahertz signal by the receiver module. If liquid is present, as in this case, the amplitude of the reflected signal is lower than expected, indicating that the inner bag is leaking.

that has leaked from the internal IV bag to the external bag. The sensors have the capability of scanning the moving bags with adjustable between 50 and 2 mm for certain resolution and sensitivity. Also, an array of sensors can cover the bottom part of the IV bag, where leaking happens, and create a 2D image of the moving IV bag.

Vega™ THz vision sensor system is employed for leaky IV bag detection with high certainty as an alternative to x-ray imaging. The system can also be used to detect fill level in plastic bottles as the THz passes through the bottle and is attenuated by the liquid inside if the bottle is full. The THz vision sensor system is a nondestructive, noncontact, and nonionizing testing technique that has the advantages of passing through plastic IV bags and seeing what liquid has leaked into the containing bag.

5.6 CONCLUSION

THz optoelectronic techniques play an essential role in future THz components and systems used for industrial application of THz technology, because of their unique capabilities in the generation and detection of coherent THz signals. From black plastics sorting to detecting leaks in IV bags, these are only some of the various industrial applications of THz technology. The continued exponential growth of THz systems and components demonstrate that this technology not only has strong implications for manufacturing industries but will also one day become available and affordable to the mass market.

REFERENCES

1. P. H. Siegel, Terahertz technology, *IEEE Transactions on Microwave Theory Techniques*, 50 (3), 910–928, 2002.
2. J. W. Fleming, High resolution submillimeter-wave Fourier-transform spectrometry of gases, *IEEE Transactions on Microwave Theory Techniques*, MTT-22, 1023–1025, December 1974.

3. D. Saeedkia and S. Safavi-Naeini, Terahertz photonics: Optoelectronic techniques for generation and detection of terahertz waves, *IEEE/OSA Journal of Lightwave Technology*, 26 (15), 2409–2423, 2008.

4. D. Grischkowsky, S. Keiding, M. Van Exter, and C. Fattinger, *Journal of the Optical Society of America B*, 7, 2006, 1990.

5. D. Saeedkia, Terahertz photoconductive antennas: Principles and applications, in *Proceedings of the Fifth European Conference on Antennas and Propagation (EUCAP)*, pp. 3326–3328, 2011.

6. D. Saeedkia, *Handbook of Terahertz Technology for Imaging, Sensing, and Communications*, Woodhead Publishing Series in Electronic and Optical Materials, No. 34, 2013.

7. M. Tonouchi, Cutting-edge terahertz technology, *Nature Photonics*, 1 (2), 97–105, 2007.

8. D. Mittleman, ed., *Sensing with Terahertz Radiation*, Vol. 85. Springer, 2013.

9. I. F. Akyildiz, J. M. Jornet, and C. Han, Full length article: Terahertz band: Next frontier for wireless communications, *Physical Communications*, 12 (September 2014), 16–32, 2014.

10. G. Fettweis, F. Guderian, and S. Krone, Entering the path towards terabit/s wireless links, in *Design, Automation & Test in Europe Conference & Exhibition (DATE)*, pp. 1–6, 14–18, March 2011.

11. D. Saeedkia, Methods and apparatus for identifying and sorting materials using terahertz waves, U.S. Patent Publication Number US20140367316 A1, filed December 11, 2012.

12. A. Maul and M. Nagel, Polymer identification with terahertz technology, In *OCM 2013-Optical Characterization of Materials-Conference Proceedings*. KIT Scientific Publishing, 2013.

13. A. Zandieh, D. M. Hailu, D. Biesty, A. Eshaghi, E. Fathi, and D. Saeedkia, Compact and reconfigurable fiber-based terahertz spectrometer at 1550 nm, in *SPIE OPTO*, International Society for Optics and Photonics, pp. 89850R–89850R, 2014.

14. E. R. Brown, F. W. Smith, and K. A. McIntosh, Coherent millimeter-wave generation by heterodyne conversion in low-temperature-grown GaAs photoconductors, *Journal of Applied Physics*, 73 (3), 1480–1484, 1993.

15. D. Saeedkia, S. Safavi-Naeini, and R. R. Mansour, The interaction of laser and photoconductor in a continuous-wave terahertz photomixer, *IEEE Journal of Quantum Electronics*, 41 (9), 1188–1196, 2005.

16. P. U. Jepsen, D. G. Cooke, and M. Koch, Terahertz spectroscopy and imaging—Modern techniques and applications, *Laser Photonics Reviews*, 5 (1), 124–166, 2011.

17. B. M. Fischer, H. Helm, and P. U. Jepsen, Chemical recognition with broadband THz spectroscopy, *Proceedings of the IEEE*, 95 (8), 1592–1604, 2007.

18. P. F. Taday, Applications of terahertz spectroscopy to pharmaceutical sciences, *Philosophical Transactions of the Royal Society of London*, A, 362, 351–364, 2004.

19. W. L. Chan, D. Jason, and D. M. Mittleman, Imaging with terahertz radiation, *Reports on Progress in Physics*, 70 (8), 1325, 2007.

20. D. Corinna, L. Koch, and P. U. Jepsen, Wall painting investigation by means of noninvasive terahertz time-domain imaging (THz-TDI): Inspection of subsurface structures buried in historical plasters, *Journal of Infrared, Millimeter, and Terahertz Waves*, 1–11, 2015.

21. W. R. Tribe, D. A. Newnham, P. F. Taday, and M. C. Kemp. Hidden object detection: Security applications of terahertz technology, in *Integrated Optoelectronic Devices 2004*, pp. 168–176, 2004.

22. A. J. Fitzgerald, B. E. Cole, and P. F. Taday, Nondestructive analysis of tablet coating thicknesses using terahertz pulsed imaging, *Journal of Pharmaceutical Sciences*, 94 (1), 177–183, 2005.

23. P. Mousavi, F. Haran, D. Jez, F. Santosa, and J. Dodge, Simultaneous composition and thickness measurement of paper using terahertz time-domain spectroscopy, *Applied Optics*, 48, 6541–6546, 2009.

24. P. H. Siegel, Terahertz technology in biology and medicine, in *Microwave Symposium Digest, 2004 IEEE MTT-S International*, Vol. 3. IEEE, 2004.

25. D. M. Hailu, I. Ehtezazi, and S. Safavi-Naeini, Terahertz imaging of biological samples, in *Antennas and Propagation Society International Symposium (APSURSI), 2010 IEEE*, pp. 1–4. IEEE, 2010.

26. K. Kawase, Y. Ogawa, Y. Watanabe, and H. Inoue, Non-destructive terahertz imaging of illicit drugs using spectral fingerprints, *Optics Express*, 11 (20), 2549–2554, 2003.

27. D. M. Hailu, H. Aziz, S. Safavi-Naeini, and D. Saeedkia, Terahertz time-domain spectroscopy of organic semiconductors, in *SPIE OPTO, International Society for Optics and Photonics*, pp. 86240C–86240C, 2013.

28. D. M. Hailu, S. Alqarni, B. Cui, and D. Saeedkia, Terahertz surface plasmon resonance sensor for material sensing, in *Photonics North 2013, International Society for Optics and Photonics*, pp. 89151G–89151G, 2013.

6 Cross-Sectional Velocity Distribution Measurement Based on Fiber-Optic Differential Laser Doppler Velocimetry

Koichi Maru

CONTENTS

6.1   Introduction 121
6.2   Differential LDV.. 122
6.3   Nonmechanical Scanning LDV 123
6.4   Directional Discrimination for Nonmechanical Scanning LDV 126
    6.4.1   LDV Using Acousto-Optic Frequency-Shifting Technique 126
    6.4.2   LDV Using Optical Serrodyne Frequency-Shifting Technique 128
6.5   Two-Dimensional Cross-Sectional Velocity Distribution Measurement..... 129
    6.5.1   Two-Dimensional Scanning Method 129
    6.5.2   Spatial Encoding Method 133
    6.5.3   Combination of Nonmechanical Scanning and Spatial Encoding.......134
6.6   Conclusions.. 136
References.. 137

6.1   INTRODUCTION

A laser Doppler velocimeter (LDV) has been a standard apparatus for measuring the velocity of a fluid flow, gas flow, or rigid object in various research fields and industries since the introduction of the concept in 1964 [1]. A differential LDV has the advantages of its noninvasive nature, small measurement volume giving excellent spatial resolution, and linear response. In many applications concerning velocity measurements, velocity distribution of the object often needs to be measured. For this purpose, various mechanical scanning techniques in differential LDVs have been proposed [2–7]. In these scanning LDVs, a moving mechanism is used in transmitting optics to scan the measurement point, for example, a movable lens [2] or rotating mirrors [3] for axial scanning (i.e., the direction of the scanning is parallel to the optical axis) or an oscillating mirror [4,5], a rotating diffraction grating [6], or a rotating transparent plate [7] for transverse scanning (i.e., the direction of the scanning is perpendicular to the optical axis). In particular, the information of cross-sectional velocity distribution, that is, the velocity distribution on the cross section perpendicular to the direction of flow, is often required in many fluid applications.

The structures of typical conventional scanning LDVs are often large. In some applications such as an industrial or biomedical use [8–11], an LDV with an easily handled probe, generally separated from a main body, has been highly desirable. When one intends to apply the conventional scanning techniques to an LDV with a probe, it is inevitable to use a moving mechanism in the probe, or otherwise the probe itself needs to be moved. However, the moving mechanism can be easily misaligned or damaged due to mechanical shock and is subject to abrasion, and therefore, the probe requires special care for handling and maintaining. The probe including a moving mechanism has also the disadvantage in terms of miniaturization. These problems should be solved to realize a scanning LDV with a compact and reliable probe.

In the conventional scanning LDVs, the direction of the scanning is limited to one dimension. The velocity distribution on a two-dimensional cross section of the flow needs to be measured in some applications such as measuring the velocity information of the fluid flow in narrow pipes. Many techniques have been developed and widely used for two-dimensional velocity distribution measurements, such as particle image velocimetry and particle tracking velocimetry [12], continuous-scan laser Doppler vibrometry [13], and Doppler global velocimetry [14]. However, these methods mainly concern velocity distribution in the front view.

In this chapter, several approaches proposed by the author's group to measuring the velocity distribution in one- or two-dimensional cross section of flow are reviewed. These techniques are based on fiber-optics and differential LDV. The use of fiber-optics enables us to realize an LDV with a compact and easily handled probe. Differential LDVs are feasible for measuring the transverse component of the velocity distributed over the cross section. In addition, these techniques involve a nonmechanical probe so that the reliability of the probe can be improved.

6.2 DIFFERENTIAL LDV

Figure 6.1 illustrates the basic structure of a differential LDV. The beam from a laser is split into two beams by a beam splitter, and each beam traverses transmitting optics. The beams are focused and incident on the measurement point. The measurement volume in the differential LDV is formed upon the crossing of the two focusing beams. By using transmitting optics with a proper design, the beams can be strongly focused and the size of the resultant measurement volume becomes small. This leads to the advantage of fine spatial resolution in the differential-type LDV. The beams scattered from the objects at the measurement point pass through receiving lenses and are detected by a photodetector (PD) such as a photodiode or photomultiplier as the beat signal whose frequency depends on the velocity of the

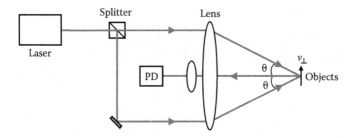

FIGURE 6.1   Basic structure of differential laser Doppler velocimeter.

objects at the measurement point. The frequency of the beat signal shifts by the Doppler frequency shift F_D is expressed as [15]

$$F_D = \frac{2v_\perp \sin\theta}{\lambda},$$ (6.1)

where
  θ is the incident angle of the beam on the objects
  $v_\perp$ is the velocity of the objects perpendicular to the bisector of the incident rays
    on the objects
  λ is the wavelength

6.3   NONMECHANICAL SCANNING LDV

Differential LDVs are basically for single-point measurements. To obtain the information of velocity distribution using the configuration of differential LDVs, techniques for scanning the measurement point, multipoint measurements, or laser Doppler profile sensing [16–18] have been proposed. In this section, a nonmechanical scanning technique is described.

The concept of the proposed fiber-optic nonmechanical scanning LDV is illustrated in Figure 6.2. Here, the term "nonmechanical" means that the LDV has no moving mechanism in its probe. In the concept proposed by the author's group [19], the measurement point is scanned by change in the wavelength of the light input to the probe, instead of using a moving mechanism in the probe. For this purpose, a tunable laser and diffraction gratings are used. Diffraction gratings have been conventionally used in LDVs for beam splitting [6,16,17,20–22], beam dividing with different wavelengths [16,17], frequency shifting [23,24], or achieving an achromatic function [20,25,26]. In contrast, in this LDV, diffraction gratings are used for beam scanning via the wavelength change. This LDV consists of a main body including the tunable laser and a probe including the gratings separated from the main body. The main body and probe are connected with optical fibers and electric cable. Optical fibers are often incorporated in velocimetry for a number of reasons such as the delivery of the beams to the flow, delivery of the scattered light to PDs, imaging applications, or delivery of high power laser beams [27]. The use of the optical fibers as the transmission paths between

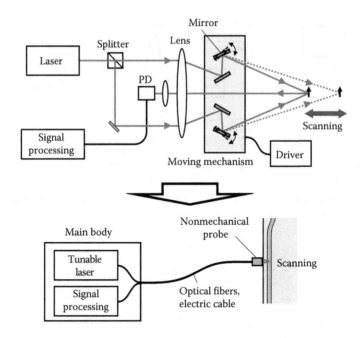

FIGURE 6.2 Concept of fiber-optic nonmechanical scanning laser Doppler velocimeter.

the main body and probe makes it possible to utilize an LDV with an easily handled probe, and many types of fiber-optic nonscanning LDVs with compact probes have been developed [9,10,28–30]. Moreover, in the proposed LDV, no moving mechanism is needed in its probe even if the tunable laser requires a moving mechanism because the tunable laser can be separated from the probe. Hence, the proposed concept brings the potential for realizing a scanning LDV with a compact and reliable probe.

Figure 6.3 illustrates the structure of a nonmechanical axial scanning LDV [19]. In this LDV, the beam from the tunable laser is transferred via a polarization-maintaining fiber (PMF). The beam is collimated and split into two beams in

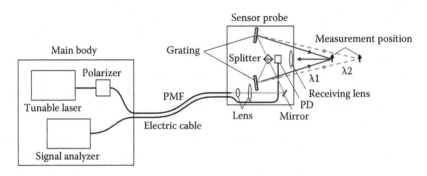

FIGURE 6.3 Structure of nonmechanical axial scanning laser Doppler velocimeter. (From Maru, K., *Opt. Express*, 19(7), 5960, 2011.)

the probe. Each beam is incident on each of two diffraction gratings. The diffracted beams from the gratings cross at the measurement point. The scattered beams are detected by the PD, and the signals are transferred to the signal analyzer in the main body. The diffraction angles of the beams change with the change in the wavelength of the beams incident on the gratings, and then the measurement point is scanned in the axial direction.

To demonstrate the scanning function of this LDV, an experiment was conducted using a setup of the probe consisting of bulk optical components [19]. The beam from a tunable laser was input to an in-line polarization beam splitter (PBS) to obtain a linearly polarized beam, and one of the beams from the PBS propagates through the PMF. The beam was launched as a vertically polarized beam into the optics of the probe setup. The beam passed through lenses for focusing and was split into two beams. Each beam was incident on each of reflection-type ruled diffraction gratings with a grating period of 1.67 μm. The first-order diffracted beams were incident on a rotating disk as a target. The beams were scattered at the surface of the rotating target. The detected beat signals were measured by a digital oscilloscope.

To investigate the scanning function in the axial direction accurately, the surface of the rotating target was angled with an angle of 57° from the vertical plane, and the target was moved in the vertical direction. The vertical position of the target was adjusted so that a peak of the beat signal appears in its spectrum. Because the distance between the measurement position and the center of rotation changed when the measurement position in the axial direction changed during the scanning, the measurement position could be estimated by monitoring the ratio of the beat frequency to the angular velocity of the target. The distance in the axial direction between each grating and the center of rotation of the target, L_z^c, was set to 273 mm. The measurement positions in the axial direction $L_z(\lambda)$ relative to L_z^c estimated from the measured beat frequency are plotted in Figure 6.4. Here, the incident angle θ on the measurement position was set to 10° for Setup (1) and 15° for Setup

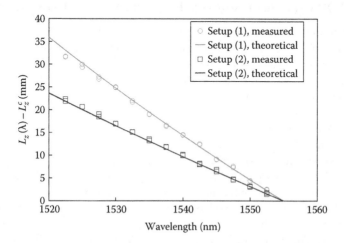

FIGURE 6.4 Relative measurement positions in the axial direction estimated from measured beat frequencies for various wavelengths. (From Maru, K., *Opt. Express*, 19(7), 5960, 2011.)

(2) at $\lambda = 1540$ nm, and $L_z(\lambda = 1555$ nm) was set to L_z^c. The measured positions agree with the theoretical values and shift almost linearly with the wavelength change. This result indicates that the axial scanning function is successfully demonstrated. The scanning range was estimated to be 29.3 mm for Setup (1) and 20.8 mm for Setup (2) over the wavelength range of 30 nm.

The measurement point can be scanned also in the transverse direction by adopting a Dove prism in transmitting optics [31]. Rearrangement of transmitting optics enables a dual-axis scanning LDV [32], in which the scanning direction is selected: in the axial direction or in the transverse direction, by changing the polarization of the beam input to the probe.

6.4 DIRECTIONAL DISCRIMINATION FOR NONMECHANICAL SCANNING LDV

In the nonmechanical scanning LDVs described in Section 6.3, the sign of the velocity component cannot be measured. To apply the proposed scanning method to most velocity measurement applications, introducing the function of directional discrimination is indispensable. Several techniques have been used to introduce the function of directional sensitivity, such as optical frequency shifting [8,9,23,24,28,33–37], quadrature mixing [38–40], or the use of a pair of PDs [41,42]. Among these techniques, optical frequency shifting has been widely used to incorporate directional sensitivity into a differential LDV, in which directional asymmetry is generated in the beat frequency by moving the interference pattern in the measurement volume in one direction. This section describes some methods based on optical frequency shifting for introducing directional sensitivity into the nonmechanical scanning LDV [43–46].

6.4.1 LDV USING ACOUSTO-OPTIC FREQUENCY-SHIFTING TECHNIQUE

The acousto-optic (AO) effect of Bragg cells is commonly used for optical frequency shifting in LDV. The structure of a nonmechanical axial scanning LDV with the function of directional discrimination [43] is illustrated in Figure 6.5a. In this LDV, the frequency of one of the beams split with a PBS is preshifted by F_0 using an acousto-optic modulator (AOM) included in the main body. Two beams, one preshifted and the other not preshifted, are transmitted separately to the probe with a pair of PMFs. The resultant beat frequency detected with the PD is based on the preshift frequency F_0 and becomes the sum of F_0 and the Doppler frequency shift F_D obtained by Equation 6.1. Because $v_\perp$ and F_D have the same sign as shown in Equation 6.1, the velocity including its directional information can be derived from F_D as the difference between the monitored beat frequency and F_0. The probe in this LDV can be kept simple and reliable because the AOM as well as the tunable laser can be separated from the probe.

In the experimental setup for the demonstration of this LDV as shown in Figure 6.5b [43], one beam traversed two AOMs with its frequency shifted before the beam was input to the probe setup. The frequency of the beam was upshifted

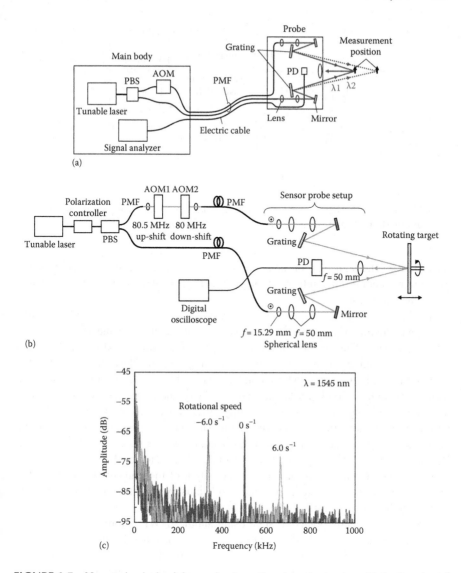

FIGURE 6.5 Nonmechanical axial scanning laser Doppler velocimeter with the function of directional discrimination: (a) structure, (b) experimental setup, and (c) measured spectra of beat signals for $\lambda = 1545$ nm. (From Maru, K. and Hata, T., *Appl. Opt.*, 51(20), 4783, 2012.)

by 80.5 MHz with the first AOM and downshifted by 80.0 MHz with the second AOM. These frequency shifts resulted in a total upshift of 0.5 MHz as the preshift frequency F_0. The spectra of beat signals measured using a rotating disk as a target are plotted in Figure 6.5c. Here, the rotational speeds of the target were set to −6.0, 0, and 6.0 s^{-1}. Each of the spectra had a sharp peak corresponding to the beat frequency. The peak was located at F_0 (0.5 MHz) when the target was not rotated and shifted with rotation. Whether the beat frequency increased or decreased from F_0

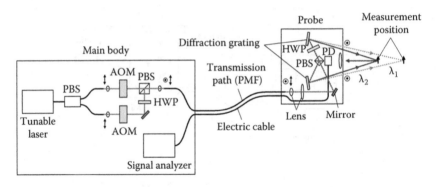

FIGURE 6.6 Nonmechanical scanning laser Doppler velocimeter using single transmission path. (From Maru, K. and Hata, T., *Optik*, 125(20), 6312, 2014.)

depended on the sign of rotational speed, that is, the direction of velocity at the measurement point. This indicates that the direction of velocity can be discriminated with this configuration.

The directionally discriminated LDV described earlier has two transmission paths between the main body and probe and, hence, is sensitive to noise due to environmental disturbances. The function of directional discrimination can also be introduced by using the configuration with a single transmission path illustrated in Figure 6.6 [45]. In this configuration, two horizontally polarized beams from a tunable laser are incident on AOMs. To obtain directional sensitivity, the frequencies of these beams are preshifted using the AOMs with slightly different frequencies. One of the beams is converted into a vertically polarized beam using a half-wave plate (HWP), and the polarization of the other beam is kept horizontal. Two beams are combined with a PBS and input to a PMF as a transmission path. The states of the two polarization modes are maintained from the input to the output of the transmission path. In the probe, the beam is collimated, split into two beams by another PBS again, and diffracted by diffraction gratings. Then, only the horizontally polarized beam is converted into a vertically polarized state using another HWP before the grating. The scanning LDV using a single PMF as its transmission path has an advantage, because two orthogonally polarized modes experience exactly the same physical conditions such as temperature or pressure and, hence, a high degree of common mode rejection of interference effects due to external conditions exists [28]. This configuration can be applied also to the two-dimensional scanning method described in Section 6.5.1 [46].

6.4.2 LDV Using Optical Serrodyne Frequency-Shifting Technique

Optical serrodyne modulation is another promising method for optical frequency shifting. A LiNbO$_3$ (LN) phase shifter that uses the electro-optic (EO) effect is suitable for optical serrodyne frequency shifting, because it has a wide bandwidth and does not require high-power signals. Optical frequency shifting with a waveguide-type LN phase shifter is suitable for introducing the function of directional discrimination into fiber-optic LDVs.

Optical serrodyne frequency shifting has been used in several types of LDVs for single-point velocity measurement with directional discrimination [34–37]. The author's group has introduced optical serrodyne frequency shifting using an LN phase shifter into a nonmechanical axial scanning LDV [44]. The configuration of this LDV is almost the same as that illustrated in Figure 6.5a, except that one beam is input to an LN phase shifter to shift its optical frequency via serrodyne modulation instead of AOMs. The frequency of one of the beams is preshifted using the LN phase shifter driven by a sawtooth voltage signal with linear ramps and a duty cycle of nominally 100%. The phase of the beam propagating the phase shifter linearly changes with the applied voltage. Serrodyne frequency shifting is accomplished when the peak-to-peak voltage of the sawtooth signal corresponds to $2n\pi$ phase shift, where n is an integer. Then, the preshift frequency F_0 is given by nF_s, where F_s is the frequency of the sawtooth signal. The higher-order harmonic of the modulated signal was sufficiently suppressed by using a waveguide-type LN phase shifter (the amplitude suppression ratio of the second-order harmonic to the fundamental harmonic was less than –29 dB for $F_s = 0.5$ MHz and –23 dB for $F_s = 1$ MHz for the wavelengths of 1525, 1545, and 1565 nm) [44].

6.5   TWO-DIMENSIONAL CROSS-SECTIONAL VELOCITY DISTRIBUTION MEASUREMENT

The LDVs described earlier are for one-dimensional velocity distribution measurements. Some applications such as measuring the fluid flow in pipes require the information of velocity distribution in a two-dimensional cross section of the flow. In this section, techniques based on differential LDV for measuring two-dimensional cross-sectional velocity distribution are described.

6.5.1   TWO-DIMENSIONAL SCANNING METHOD

The concept of the two-dimensional nonmechanical scanning LDV [47] is shown in Figure 6.7. The main body of this LDV includes an optical switch, a tunable laser, and a signal analyzer. The main body and probe are connected with a fiber array and electric cable. The beam from the tunable laser is input to the optical switch and input to the probe via one of the fibers in the fiber array. The beam is collimated and incident on the diffraction grating in the probe. Here, the angle of the incident on the grating is set to 0° to enhance the axial scanning range [48]. The beam is diffracted symmetrically, and the +first- and −first-order diffracted beams cross each other at the measurement point.

In this concept, a change in the input position using the optical switch and fiber array and nonmechanical scanning using the tunable laser and diffraction grating are combined to scan the measurement point two-dimensionally. The measurement point is scanned in the axial direction by changing the wavelength of the light input to the probe and in the transverse direction by changing the port of the fiber array. The measurement position in the transverse direction depends on the input position of the beam to the probe. Hence, the transverse measurement position can be controlled by

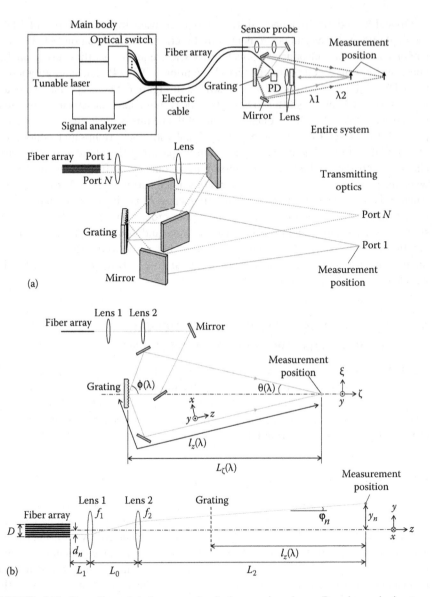

FIGURE 6.7 Two-dimensional nonmechanical scanning laser Doppler velocimeter: (a) structure and (b) model of transmitting optics. (From Maru, K. and Hata, T., *Appl. Opt.*, 51(34), 8177, 2012.)

selecting the input port of the fiber array with the optical switch. Because the optical switch as well as the tunable laser can be separated from the probe, the probe can be kept simple and reliable.

The scanning range can be theoretically derived by applying ray tracing to the transmitting optics [47]. The model of the transmitting optics is illustrated in Figure 6.7b. The output ports of the fiber array are arranged along the *y* axis. Lenses 1 and 2 are put after the fiber array to focus the beam and adjust the transverse measurement position. The position and direction of the end facet of the fiber array are set so that the beam array lies in a meridional plane of the lens system included in the yz plane. The beam from the nth port of the fiber array is first collimated with Lens 1 with a focal length f_1 and next focused with Lens 2 with a focal length f_2. The beam is diffracted at the grating whose rulings are parallel to the y axis. On the basis of the behavior of the ray in conical diffraction [49,50], the elevation angle of the beam in the yz plane, φ_n, is kept constant before and after the diffraction because the direction cosine of the ray along the rulings does not change before and after the diffraction. Under the paraxial approximation, the position along the y axis, y_n, and the elevation angle φ_n at the measurement point are expressed using ray-transfer matrices for two thin lenses as [26]

$$\begin{pmatrix} y_n \\ \varphi_n \end{pmatrix} = d_n \begin{pmatrix} 1 - L_0 f_1^{-1} + L_2 \left(L_0 f_1^{-1} f_2^{-1} - f_1^{-1} - f_2^{-1} \right) \\ L_0 f_1^{-1} f_2^{-1} - f_1^{-1} - f_2^{-1} \end{pmatrix}, \qquad (6.2)$$

where
   d_n is the displacement of the nth core in the fiber array in the y direction from the z axis
   L_1 is the distance between the end facet of the fiber array and Lens 1
   L_0 is the distance between Lenses 1 and 2
   L_2 is the distance between Lens 2 and the measurement point

Note that L_2 changes according to the axial scan and hence depends on the wavelength. When the beam incident on Lens 2 is collimated, the beam is focused at $L_2 = f_2$. In this case, the position y_n at the focusing point is independent of L_0 and given by

$$y_n = -d_n \frac{f_2}{f_1}. \qquad (6.3)$$

Accordingly, the beam is almost focused at the measurement point by placing the measurement point at the distance f_2 from Lens 2. In this case, letting D be the distance between the cores of the marginal ports in the fiber array, the transverse scanning range is expressed as Df_2/f_1. Here, y_n is approximated to $-d_n f_2/f_1$ over the scanning range provided that f_1 and f_2 are much larger than d_n so that φ_n becomes small.

The measurement position is derived as follows. Suppose that the ray is incident on the grating at the angle φ_n in the yz plane and $0°$ in the $\xi\zeta$ plane. The diffraction angle of the $\pm$first-order diffracted beams projected onto the $\xi\zeta$ plane, $\phi(\lambda)$, is expressed on the basis of the grating equation for oblique incidence [49,50] as

$$\phi(\lambda) = \sin^{-1} \left(\frac{\lambda}{\Lambda \cos \varphi_n} \right), \qquad (6.4)$$

where Λ is the grating period. The axial measurement position for a wavelength λ, $L_\zeta(\lambda)$, is derived as the relative value from the position for $\lambda = \lambda_0$, $L_\zeta(\lambda_0)$:

$$L_\zeta(\lambda) - L_\zeta(\lambda_0) = l_z(\lambda_0)\sin(\theta(\lambda_0)) \left[\frac{1}{\tan(\theta(\lambda_0) + \phi(\lambda) - \phi(\lambda_0))} - \frac{1}{\tan(\theta(\lambda_0))} \right], \quad (6.5)$$

where

$\theta(\lambda)$ is the incident angle on the measurement point projected on the $\xi\zeta$ plane

$l_z(\lambda_0)$ is the path length of the ray from the grating to the measurement point along the z axis for $\lambda = \lambda_0$

$l_z(\lambda)$ for a general value of λ is expressed as

$$l_z(\lambda) = \frac{l_z(\lambda_0)\sin(\theta(\lambda_0))}{\sin(\theta(\lambda_0) + \phi(\lambda) - \phi(\lambda_0))}. \quad (6.6)$$

Then, from Equation 6.2, the transverse measurement position y_n is determined by letting L_2 be the sum of $l_z(\lambda)$ and the distance between Lens 2 and the grating along the z axis.

The incident angle on the measurement point, γ, is given by $\gamma = \sin^{-1}(\sin(\theta(\lambda)) \cos \varphi_n)$; that is, γ slightly deviates from $\theta(\lambda)$ when φ_n is not zero.

A proof-of-concept experiment using a commercial 24-channel PMF array was performed to demonstrate the two-dimensional scanning function [47]. In this experiment, switching of the port was done manually by detaching and reattaching FC/PC connectors. The design parameters were $f_1 = 50$ mm, $f_2 = 500$ mm, $D = 2.92$ mm, $L_0 = 100$ mm, $\lambda_0 = 1545$ nm, $\Lambda = 1.67$ μm, $\theta(\lambda_0) = 15°$, and $l_z(\lambda_0) = 349$ mm. The core pitch of the PM fiber array was 127 μm at the end facet on the probe side. For receiving optics in the probe setup, two cylindrical lenses with different focal lengths were combined to expand the beam spot in the vertical direction at the detected surface of the PD so that sufficient optical power could be received from any measurement point. Two-dimensional measurement positions for seven wavelengths between 1536 and 1554 nm and eight ports in the fiber array are plotted in Figure 6.8. The intersections of the dotted lines indicate theoretical values. The result reveals that the measurement point can be scanned in two dimensions by using this configuration. The interval in the transverse direction between adjacent ports increased with the increase in the axial position because the elevation angle of the ray φ_n was positive for smaller port numbers and negative for larger port numbers around the measurement points. The scanning range was 39.7 mm in the axial direction and 26.1 mm in the transverse direction.

In this experiment, the wavelength was not swept but set at intervals. Moreover, the port was switched by the detachment and reattachment of the connectors instead of the use of an optical switch. The two-dimensional scanning technique described in this section can be adapted for practical applications that require fast two-dimensional scanning such as the measurement in an unsteady flow or pulsatile flow by using a high-speed multichannel optical switch such as those reported in [51–53] and a fast wavelength-swept laser as those reported in [54,55].

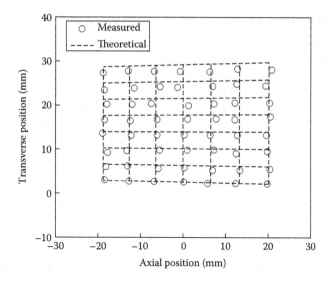

FIGURE 6.8 Two-dimensional measurement positions for seven wavelengths over 1536–1554 nm with a period of 3 nm for ports 3, 6, 9, 12, 15, 18, 21, and 24 of the fiber array. (From Maru, K. and Hata, T., *Appl. Opt.*, 51(34), 8177, 2012.) The intersections of the dotted lines indicate theoretical values.

6.5.2 SPATIAL ENCODING METHOD

Simultaneous multipoint measurement, in which the velocities at multiple measurement points are measured simultaneously, is another approach to measure a velocity distribution nonmechanically. Many types of differential LDVs have been reported for simultaneous multipoint measurement, for example, an LDV using multiple probes [56] or using an optical fiber array and multiple PDs in receiving optics [57–59].

Spatial encoding [60–62] is one of the promising techniques to obtain multiple measurement points. This method has an advantage, that is, the velocity distribution can be measured with a single PD. Li et al. [60] reported a method for measuring three-point velocity distribution in one dimension. Fu et al. [61,62] reported reference-type LDVs for two-dimensional vibration measurement using a 2 × 5 or 5 × 4 beam array. The spatially encoding technique uses multiple measurement points encoded with different carrier frequencies generated by an optical frequency shifting technique such as those described in Section 6.4, and the velocity information from the different measurement points is extracted by separating the signals in a spectral manner. The frequency of one of the rays forming the nth measurement point is shifted by F_n. Then, on the basis of heterodyne interferometry, the beat frequency of the rays scattered at this point is preshifted by F_n. Here, the preshift frequencies for all the measurement points are set to be different from one another so that each measurement point is encoded with the corresponding preshift frequency. Then, provided that the Doppler frequency shift is limited to less than half the spacing of the adjacent preshift frequencies, the velocity information of the nth measurement point can be extracted as the spectral peak with the beat frequency around F_n.

6.5.3 COMBINATION OF NONMECHANICAL SCANNING AND SPATIAL ENCODING

Simultaneous multipoint measurement using frequency shifting requires many frequency shifters for dense velocity distribution measurement. In the conventional multipoint LDVs, AOMs was used for optical frequency shifting to generate spatially encoded measurement points. However, the typical available AOMs are bulky and the configuration of the transmitting optics using AOMs becomes complicated as the number of measurement points increases. On the other hand, the use of optical serrodyne frequency shifting (described in Section 6.4.2) with a waveguide-type LN phase shifter simplifies the structure of the system because many phase shifters can be integrated in one chip and a beam array can be easily input and output using fiber arrays. However, if one intends to apply the multipoint method to dense two-dimensional velocity distribution measurement, many frequency/phase shifters are required for frequency shifting because a large number of beams with different carrier frequencies should be prepared to cover a two-dimensional area.

To overcome this problem, the author's group has proposed a technique combining nonmechanical scanning and simultaneous multipoint measurement using spatial encoding by multichannel optical serrodyne frequency shifting for two-dimensional cross-sectional velocity distribution measurement [63,64]. The structure is illustrated in Figure 6.9. The main body of this LDV includes an LN phase-shifter array as well as a tunable laser. The beam from the tunable laser is divided into two beam arrays, each of which includes N beams. The beams composing one of the beam arrays are modulated with serrodyne modulation by the LN phase-shifter array to generate N spatially encoded measurement points. The beam arrays are launched into the probe via two PM fiber arrays, and the beams are collimated, diffracted on the grating, and focused while traversing the transmitting optics. Here, cylindrical lenses are inserted into the transmitting optics to reduce the astigmatism due to the diffraction grating. The two beam arrays intersect and generate N spatially encoded measurement points aligned along the transverse direction. The measurement points are scanned in the axial direction in the nonmechanical manner by changing the wavelength. This LDV requires only a moderate number of phase shifters because the simultaneous multipoint measurement is performed only in one dimension and velocity distribution in another dimension is measured by employing the nonmechanical scanning. The use of the LN phase-shifter array makes the optical system for generating spatially encoded measurement points simpler than that using AOMs.

In the experiment using a six-channel spatially encoded beam array [63], one of the beam arrays was input to the LN phase-shifter array that consisted of straight waveguides arranged in parallel on a z-cut LN substrate. Here, the polarization of each input beam was adjusted to be coupled to the transverse magnetic mode of each waveguide. The number of measurement points can be increased by employing a series of phase shifters [64] or introducing an asymmetrical push–pull configuration. In this experiment, six spatially encoded measurement points with preshift frequencies $F_1, \ldots, F_6$ over 0.25–1.50 MHz at an interval of 0.25 MHz were generated. Sawtooth signals were applied to the electrodes on the waveguides by signal generators. The bias voltages of the sawtooth signals were set to zero to avoid DC drift. The measurement positions in two dimensions for all six ports and seven

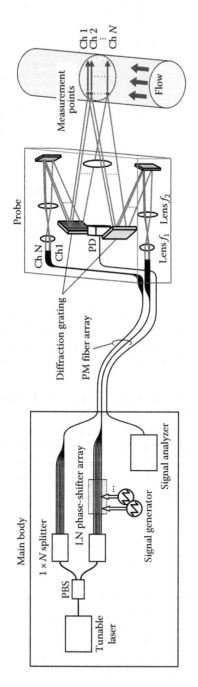

FIGURE 6.9 Cross-sectional laser Doppler velocimeter combining nonmechanical scanning and simultaneous multipoint measurement using spatial encoding.

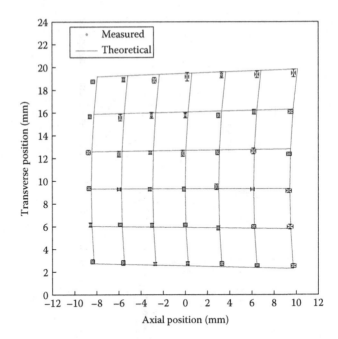

FIGURE 6.10 Measurement positions in two dimensions for seven wavelengths over 1530–1560 nm with a period of 5 nm for six ports. (From Maru, K. and Watanabe, K., *Opt. Lett.*, 39(1), 135, 2014.) Each point represents the average of six measurements in each direction. Error bars represent one standard deviation. The intersections of the dotted lines indicate theoretical values.

wavelengths over 1530–1560 nm are plotted in Figure 6.10. The intersections of the dotted lines indicate the theoretical values. Here, the distance between the gratings and measurement points was set to 167 mm along the axial direction for the beams with λ = 1545 nm. The measurement positions were distributed two-dimensionally and agreed well with the theoretical ones. The estimated measurement area was 18.5 mm in the axial direction over the wavelengths of 1530–1560 nm and 16.9 mm in the transverse direction for the six channels.

6.6 CONCLUSIONS

Several approaches based on differential LDV to the measurements of one- or two-dimensional cross-sectional velocity distribution are described. The concept of combining fiber optics and nonmechanical scanning makes it possible to realize velocity distribution measurement using an apparatus with a compact, reliable, and easily handled probe. The measurement point can be scanned in the axial or transverse direction without any moving mechanism in the probe by using a tunable laser and gratings. The function of directional discrimination is introduced into the nonmechanical scanning LDVs by using optical frequency shifting via the AO or EO effect. Two-dimensional velocity distribution can be measured by using an optical switch and fiber arrays as transmission paths between the main body and the probe. The combination of nonmechanical scanning and spatial encoding enables us to realize the method that can measure more dense two-dimensional velocity distribution.

The proposed concept can also be adopted to miniaturize the entire system of the LDV by using an integrated tunable laser such as the combination of a semiconductor optical amplifier and tunable waveguide ring resonators [65,66]. Applying integrated waveguide technologies to laser velocimetry has a potential for realizing ultrasmall devices for velocity distribution measurements.

REFERENCES

1. Y. Yeh and H. Z. Cummins, Localized fluid flow measurements with an He-Ne laser spectrometer, *Appl. Phys. Lett.*, 4(10), 176–178, 1964.
2. G. R. Grant and K. L. Orloff, Two-color dual-beam backscatter laser Doppler velocimeter, *Appl. Opt.*, 12, 2913–2916, 1973.
3. M. Uchiyama and K. Hakomori, A beam scanning LDV to measure velocity profile of unsteady flow, *Seimitsu Kikai (Journal of the Japan Society of Precision Engineering)*, 48, 939–944, 1982 (in Japanese).
4. F. Durst, B. Lehmann, and C. Tropea, Laser-Doppler system for rapid scanning of flow fields, *Rev. Sci. Instrum.*, 52, 1676–1681, 1981.
5. P. Sriram, S. Hanagud, J. Craig, and N. M. Komerath, Scanning laser Doppler technique for velocity profile sensing on a moving surface, *Appl. Opt.*, 29, 2409–2417, 1990.
6. N. Nakatani, T. Nishikawa, Y. Yoneda, Y. Nakano, and T. Yamada, Space-correlation measurement of attaching jets by the new scanning laser Doppler velocimeter using a diffraction grating, in *Proceedings of Seventh Symposium on Turbulence*, pp. 380–389, University of Missouri-Rolla, Rolla, MO, 1981.
7. E. B. Li, A. K. Tieu, and W. Y. D. Yuen, Measurements of velocity distributions in the deformation zone in cold rolling by a scanning LDV, *Opt. Lasers Eng.*, 35, 41–49, 2001.
8. T. Eiju, K. Matsuda, J. Ohtsubo, K. Honma, and K. Shimizu, A frequency shifting of LDV for blood velocity measurement by a moving wedged glass, *Appl. Opt.*, 20, 3833–3837, 1981.
9. H. Nishihara, J. Koyama, N. Hoki, F. Kajiya, M. Hironaga, and M. Kano, Optical-fiber laser Doppler velocimeter for high-resolution measurement of pulsatile blood flows, *Appl. Opt.*, 21, 1785–1790, 1982.
10. M. H. Koelink, F. F. M. de Mul, J. Greve, R. Graaff, A. C. M. Dassel, and J. G. Aarnoudse, Laser Doppler blood flowmetry using two wavelengths: Monte Carlo simulations and measurements, *Appl. Opt.*, 33, 3549–3558, 1994.
11. W. N. Sharpe, Jr., *Springer Handbook of Experimental Solid Mechanics*, Section 29.6, Springer, New York, 2008.
12. M. Stanislas, K. Okamoto, and C. Kähler, Main results of the First International PIV Challenge, *Meas. Sci. Technol.*, 14(10), R63–R89, 2003.
13. M. S. Allen and M. W. Sracic, A new method for processing impact excited continuous-scan laser Doppler vibrometer measurements, *Mech. Syst. Signal Proc.*, 24(3), 721–735, 2010.
14. J. F. Meyers, Development of Doppler global velocimetry as a flow diagnostics tool, *Meas. Sci. Technol.*, 6(6), 769–783, 1995.
15. H.-E. Albrecht, M. Borys, N. Damaschke, and C. Tropea, *Laser Doppler and Phase Doppler Measurement Techniques*, Section 2.1, Springer, Berlin, Germany, 2003.
16. J. Czarske, L. Büttner, T. Razik, and H. Müller, Boundary layer velocity measurements by a laser Doppler profile sensor with micrometre spatial resolution, *Meas. Sci. Technol.*, 13, 1979–1989, 2002.

17. L. Büttner, J. Czarske, and H. Knuppertz, Laser-Doppler velocity profile sensor with submicrometer spatial resolution that employs fiber optics and a diffractive lens, *Appl. Opt.*, 44(12), 2274–2280, 2005.

18. T. Pfister, L. Büttner, K. Shirai, and J. Czarske, Monochromatic heterodyne fiber-optic profile sensor for spatially resolved velocity measurements with frequency division multiplexing, *Appl. Opt.*, 44(13), 2501–2510, 2005.

19. K. Maru, Axial scanning laser Doppler velocimeter using wavelength change without moving mechanism in sensor probe, *Opt. Express*, 19(7), 5960–5969, 2011.

20. J. Schmidt, R. Völkel, W. Stork, J. T. Sheridan, J. Schwider, N. Streibl, and F. Durst, Diffractive beam splitter for laser Doppler velocimetry, *Opt. Lett.*, 17(17), 1240–1242, 1992.

21. H. W. Jentink, J. A. J. van Beurden, M. A. Helsdingen, F. F. M. de Mul, H. E. Suichies, J. G. Aarnoudse, and J. Greve, A compact differential laser Doppler velocimeter using a semiconductor laser, *J. Phys. E Sci. Instrum.*, 20, 1281–1283, 1987.

22. F. Onofri, Three interfering beams in laser Doppler velocimetry for particle position and microflow velocity profile measurements, *Appl. Opt.*, 45, 3317–3324, 2006.

23. J. Oldengarm and P. Venkatesh, A simple two-component laser Doppler anemometer using a rotating radial diffraction grating, *J. Phys. E Sci. Instrum.*, 9, 1009–1012, 1976.

24. J. P. Sharpe, A phase-stepped grating technique for frequency shifting in laser Doppler velocimetry, *Opt. Lasers Eng.*, 45, 1067–1070, 2007.

25. R. Sawada, K. Hane, and E. Higurashi, *Hikari Maikuromashin (Optical Micro Electro Mechanical Systems)*, Section 5.2, Ohmsha, Tokyo, 2002 (in Japanese).

26. H.-E. Albrecht, M. Borys, N. Damaschke, and C. Tropea, *Laser Doppler and Phase Doppler Measurement Techniques*, Section 7.2, Springer, Berlin, Germany, 2003.

27. T. O. H. Charrett, S. W. James, and R. P. Tatam, Optical fibre laser velocimetry: A review, *Meas. Sci. Technol.*, 23(3), 032001, 2012.

28. J. Knuhtsen, E. Olldag, and P. Buchhave, Fibre-optic laser Doppler anemometer with Bragg frequency shift utilising polarization-preserving single-mode fibre, *J. Phys. E Sci. Instrum.*, 15, 1188–1191, 1982.

29. M. Stiegimeier and C. Tropea, Mobile fiber-optic laser Doppler anemometer, *Appl. Opt.*, 31, 4096–4105, 1992.

30. H.-E. Albrecht, M. Borys, N. Damaschke, and C. Tropea, *Laser Doppler and Phase Doppler Measurement Techniques*, Section 7.3, Springer, Berlin, Germany, 2003.

31. K. Maru, T. Fujiwara, and R. Ikeuchi, Nonmechanical transverse scanning laser Doppler velocimeter using wavelength change, *Appl. Opt.*, 50(32), 6121–6127, 2011.

32. K. Maru, Non-mechanical dual-axis scanning laser Doppler velocimeter, *IEEE Sens. J.*, 12(8), 2648–2652, 2012.

33. X. Shen, J. Zhang, Z. Wang, and H. Yu, Two component LDV system with dual-differential acousto-optical frequency shift and its applications, *Acta Mechanica Sinica*, 2(1), 81–92, 1986.

34. H. Toda, M. Haruna, and H. Nishihara, Optical integrated circuit for a fiber laser Doppler velocimeter, *J. Lightwave Technol.*, 5(7), 901–905, 1987.

35. H. Toda, K. Kasazumi, M. Haruna, and H. Nishihara, An optical integrated circuit for time-division 2-D velocity measurement, *J. Lightwave Technol.*, 7(2), 364–367, 1989.

36. Y. Li, S. Meersman, and R. Baets, Realization of fiber-based laser Doppler vibrometer with serrodyne frequency shifting, *Appl. Opt.*, 50(17), 2809–2814, 2011.

37. Y. Li, S. Verstuyft, G. Yurtsever, S. Keyvaninia, G. Roelkens, D. Van Thourhout, and R. Baets, Heterodyne laser Doppler vibrometers integrated on silicon-on-insulator based on serrodyne thermo-optic frequency shifters, *Appl. Opt.*, 52(10), 2145–2152, 2013.

38. D. O. Hogenboom and C. A. DiMarzio, Quadrature detection of a Doppler signal, *Appl. Opt.*, 37(13), 2569–2572, 1998.

39. L. Büttner and J. Czarske, Passive directional discrimination in laser-Doppler ane-mometry by the two-wavelength quadrature homodyne technique, *Appl. Opt.*, 42(19), 3843–3852, 2003.
40. Y. Li and R. Baets, Homodyne laser Doppler vibrometer on silicon-on-insulator with integrated 90 degree optical hybrids, *Opt. Express*, 21(11), 13342–13350, 2013.
41. T. Ito, R. Sawada, and E. Higurashi, Integrated microlaser Doppler velocimeter, *J. Lightwave Technol.*, 17(1), 30–34, 1999.
42. K. Plamann, H. Zellmer, J. Czarske, and A. Tünnermann, Directional discrimination in laser Doppler anemometry (LDA) without frequency shifting using twinned optical fibres in the receiving optics, *Meas. Sci. Technol.*, 9, 1840–1846, 1998.
43. K. Maru and T. Hata, Nonmechanical axial scanning laser Doppler velocimeter with directional discrimination, *Appl. Opt.*, 51(20), 4783–4787, 2012.
44. K. Maru and K. Watanabe, Non-mechanical scanning laser Doppler velocimetry with sensitivity to direction of transverse velocity component using optical serrodyne fre-quency shifting, *Opt. Commun.*, 319, 80–84, 2014.
45. K. Maru and T. Hata, Directional discrimination for fiber-optic non-mechanical scanning laser Doppler velocimeter using single transmission path, *Optik*, 125(20), 6312–6314, 2014.
46. K. Maru and T. Hata, Nonmechanical cross-sectional scanning laser Doppler velocim-eter with directional discrimination of transverse velocity component, *Opt. Eng.*, 54(1), 017102, 2015.
47. K. Maru and T. Hata, Nonmechanical scanning laser Doppler velocimeter for cross-sectional two-dimensional velocity measurement, *Appl. Opt.*, 51(34), 8177–8183, 2012.
48. K. Maru and T. Hata, Axial non-mechanical scan in laser Doppler velocimeter using single diffraction grating, *Opt. Rev.*, 20(2), 137–140, 2013.
49. J. E. Harvey and C. L. Vernold, Description of diffraction grating behavior in direction cosine space, *Appl. Opt.*, 37(34), 8158–8160, 1998.
50. M. Pascolini, S. Bonora, A. Giglia, N. Mahne, S. Nannarone, and L. Poletto, Gratings in a conical diffraction mounting for an extreme-ultraviolet time-delay-compensated monochromator, *Appl. Opt.*, 45(13), 3253–3262, 2006.
51. Y. Kai, Y. Takita, Y. Aoki, A. Sugama, S. Aoki, and H. Onaka, 4×4 high-speed switching subsystem with VOA (<10 µs) using PLZT beam-deflector for optical burst switching, in *Optical Fiber Communication and the National Fiber Optic Engineers Conference 2006*, paper OFJ7, Anaheim, CA, 2006.
52. I. M. Soganci, T. Tanemura, K. A. Williams, N. Calabretta, T. de Vries, E. Smalbrugge, M. K. Smit, H. J. S. Dorren, and Y. Nakano, High-speed 1×16 optical switch monolithi-cally integrated on InP, in *35th European Conference on Optical Communication 2009 (ECOC 2009)*, paper 1.2.1, Vienna, Austria, 2009.
53. K. Nashimoto, D. Kudzuma, and H. Han, High-speed switching and filtering using PLZT waveguide devices, in *15th OptoElectronics and Communications Conference (OECC2010)*, paper 8E1-1, Sapporo, Japan, 2010.
54. S. H. Yun, C. Boudoux, G. J. Tearney, and B. E. Bouma, High-speed wavelength-swept semiconductor laser with a polygon-scanner-based wavelength filter, *Opt. Lett.*, 28(20), 1981–1983, 2003.
55. N. Fujiwara, R. Yoshimura, K. Kato, H. Ishii, F. Kano, Y. Kawaguchi, Y. Kondo, K. Ohbayashi, and H. Oohashi, 140-nm quasi-continuous fast sweep using SSG-DBR lasers, *IEEE Photon. Technol. Lett.*, 20(12), 1015–1017, 2008.
56. S. Kato, T. Ichikawa, H. Ito, M. Matsuda, and N. Takahashi, Multipoint sensing laser Doppler velocimetry based on laser diode frequency modulation, in *International Conference on Optical Fiber Sensors*, paper Th3-47, Sapporo, Japan, 1996.

57. T. Hachiga, N. Furuichi, J. Mimatsu, K. Hishida, and M. Kumada, Development of a multi-point LDV by using semiconductor laser with FFT-based multi-channel signal processing, *Exp. Fluids*, 24(1), 70–76, 1998.

58. H. Ishida, H. Shirakawa, T. Andoh, S. Akiguchi, D. Kobayashi, K. Ueyama, Y. Kuraishi, and T. Hachiga, Three-dimensional imaging techniques for microvessels using multi-point laser Doppler velocimeter, *J. Appl. Phys.*, 106, 054701, 2009.

59. T. Kyoden, Y. Yasue, H. Ishida, S. Akiguchi, T. Andoh, Y. Takada, T. Teranishi, and T. Hachiga, Multi-channel laser Doppler velocimetry using a two-dimensional optical fiber array for obtaining instantaneous velocity distribution characteristics, *Jpn. J. Appl. Phys.*, 54, 012501, 2015.

60. E. B. Li, J. Xi, J. F. Chicharo, J. Q. Yao, and D. Y. Yu, Multi-point laser Doppler velocimeter, *Opt. Commun.*, 245, 309–313, 2005.

61. Y. Fu, M. Guo, and P. B. Phua, Spatially encoded multibeam laser Doppler vibrometry using a single photodetector, *Opt. Lett.*, 35, 1356–1358, 2010.

62. Y. Fu, M. Guo, and P. B. Phua, Multipoint laser Doppler vibrometry with single detector: Principles, implementations, and signal analyses, *Appl. Opt.*, 50, 1280–1288, 2011.

63. K. Maru and K. Watanabe, Cross-sectional laser Doppler velocimetry with nonmechanical scanning of points spatially encoded by multichannel serrodyne frequency shifting, *Opt. Lett.*, 39(1), 135–138, 2014.

64. K. Maru and K. Watanabe, Fiber-optic laser Doppler velocimeter with non-mechanical scanning of spatially encoded points for cross-sectional velocity distribution measurement, *Proc. SPIE*, 9203, 920314, 2014.

65. M. Takahashi, S. Watanabe, M. Kurihara, T. Takeuchi, Y. Deki, S. Takaesu, M. Horie et al. Tunable lasers based on silica waveguide ring resonators, in *Optical Fiber Communication and the National Fiber Optic Engineers Conference 2007*, paper OWJ1, Anaheim, CA, 2007.

66. D. G. Rabus, Z. Bian, and A. Shakouri, Ring resonator lasers using passive waveguides and integrated semiconductor optical amplifiers, *IEEE J. Sel. Top. Quantum Electron.*, 13, 1249–1256, 2007.

7 Integrated CMOS Optical Biosensor Systems

Sameer Sonkusale and Jian Guo

CONTENTS

7.1 Introduction 141
7.2 CMOS Photodetectors for Optical Biosensing 142
    7.2.1 Linear-Mode P–N Junction Photodetectors 143
    7.2.2 Avalanche Photodiodes 147
7.3 Applications 150
    7.3.1 CMOS Imagers for Luminescence Spectroscopy 150
        7.3.1.1 Intensity-Based Luminescence Imaging 150
        7.3.1.2 Time-Resolved Luminescence Spectroscopy 153
    7.3.2 Integrated CMOS Image Sensors for Spectroscopy 160
    7.3.3 Biosensors with Commercial Off-the-Shelf CMOS Camera 162
        7.3.3.1 Lensless Ultrawide-Field-of-View Cell Monitoring 162
7.4 Summary 164
References 164

7.1 INTRODUCTION

Biosensors are essential for biomedical diagnostics and to perform fundamental investigations in biology and life sciences. Of many different approaches to biosensing, optical methods are highly desirable because they provide noninvasive remote and possibly multiplexed measurements over a wide field of view that can resolve fast processes underlying their optical responses (e.g., kinetics of biochemical interaction). Optical methods are primarily based on luminescence, which is a result of photoemission such as phosphorescence, fluorescence, bioluminescence, or chemiluminescence from an optical marker attached to the desired target, or it is due to intrinsic material response to light from scattering, absorption, change in polarization, nonlinear interaction, or heating. Some well-known examples of optical biosensors include imaging of pathogens (viruses and bacteria) in cell cultures, infrared (IR) *in vivo* imaging of tissues for tumor identification, metabolic assay for imaging glucose, oxygen consumption in cells and tissues, and high-throughput DNA and genome sequencing and protein detection. However, many traditional optical biosensors require extensive sample extraction and preparation that consumes lots of reagents and also requires large amount of input sample for reliable measurements. Instrumentation for such biosensors is largely based on discrete optical and electronic components such as CCD cameras, photomultiplier tubes (PMTs), lenses, objectives, beam splitters, and readout amplifiers that result in large benchtop instruments. While sophisticated optical instrumentation may be required for extreme measurements (nanometer spatial resolution or femtosecond time resolution), they are, however, very expensive and bulky for routine biological investigations and medical diagnostics. Miniaturization and cost reduction of optical biosensors are therefore essential. Mainstream complementary metal-oxide semiconductor (CMOS) technology provides an ideal platform for low-cost, portable imaging platforms for biosensing.

CMOS technology is a multibillion dollar industry for making analog and digital circuits; these chips can be found in digital computers, wireless communication devices, and consumer electronic products like TV. CMOS image sensors utilize this mainstream technology to make digital imagers and are now widely used in smartphones and mobile PCs. CMOS-based scientific imagers are reaching the quality and resolution of their much more expensive counterpart CCD imagers, and are now beginning to replace them in high-quality microscopes and spectroscopy systems. CMOS imager integrates both photodetector array and the mixed-signal readout circuits on the same silicon substrate, which allows for flexible signal-processing algorithms at pixel- or chip-level, low-voltage operation, low power consumption, and high frame rate. Although early generations of CMOS image sensors suffered from poor noise performance and device mismatch from readout transistors, recent developments of the active pixel sensor (APS) [34] and methods to mitigate fixed-pattern noise (FPN) (e.g., correlated double sampling [CDS]) have pushed the CMOS imagers' noise and linearity performance to rival that of the high-end CCDs. It is without doubt that CMOS image sensors have become the most widely used functional modules in developing low-cost and low-power lab-on-a-chip (LOC) devices for biochemical sensing applications in medical diagnostics, life science, biology, and environmental and homeland security.

This chapter is organized as follows: Section 7.2 presents a brief introduction to CMOS-based photodetectors for optical biosensing applications. In this section, various P–N junction photodiodes and their related pixel structures are presented; Section 7.3 presents three distinct application areas commonly used in optical biosensing applications. For each approach, the basic operation principles and CMOS-based single-chip implementations are presented; the objective is to introduce the main concepts and it is not intended to be a comprehensive review of all the approaches in the literature. Section 7.4 concludes this chapter.

7.2 CMOS PHOTODETECTORS FOR OPTICAL BIOSENSING

Solid-state optical detectors based on the metal-oxide semiconductor (MOS) technology were first introduced in the mid-1960s [19,47], where they utilized the inherent optical response of silicon P–N junction diode. Nowadays, the advancement of CMOS technology has made CMOS-based integrated image sensor systems popular for many applications in scientific research, industry, entertainment, consumer electronics, security surveillance, and portable media devices. CMOS image sensors enjoy a wide range of advantages such as high-level system integration with focal-plane or pixel-level image processing, high-speed and flexible readout schemes, and ultralow power operation. In this section, we will discuss traditional linear-mode P–N junction photodetectors and avalanche photodiodes (APDs) implemented in CMOS process.

7.2.1 Linear-Mode P–N Junction Photodetectors

Figure 7.1a shows a simplified cross-sectional view and operation diagram of a conventional solid-state P–N junction photodiode, which is used in both CMOS imagers and CCDs. The photodiode is implemented using a reverse-biased N-type/P-type substrate junction diode with N-type layer biased at a positive potential and P-type substrate biased at ground. The N-type layer can be either an N+ diffusion layer or a deeper N-well, depending on the wavelength of interest and other performance requirements (i.e., pixel size) of the photodiode. Because of the reverse-biased voltage, a depletion region is formed between the N-type layer (N-well or N+) and the P-type substrate. When a photon with an energy level $h\upsilon$ (where h is the Planck constant and υ is the photon's frequency) greater than the bandgap of the silicon (1.11 eV) gets absorbed in the depletion region, an electron–hole pair is generated. Due to the strong electric field, both the electrons and holes will be swept out of the depletion region. The electrons are collected by the N+ contact and the hole are collected at the P-type substrate. If a continuous flux of photons gets absorbed in the depletion region, it will generate electron–hole pairs that will result in a photon-generated current, which on the average is linearly proportional to the optical intensity. Usually, the photon-generated charges are stored in the parasitic capacitor of the photodiode as charge packets, which are later read out using charge transfer technique as in CCDs or converted to a voltage signal via readout transistors as in the MOS imagers. Normally, the photocurrent is generated from three sources: the current generated by the electron–hole pairs from the depletion region, the current generated by the free-space holes in the quasi-neutral N-type layer that randomly get absorbed into the depletion region, and the current generated by the free-space electrons from the quasi-neutral P-type substrate that randomly get absorbed into the depletion region [16]. Figure 7.1b shows a simplified R-C model of a P–N junction photodiode, where I_{ph} is the photon-generated current, $R_{j,PD}$ is the P–N junction diode parasitic resistor, R_s is the series resistor resulting from metal contacts and signal routing layer's parasitic resistors if the photodiode is connected to readout circuits, and C_{PD} is the photodiode parasitic capacitor. The values of these circuit models depend greatly on the geometry and nature of the diode and the technology parameters.

Here we briefly review the operation principles of several commonly used photodetector structures and pixel architectures. Figure 7.2 shows four commonly used CMOS-based pixel structures, including the passive pixel sensor (PPS; Figure 7.2a), APS (Figure 7.2b and c), and pinned photodiode (PPD; Figure 7.2d).

PPS represents a category of CMOS image sensors whose pixel functions as a passive device. Figure 7.2a shows an example of PPS pixel that consists of an

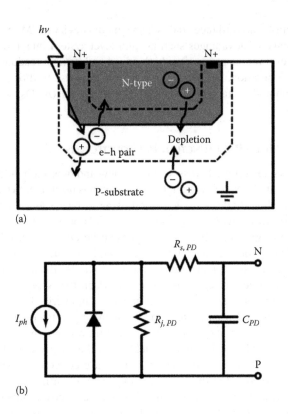

(a)

(b)

FIGURE 7.1   Typical implementation of a P–N junction photodiode. (a) Illustration of a P–N junction photodiode operation. (b) A simplified R-C photodiode model. (b: Revised from Yadid-Pecht, O. and Etienne-Cummings, R., *CMOS Imagers: From Phototransduction to Image Processing*, Kluwer Academic Publishers, Dordrecht, the Netherlands, 2004.)

N-well/P-type substrate photodiode and a *select* transistor. The readout of the photon-generated charge is carried out by a charge integrator. The charge-integration readout integrates the photon-generated current onto an integration capacitor C_{int} and utilizes a *reset* switch at the end of each integration process to reset the readout circuit. The PPS pixel allows for small pixel size and high fill factor implementations, leading to high quantum efficiency (QE). However, the large parasitics at the input of the readout circuits, which is shared among the entire pixel array, result in slow readout speed and very limited dynamic range (DR) performance.

   Figure 7.2b shows a typical three-transistor (3-T) photodiode pixel structure that consists of an N-well/P- type substrate photodiode: a *reset* transistor $N1$, a source follower input transistor $N2$, and a *select* transistor $N3$. The APS structure was first introduced in Reference 35 and it has been significantly improved over the years to achieve comparable performance as CCDs [34]. The main advantage of APS pixel compared to PPS pixel is that it has a nondestructive readout, during which the integrated photodiode output voltage is buffered by a source follower.

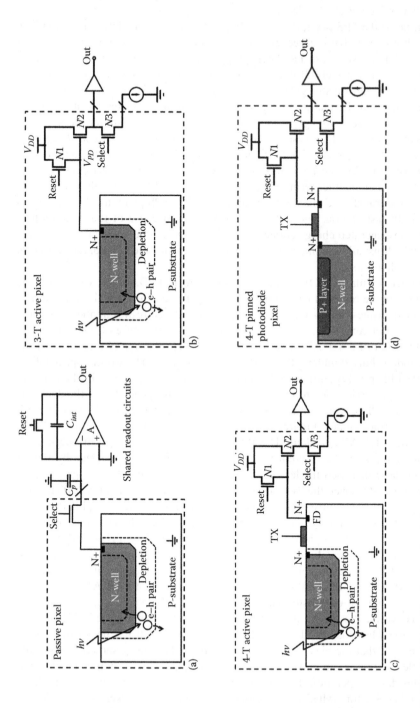

FIGURE 7.2 Typical implementations of complementary metal-oxide semiconductor–based pixels. (a) Passive pixel. (b) Three-transistor active pixel sensor (APS) pixel. (c) Four-transistor APS pixel. (d) Pinned photodiode pixel.

Moreover, the readout speed is faster because the charge-to-voltage conversion is directly carried out at pixel level.

Similar to the PPS pixel, the 3-T APS pixel suffers from a low conversion gain if the photodiode size is large. In order to overcome this issue, a four-transistor (4-T) APS pixel structure has been proposed by Mikio Kyomasu [25] in 1991 and was later extensively discussed in Reference 46. The 4-T APS pixel adds an additional charge transfer gate TX to separate the photon-generated charge collection node and the charge-to-voltage conversion node. Moreover, it maintains a relatively constant voltage at the photodiode. Figure 7.2c shows the 4-T APS pixel schematic, which consists of a photodiode, a charge transfer gate TX, a reset transistor $N1$, a source follower input transistor $N2$, and a select transistor $N3$. The charge transfer gate TX shares one of its terminals with the N+ contact of the photodiode, and its other terminal serves as a floating diffusion node to temporarily store the transferred charge. Compared with 3-T APS pixel, the 4-T APS pixel offers both fast readout and high conversion gain, because of the separation of the photon-generated charge collection node and the charge-to-voltage conversion node, which usually has much smaller parasitic capacitance compared to that of the large photodiode.

Another popular CMOS-based pixel structure is PPD, which was originally developed in the research laboratory at Eastman Kodak Company by B.C. Burkey et al. for interline-transfer CCD in 1984 [4]. Such structure was applied to CMOS imagers in References 12, 15, 21, 22, 30, and 37. The structure of APS PPD pixel resembles that of a 4-T APS photodiode pixel except that the N-type layer of the photodiode is buried under a P-type diffusion layer (P+). The working principle of the APS PPD pixel is similar to that of a 4-T APS photodiode pixel. The majority of the charge carriers due to incoming photons are still generated in the depletion region of the PPD, which now consists of two diodes, namely, the P+-well/N-well/ P-type substrate diodes. Figure 7.2d shows a typical implementation of a CMOS PPD with a 4-T APS readout. The PPD pixel has low dark current because the photon-charge collection region is separated from the silicon surface by the P+ layer, because of which the leakage current due to surface defects is greatly suppressed [27]. Moreover, the P+ layer is transparent to visible light under 500 nm and it also passes through most of the light over the range of 500–750 nm [48]; therefore, it does not degrade the QE performance. Moreover, the light sensitivity is higher since the width of the depletion range extends all the way from N-well/P-type substrate to very close to silicon interface. Moreover, the double junction (P+/N-well and N-well/P-type substrate) means higher charge storage capacitance resulting in higher DR. The 4-T PPD APS also has all the advantages of a 4-T APS photodiode pixel, such as high conversion gain, high SNR, and reduced reset noise through CDS. However, the design and fabrication of PPD is complicated [27] because an additional layer is added to the photodiode structure. It has been discussed in Reference 27 that the bias voltage applied to the N-type layer should be carefully controlled so that it can make the PPD fully depleted while leaving enough headroom for device operation. But the advantages of this approach have made this a preferred choice for CMOS image sensor pixels in today's market.

7.2.2 Avalanche Photodiodes

While linear-mode P–N junction photodiodes are commonly used for intensity-based optical biosensing applications, the nonlinear-mode APDs with single-photon detection sensitivity play an important role in time-resolved optical biosensing applications. The APDs are P–N junction diodes reverse biased at a high-voltage close to the diode breakdown region [7,23,24]. The high bias voltage produces a depletion region with a very strong electric field, and when a photon with sufficient energy (≥ 1.1 eV [1]) gets absorbed in the depletion region, an electron–hole pair will be generated. Due to the high electric fields, the generated electrons or holes can be accelerated to an energy level that is high enough to collide with other electron–holes pairs, generating more electrons and holes. As this process continues, an *avalanche* effect will occur, generating a large amount of electrons and holes in the depletion regions, which eventually forms a photocurrent that can be read out using external electronic circuits. If the P–N junction diode is biased below the diode breakdown region, the APD is working at linear avalanche region, where all the photon-generated electrons or holes can exit the depletion region and get collected at N- or P-terminal. In linear mode, the APD has an internal multiplication gain that is defined as the number of electron–hole pairs generated by a single-photon absorption. If the P–N junction diode is biased at or above the diode breakdown region, then the APD works in *Geiger mode*, where the avalanche effect continues to generate more electron–hole pairs in the depletion region unless an external circuit is used to reduce the APD's bias voltage and eventually turns off the APD so that no more electron–hole pairs will be generated. The process of shutting down the APD is called "quenching," and nowadays, most silicon-based APDs have integrated quenching circuits for multiple detections. APDs operating in *Geiger mode* for single-photon detection are also called single-photon avalanche diodes (SPADs).

Silicon-based APDs have been extensively studied and developed [8,29]. Early implementation of APDs usually required high-voltage operation (≥ 100 V), and the photodiodes were fabricated in various materials such as silicon or InGaAs/InP, and only recently have there been low-voltage APDs developed using commercially available CMOS technologies such as 0.35 μm CMOS [40] and 0.13 μm CMOS [32]. Figure 7.3 shows an example implementation of a SPAD structure fabricated using a commercial 0.35 μm CMOS technology. Figure 7.3a shows the cross-sectional view of the CMOS-based SPAD. A depletion region is formed between the P+ layer and deep N-well layer for photon absorption.

Since the diode output N+ needs to interface with conventional CMOS circuits, the N-well potential is usually kept at below CMOS compatible 5 V, and P+ layer is biased at a high negative voltage (usually ≤ -10 V depending on the specific technology used) so that the overall reverse-bias voltage across P+/N-well junction diode is higher than the diode breakdown voltage for *Geiger-mode* operation. A guard ring using lightly doped P-well is formed around the active region of the SPAD to prevent the diode from lateral breakdown. The SPAD structure is highly dependent on the technology used; other guard ring configurations have been proposed in different technologies [13]. Figure 7.4a shows a typical SPAD pixel schematic with passive quenching and passive recharge circuit. The P+ is the anode of the SPAD and is

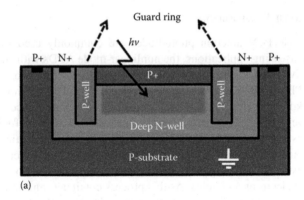

(a)

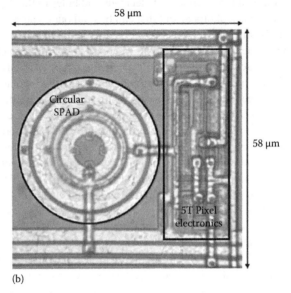

(b)

FIGURE 7.3   An example single-photon avalanche diode (SPAD) implemented in a 0.35 μm complementary metal-oxide semiconductor (CMOS) technology. (a) Cross-sectional view of a CMOS SPAD. (Revised from Niclass, C. et al., *IEEE J. Solid State Circ.*, 40(9), 1847, September 2005.) (b) An example layout of a CMOS SPAD. (From Niclass, C. et al., *IEEE J. Solid State Circ.*, 40(9), 1847, September 2005.)

biased at $V_{DD} - (V_{bd} + V_e)$ so that the reverse-bias voltage of P–N junction is $V_{bd} + V_e$, which is the diode breakdown voltage V_{bd} plus an overdrive voltage V_e. A V_e greater than 0 V ensures that the diode is operating in *Geiger mode*.

Figure 7.4b shows a simplified timing diagram for the SPAD pixel operation. Initially the SPAD is reversed biased above breakdown region and no current is flowing through the p-type MOSFET (PMOS) transistor; the voltage V_n is at the same potential as V_{DD}. Upon the absorption of a photon at t_1, the SPAD is triggered into *Geiger-mode* operation and a large photocurrent I_{ph} generated by

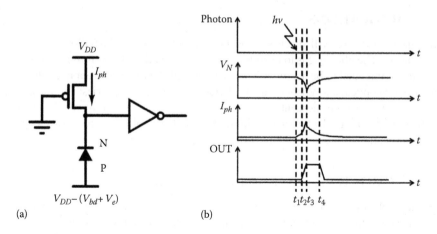

(a)   (b)

FIGURE 7.4   A single-photon avalanche diode (SPAD) pixel implemented in a 0.35 μm CMOS technology. (a) An example SPAD pixel schematic with passive quenching circuits. (Revised from Niclass, C. et al., *IEEE J. Solid State Circ.*, 40(9), 1847, September 2005.) (b) A simplified time diagram of SPAD pixel operation. (Revised from Schwartz, D. et al., *IEEE J. Solid State Circ.*, 43 (11), 2546, 2008.)

impact ionization starts to flow through the SPAD. Due to the finite on-resistance $R_{on, p}$ of the PMOS transistor, the voltage V_n starts to drop as the photocurrent I_{ph} increases. At t_2, the SPAD output starts to increase and eventually reaches a digital high level. As the reverse bias voltage of SPAD decreases below the breakdown region, avalanche effect dies out at t_3, causing the I_{ph} to decrease, and eventually the SPAD is shut down. At the output of the SPAD, an inverter converts the transient response into a digital pulse representing an event trigger signal. A more detailed SPAD transient response can be found in Reference 39, and other quenching techniques such as passive quenching/active recharge or active quenching/active recharge that can generate faster transient response of SPAD pixel have been extensively discussed in References 6 and 39.

To sum up this section, the introduction of CMOS APS [35] and the subsequent improvements contributed by numerous research groups have proven that CMOS image sensors can offer comparable performance as compared to CCDs [14,34] in terms of noise and FPN. Its ability to integrate analog and mixed-signal circuits on the same substrate, high degree of flexibility and compatibility to a wide range of applications, and ultralow power consumption make them competitive candidates for future low-cost and ultracompact scientific imaging applications. CMOS-based photodetectors have become excellent candidates for high-performance scientific imaging. For example, integration with microfluidics can result in an integrated LOC device for many portable sensing applications [28] as a replacement to bulky sensing platforms that need expensive imagers and instruments.

7.3 APPLICATIONS

In this section, we will discuss some basic applications of CMOS-based optical biosensors. The objective is to introduce the fundamental concepts through examples and is not intended to be a comprehensive review of the topic. The application space for CMOS optical biosensors can be divided into three categories: luminescence imaging, where a luminescent optical marker is used; spectroscopy, where intrinsic optical properties of the target are captured through excitation at given wavelengths; and biosensing with commercial off-the-shelf CMOS camera.

7.3.1 CMOS IMAGERS FOR LUMINESCENCE SPECTROSCOPY

In the application of luminescence-based biochemical sensing, the target analytes are labeled with luminescent markers, which are illuminated using short-wavelength excitation light. Depending on the surrounding biochemical environment, the labeled analytes generate luminescence emission at a different wavelength that is detected by CMOS imagers equipped with appropriate excitation filters to block the background excitation. Postsilicon processing on a CMOS chip is performed utilizing soft lithography or chemical etching techniques, in order to prepare the chip for immobilization of the luminescently labeled optical biomarkers on top of the CMOS chips. There are two major techniques associated with luminescence spectroscopy: (1) steady-state luminescence imaging that relies on the monitoring of the luminescence intensity levels at its emission wavelength and (2) time-resolved luminescence imaging that extracts the intrinsic luminescence time decay. Here, we discuss the two methods in more detail in the following subsections.

7.3.1.1 Intensity-Based Luminescence Imaging

Biosensing based on the intensity of luminescence has been widely used for many applications in chemical and biological sensing. DNA microarray is one such powerful platform for DNA-based detection of cancerous genes and screening for infectious diseases; low-cost implementations will bring this powerful tool for bedside point-of-care diagnostics. CMOS-based optical sensor arrays provide an ideal high-throughput platform for DNA microarrays with integrated photodetectors and electronics for readout, signal processing, and smart computation. CMOS chips facilitate "contact imaging" that do not require bulky optical components like objectives, lenses, and mirrors; it is a "lensless" imaging paradigm, which lends itself very well to portable implementation. The conceptual implementation of one such version of a DNA detection device utilizing CMOS chips in lensless imaging paradigm is shown in Figure 7.5. In this device, different DNA probe sequences are immobilized using robotic printing on a glass slide or a fiber-optic faceplate, creating a microarray. The target DNA from diseased cells is extracted and amplified and furthermore tagged with a fluorescent reporter and allowed to hybridize with sequences on the microarray. Perfect hybridization occurs only when target DNA finds a complementary sequence in this array. Washing the microarray will release any partial or unhybridized target DNA sequences. In the presence of short-wavelength excitation, the spots on the microarray that were hybridized will generate fluorescence, which is

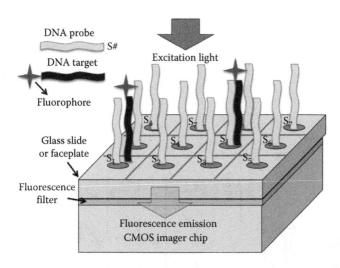

DNA probe

S#

DNA target

Excitation light

Fluorophore

Glass slide
or faceplate

Fluorescence
filter

Fluorescence emission
CMOS imager chip

FIGURE 7.5 Conceptual implementation of complementary metal-oxide semiconductor–based DNA detection based on luminescence.

then detected by the underlying CMOS imager. A fluorescence filter is needed to suppress background excitation light and to filter out the fluorescence signal. In differential gene expression studies, one normally also supplies a target DNA from normal cells (for control) tagged with a different fluorescent reporter at another wavelength. The underlying CMOS imager is expected to capture the intensity of the fluorescence, which is usually weak; therefore, high-sensitivity photodetector and high-performance readout circuits are desired. Several techniques have been used to reduce the readout noise and increase the photosensitivity performance. In Reference 11, a differential pixel structure was proposed to reduce the common mode noise associated with traditional source follower readout circuits. Moreover, the photodiode utilizes a P+/N-well/P-type substrate structure to reduce the dark current by isolating the light-sensitive N-well from the silicon surface. Moreover, high sensitivity is achieved through correlated multiple sampling with multiple capture and averaging to suppress offsets, FPN, reset noise, and flicker noise, all performed in digital domain with a high-resolution analog-to-digital conversion (ADC) on chip. The ADC is a two-step ADC using a single-slope ADC architecture. The differential pixel schematic, ADC architecture, and the operation of correlated multiple sampling are shown in Figure 7.6. Several dark frames with the same integration time are used to accurately estimate the background. Light emission is then triggered by the addition of chemical reagent (or target DNA) and several different types of sampling schemes can be employed. The figure shows six different sampling schemes possible with this imager, namely, (a) regular video sampling where sample is acquired at the end of the ramp; (b) delta-reset sampling where sample is acquired at the end of the ramp right after reset to suppress 1/f noise and to cancel offset FPN but results in increased reset and read noise; (c) CDS where sample is taken right after reset, at the beginning of integration time, and after integration time, which eliminates reset noise and offset FPN but does not provide an effective $1/f$ noise cancellation;

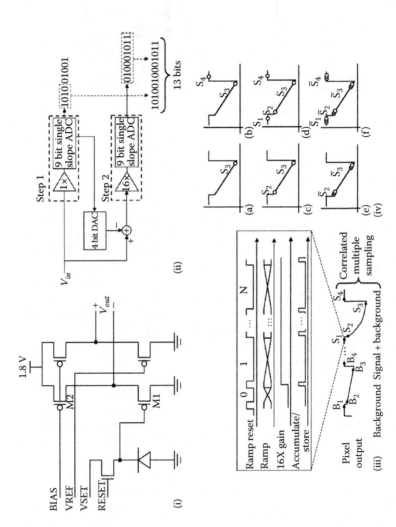

FIGURE 7.6 Luminescence intensity imager implemented in Reference 11: (i) pseudo-differential pixel schematic, (ii) two-step single-slope analog-to-digital conversion, (iii) conceptual timing diagram showing multiple correlated sampling, and (iv) six schemes for sampling such as (a) regular sampling, (b) delta-reset sampling, (c) correlated double sampling, (d) correlated triple sampling, (e) correlated multiple sampling, and (f) another correlated multiple sampling.

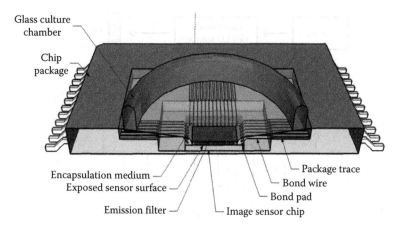

Glass culture
chamber

Chip
package

Encapsulation medium
Exposed sensor surface
Emission filter

Package trace
Bond wire
Bond pad
Image sensor chip

FIGURE 7.7 Cross-sectional diagram of the chip package, glass culture chamber, and mounted emission filter as presented in Reference 3.

(d) correlated triple sampling, which combines the benefits of delta-reset sampling and CDS; and (e) multiple correlated sampling, which reduces even the read noise through averaging by multiple sampling. The chip was built and it showed sensitivity to emission rates below 10^{-6} lux [11].

The luminescence imaging also requires an optical band-pass or longpass filter to prevent excitation light from reaching the image sensor chip, saturating the weak luminescence signal. In Reference 3, a longpass filter with a thickness of only 98 μm has been fabricated in-house by doping SUDAN-II blue dye in PDMS. It was mounted directly on CMOS and packaged with glass culture chamber. See Figure 7.7. Packaging and encapsulation of the CMOS chips for biological compatibility and also to prevent chip from damage for biosensing is a fundamental requirement. Some type of an epoxy or sealant is used to prevent the exposed wirings and bond pads from biological environment. The passivation of the CMOS chip is made from silicon dioxide and silicon nitride and is itself quite biocompatible. A glass culture chamber can be mounted directly on this chip using the epoxy. Packaging of CMOS chips discussed here is universally applicable to all types of CMOS-based integrated sensors for biomedical applications.

7.3.1.2 Time-Resolved Luminescence Spectroscopy

Traditional optical sensors based on luminescence are largely based on measuring intensity of luminescence at their characteristic wavelength. Beyond intensity, other characteristics of the emission, such as lifetime of emission measured as time to decay after turning off the excitation, add diversity to measurement by capturing the local physiochemical dependencies of luminescence. This measurement modality of capturing lifetime in case of fluorescence is called fluorescence lifetime imaging microscopy (FLIM). The time-domain luminescence-sensing technique offers many advantages over the intensity-based approach, such as high sensitivity and selectivity, less sensitive to the photobleaching effect, and less dependent on the luminophore

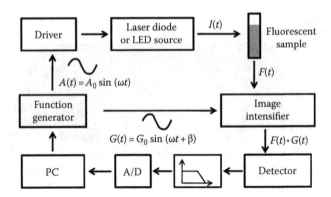

FIGURE 7.8   A typical system setup for frequency-domain fluorescence lifetime imaging microscopy.

concentration. Nowadays, the luminescence lifetime imaging has been widely used in chemical and biological applications, such as protein identification [10], tumor detection [38], and pH measurement [2]. Figure 7.8 shows a typical system setup for FLIM using frequency-domain phase/modulation technique. The intensity-modulated excitation signal is generated by a laser diode, which is electronically modulated by a function generator (or a driver circuit triggered by a frequency synthesizer). The emitted fluorescence is also intensity modulated at the same frequency and, at the same time, exhibits a phase-shift α and a demodulation index k_F. An image intensifier is usually used as a variable gain stage for cross-correlation operation, and the intensifier gain is also intensity modulated at the same frequency ω. The photodetector, usually a cooled CCD or a PMT, captures the modulated fluorescence signal (as power detectors). A low-pass filter (LPF) performs demodulation and extracts the fluorescence lifetime information. Finally, an ADC quantizes the analog signal at the LPF output and a data acquisition/signal-processing program running on a PC performs the phase calculation and fluorescence lifetime reconstruction. However, such traditional optical detectors used in luminescence imaging systems such as PMTs or CCDs are too bulky and too power consuming. An alternative solution is CMOS image sensor technology that has been extensively used for reasons of low cost due to mass production. There are two primary methods for implementing hardware for time-resolved luminescence imaging using CMOS-based imagers. They are classified into time-domain methods such as time-correlated single-photon counting (TCSPC) [2,9,41] or frequency-domain phase-modulation [26] techniques. Figure 7.9a illustrates the operation of TCSPC, where it extracts the luminescence lifetime by reconstructing the luminescence impulse response. An ultrafast light pulse excites the luminescence-labeled biosensor, which illuminates luminescence with an exponentially decayed intensity. The time interval between the light pulse excitation and the time the emitted photons is detected is recorded as the time-of-arrival data. Such process is repeated for many times (millions of repetitions), therefore reconstructing the decay curve as shown in the figure.

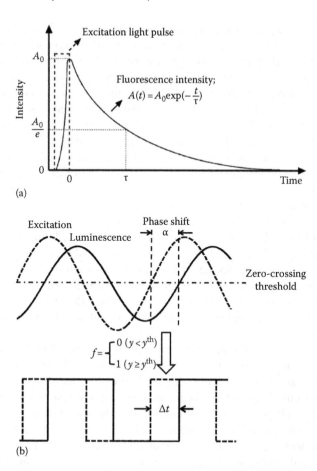

(a)

(b)

FIGURE 7.9 Operation principles of time-resolved luminescence spectroscopy. (a) Time-correlated single-photon counting. (From Yoon, H.-J. et al., *IEEE Trans. Electron Dev.*, 56(2), 214, February 2009.) (b) Frequency modulation with zero-crossing detection operation.

A better alternative to the time-domain luminescence lifetime imaging is the frequency-domain modulation method, which utilizes an intensity-modulated excitation light source to excite the luminescent sensors and subsequently converts the lifetime time information into a frequency-domain phase shift that can be extracted using various demodulation techniques. In essence, the emitted luminescence is also an intensity-modulated signal at the same modulation frequency. The phase shift between the excitation light and emitted luminescence is a function of the luminescence lifetime. Figure 7.9b represents one approach to measuring such lifetime. It measures the phase shift between reference and phase-modulated signal by monitoring time difference between zero crossings, which can be subsequently measured using a timing counter or time-to-digital converter (TDC) [17]. A lock-in amplifier can also be used to capture the phase information in analog domain; a lock-in amplifier works in a manner similar to direct-conversion radio-frequency (RF) receivers where the modulated signal is multiplied by in-phase and quadrature-phase reference signals and the result is low-pass filtered to capture the in-phase and quadrature components. One example of CMOS-based lock-in amplifier for optoelectronic sensor arrays can be found in Reference 20. Both of these implementations for frequency-domain luminescence lifetime measurement allow for use of lower-cost light-emitting diodes (LEDs) to operate as excitation sources, thus reducing the overall system cost and power consumption. Moreover, the luminescence lifetime reconstruction algorithms are less complicated and are potentially faster compared to those of time-domain measurement approaches. Therefore, the frame rate can be greatly improved. Nowadays, the frequency-domain technique can achieve subnanosecond resolution using precision RF analog electronic readout circuits and wide DR by modulating LEDs or LDs at different frequencies to measure both fluorescence and phosphorescence lifetime [17,18]. On the other hand, TCSPC is limited to measurement of short fluorescence lifetime for low-light-level signals due to the pulse pileup problem; pulse pileup happens when photon arrives faster than the timing resolution or more than one photon strikes the detector where the instrument is unable to discriminate between multiple pulses, making it difficult to extract accurate lifetime from waveforms.

There have been many implementations of TCSPC-based CMOS image sensor for fluorescence lifetime imaging with fine temporal resolution and high-level system integration. In Reference 42, an array of CMOS-based APDs in a submicron CMOS technology and a subnanosecond temporal resolution are reported. The system has a high power consumption due to the operation of the APD array, which needs to be biased beyond diode breakdown voltage (above 10 V) for single-photon detection. The frame rate is extremely low at 3.9 Hz because of the complicated and time-consuming postsignal processing. In another chip [44], a SPAD-based pixel array for the analysis of fluorescence phenomena is presented (see Figure 7.10). Each 180 × 150 μm² pixel integrates a single-photon detector combined with an active quenching circuit and a 17-bit digital event counter (Figure 7.10b). The operation of quenching circuit was explained in an earlier section. The pixel exhibits an average dark count rate of 3 kcps and a DR of over 120 dB in time-uncorrelated operation. The chip shows time-resolved fluorescence measurements of fluorescence decay of quantum-dot samples without the aid of any optical filters for excitation laser light cutoff. Although both chips demonstrate big improvement over the traditional lab-based luminescence lifetime imaging systems in terms of device integration, they still are unsuitable for portable implementations due to high power consumption (≥100 mW). Moreover, both CMOS imagers require a high-speed laser source for luminescence excitation, which dramatically increases the overall device power and cost.

In References 5 and 49, a CMOS-based fluorometer based on frequency-domain approach with an integrated 16 × 16 phototransistor array, photocurrent readout circuits, analog filters, and phase detector has been designed and fabricated in a 1.5 μm CMOS technology. A LED is used as the excitation source; the output is a DC voltage that corresponds to the detected fluorescence phase shift. With a 14 kHz

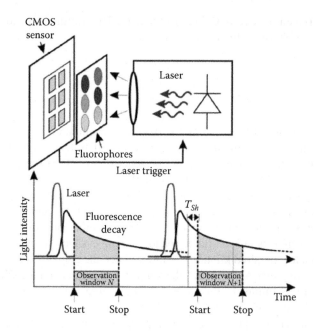

(a)

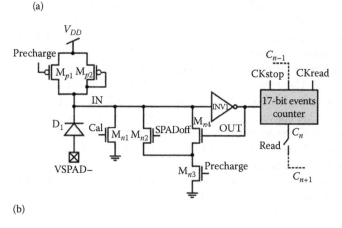

(b)

FIGURE 7.10 Complementary metal-oxide semiconductor (CMOS) avalanche photodiode (APD) array for time-resolved luminescence imaging using time-correlated single-photon counting (TCSPC). (a) Measuring lifetime using 2D CMOS APD array with time gating. (b) Schematic of an APD pixel for TCSPC application. (From Stoppa, D. et al., *IEEE Sens. J.*, 9(9), 1084, September 2009.)

modulation frequency, the entire system including driving the LED consumes 80 mW of average power. The chip was used as a sensor to measure oxygen (O_2) concentration using sol-gel-derived xerogel thin-film sensor elements. The sensor system determines analyte concentrations using the excited-state lifetime measurements of an O_2-sensitive luminophore (*tris*(4,7-diphenyl-1,10-phenanthroline) ruthenium (II)) embedded in the xerogel matrix. The lifetime dependency can be captured by the Stern–Volmer equation:

$$\frac{\tau_0}{\tau} = 1 + k_q[O_2] \tag{7.1}$$

In this equation, $[O_2]$ represents the oxygen concentration in mol/L, k_q is the quenching constant, τ_0 is the inherent fluorescence lifetime of the ruthenium complex dye in the water solution without the presence of quencher (in this case oxygen), and τ is the expected fluorescence lifetime in the presence of dissolved oxygen. The CMOS chip gave an overall sensitivity of 3.5 mV/% O_2 concentration. It should be noted that this sensitivity is influenced by the temporal resolution of the chip to resolve phase shifts. Sensitivity was limited due to inherent speed limitations of phototransistor and its current-mode readout circuits, which limit the modulation frequency to a few hundred kilohertz or lower. Since the chip did not implement the back-end circuitry such as the ADC, the temporal resolution and the DR of the system will be further limited by this external ADC. Figure 7.11 shows the schematic of the fabricated CMOS fluorometer.

An alternate chip implementation that provides direct phase-to-digital conversion was recently demonstrated by the authors of this chapter for oxygen sensing application [17,18]. The zero-crossing detection approach utilized in this chip was shown conceptually in Figure 7.9b. Here, both the excitation light (dotted line) and the emitted fluorescence (solid line) are sinusoidally modulated, and the phase-shift α is a function of fluorescence lifetime, τ: $\alpha = \tan^{-1}(\omega\tau)$ where ω is the modulation frequency. The zero-crossing detection (ZCD) is carried out by performing a threshold detection on both the excitation signal and the fluorescence. Therefore, the sine waves are converted into digital square pulses. The resulting rising and falling edges of ZCD output pulses are determined by the relative locations of the DC common mode levels with respect to the entire sine waveforms. As a result of the ZCD operation, the frequency-domain phase-shift α can be represented as a time-domain delay signal Δt. Let us consider the case where the modulation frequency is low in the submegahertz range so that ωt becomes relatively small ($\omega\tau < 0.01$), then we can approximate the phase shift by $\alpha \approx \omega\tau$, which means that the converted time delay $\Delta t = \alpha/\omega = \tau$. In other words, if the fluorescence is modulated at a very low frequency, then the phase-shift α can be approximated as a linear function of the fluorescence lifetime τ, independent of the modulation frequency. The converted time delay $\tau = \delta t$ can be accurately measured using a high-resolution TDC, whose output codes are direct digital representations of the phase shifts. The use of low modulation frequency also relieves both the system power consumption and cost because lower-speed LEDs can be used.

Figure 7.12 shows the architecture of the digital phase imager. The implemented chip consists of 32 × 32 pixels with a P+/N-well/P-type substrate photodiode, a phase readout circuit with integrated ZCD to extract the frequency-domain phase shift into a time-domain delay, and a high-resolution TDC for digital quantization of the converted time delay [17]. This chip was implemented in a 65 nm CMOS technology.

The operation of this chip can be explained briefly as follows. A P+/N-well/P-type substrate photodiode detects the intensity-modulated fluorescence signal, and its photon-generated current is converted to a voltage output by a high-gain transimpedance amplifier (TIA). The TIA is implemented utilizing a T-network feedback with 100 M conversion gain. After the TIA, a high-speed comparator converts this output

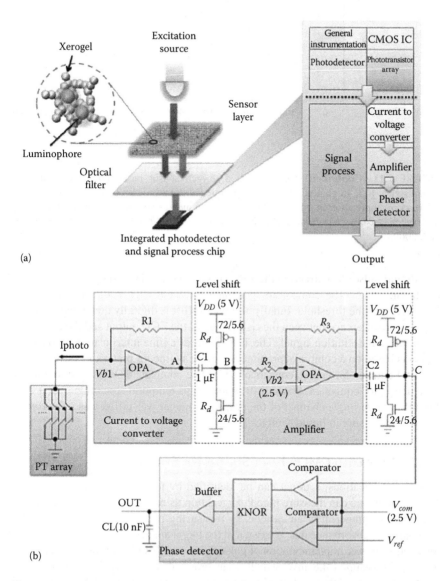

FIGURE 7.11 Xerogel-based oxygen sensor based on lifetime measurement using a 16 × 16 photodiode array for frequency-domain fluorescence lifetime imaging microscopy application in a 1.5 μm complementary metal-oxide semiconductor technology. (a) Overall sensor platform. (b) Circuit schematic. *(Continued)*

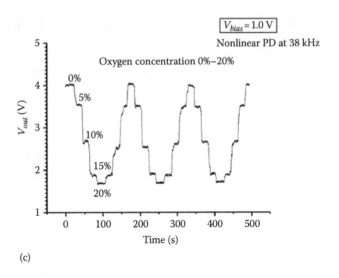

(c)

FIGURE 7.11 (*Continued*)   Xerogel-based oxygen sensor based on lifetime measurement using a 16 × 16 photodiode array for frequency-domain fluorescence lifetime imaging microscopy application in a 1.5 μm complementary metal-oxide semiconductor technology. (c) Oxygen sensing measurement. (From Yao, L. et al., *IEEE Trans. Biomed. Circ. Syst.*, 3(5), 304, 2009.)

into digital pulses by extracting the zero-crossing points defined by a threshold from a reference TIA. The comparator generates an event trigger signal whenever the signal crosses the threshold. Finally, the phase shift is digitally quantized by a TDC as the time difference between the rising edge of the ZCD output and the rising edge of an external excitation signal. The TDC features a time interpolation architecture, which is based on a combined architecture of delay line for precision conversion and a digital counter for coarse conversion [17]. The complete TDC is able to quantize time delays (or equivalent phase shift) with 110 ps temporal resolution over a 22 bit DR (414 μs). A single channel of the proposed imager was utilized as a fluorometer for oxygen sensing in liquid samples [18].

7.3.2   Integrated CMOS Image Sensors for Spectroscopy

In many applications, a target can be identified by its intrinsic optical properties. For example, one can measure blood oxygenation using an instrument called a pulse oximeter at a fingertip or an earlobe, by measuring intensity fluctuation caused by the change in absorbance due to oxygen in the blood at near-IR wavelengths. In 2010, an ultralow power implementation of pulse oximeter CMOS chip was presented [45]. The chip consumed 4.8 mW, which is an order of magnitude below commercial implementations with smart circuit techniques like logarithmic TIA and automatic gain control circuitry.

The near-IR region of the spectrum is highly suitable for IR spectroscopy due to nonionizing nature of the IR radiation and reduced scattering from tissues, allowing

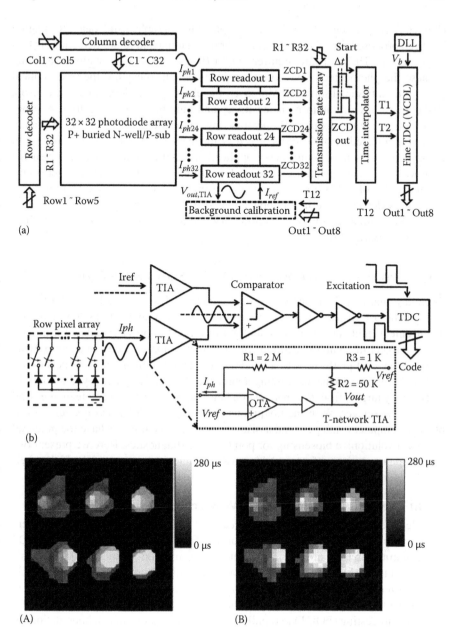

FIGURE 7.12 A digital phase imager in 65 nm complementary metal-oxide semiconductor technology with built-in time-to-digital converter for fluorescence lifetime imaging applications [17]. Application for oxygen sensing has been demonstrated. (a) Architecture of the digital phase imager. (b) Row-level readout schematic. (c) (A) Original phase profile and (B) reconstructed phase profile at 1 KHz modulation frequency. (From Guo, J. et al., *IEEE Sens. J.*, 12(7), 2506, 2012.)

characterization of deep tissues. Diffuse optical tomography using near-IR imaging has been proposed for functional near-IR brain imaging and tumor detection such as in the breasts. The hardware requirement for IR spectroscopy is usually demanding because the photodetector has to be sensitive enough to capture the weak optical signal due to tissue absorption. Moreover, the electronic readout circuits are also a challenge to design due to high-sensitivity and high-bandwidth requirement especially for frequency-domain IR spectroscopy application, where the IR excitation light is modulated at high frequency. It is for these reasons that PMTs and gated intensified CCDs have found many applications in this area. However, given the improved noise and sensitivity metrics of emerging scientific CMOS imagers comparable to that of CCDs, the gap is finally closing to the point where CMOS imagers will find widespread use in all applications of IR spectroscopy. Two recent works using CMOS imagers, although preliminary and not up to the levels of the CCDs, have been reported [51].

7.3.3 BIOSENSORS WITH COMMERCIAL OFF-THE-SHELF CMOS CAMERA

There are many applications where one is interested in sensing biological targets without the use of any luminescent reporters. In such applications, one may use a convenient light source such as sunlight or a LED to provide ambient illumination of targets in complex medium. In such applications, CMOS imagers provide an ideal low-cost platform with a potential for "lensless imaging" of targets over a wide field of view. Image processing approaches are typically employed to identify and classify targets from the shadow or a holographic effect they create on the focal plane of the imager. With the ubiquitous presence of such CMOS imagers in cell phones, computers, and personal assistant devices, these imagers have the potential to truly revolutionize biosensing for point-of-care diagnostics. Here we present two such implementations of optical biosensors employing commercial off-the-shelf CMOS cameras.

7.3.3.1 Lensless Ultrawide-Field-of-View Cell Monitoring

A wide-field-of-view imaging may be required to monitor and count cells where resolution of the image may not be important. One such application is that of CD4 cell counting for HIV monitoring in patients from blood samples [31]. The basic premise of the technology is to record shadows (or diffraction pattern) of microscale objects directly on CMOS chip and utilize inverse computation to detect cells from this diffraction [36] See Figure 7.13. The integrated platform is ideally suited for point-of-care testing (POCT) to rapidly capture, image, and count subpopulations of cells from blood samples in an automated matter.

Another wide-field-of-view imaging utilizes a holographic imaging where a monochromatic light source is employed to illuminate the biological targets [43]. This light from the source interferes with the scattered light from microscale objects, generating a hologram that is recorded by the underlying CMOS imager in two dimensions. Subsequent digital image processing is employed to extract the 3D view of the object from 2D hologram. For high-throughput detection and real-time enumeration of objects in complex media, the recorded 2D image is typically compared with the library of on-chip images without reconstruction. This approach is shown in Figure 7.13 with some representative results. It was used to identify red blood cells, yeast cells, *Escherichia coli*, and various microparticles over a wide field of view. A related work by the same group implemented a lens-free microscope, utilizing a cell phone camera in a truly portable platform [52]. See Figure 7.14. It utilizes a simple incoherent LED light source, which maintains spatial coherence for holography to work at the CMOS image sensor plane. This lens-free microscope on a cell phone was used to image various sized microparticles, red blood cells, white blood cells, platelets, and a waterborne parasite (*Giardia lamblia*).

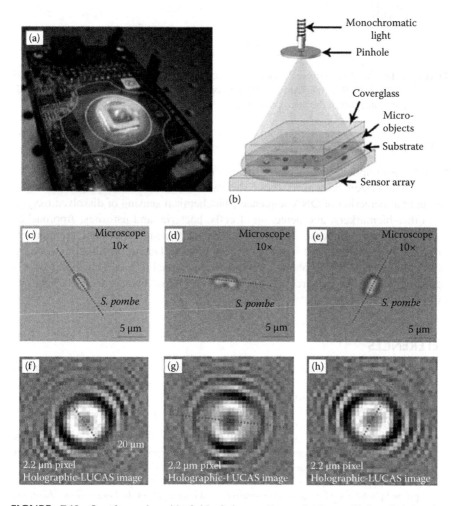

FIGURE 7.13 Lensless ultrawide-field-of-view cell monitoring utilizing holography. (a) Platform, (b) operation, (c–e) microscope images of asymmetric *Schizosaccharomyces pombe* yeast cells (10×), (f–h) detection of the 2D orientation of the cells using the proposed platform [43].

(a)

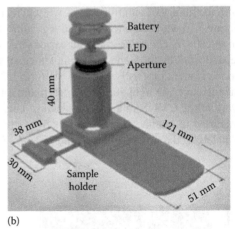
(b)

FIGURE 7.14   (a) A lens-free cell phone microscope. The samples to be imaged are loaded from the side through a mechanical sample holder. An incoherent LED source is used. (b) Schematic diagram of the microscope [52].

7.4   SUMMARY

CMOS-based optical biosensors utilize the mainstream CMOS foundry to implement miniaturized biosensors for a variety of applications ranging from high-throughput detection of DNA sequences, biochemical sensing of dissolved oxygen and other biomarkers, and detection of cells, bacteria, and parasites. Approaches for detection rely on measuring intensity either of lifetime of luminescent markers or on imaging scattering of light by targeted objects. Due to the ubiquitous nature of CMOS cameras in cell phones, laptops, and personal assistant devices, CMOS-based optical biosensors are expected to be ever more present for POCT in our everyday lives.

REFERENCES

1. B. F. Aull, A. H. Loomis, D. J. Young, R. M. Heinrichs, B. J. Felton, P. J. Daniels, and D. J. Landers. Geiger-mode avalanche photodiodes for three-dimensional imaging. *Lincoln Laboratory Journal*, 13(2):335–350, 2002.
2. W. Becker, Wolfgang, and A. Bergmann. Lifetime imaging techniques for optical microscopy. *Technical report, Becker & Hickl GmbH*, 2003.
3. M. Beiderman, T. Tam, A. Fish, G. Jullien, and O. Yadid-Pecht. A low-light CMOS contact imager with an emission filter for biosensing applications. *IEEE Transactions on Biomedical Circuits and Systems*, 2(3):193–203, 2008.
4. B. Burkey, W. Chang, J. Littlehale, T. Lee, T. Tredwell, J. Lavine, and E. Trabka. The pinned photodiode for an interline-transfer ccd image sensor. In *International Electron Devices Meeting 1984*, San Francisco, CA, vol. 30, pp. 28–31, 1984.
5. V. P. Chodavarapu, D. O. Shubin, R. M. Bukowski, A. H. Titus, A. N. Cartwright, and F. V. Bright. CMOS-based phase fluorometric oxygen sensor system. *IEEE Transactions on Circuits and Systems—I*, 54:111–118, 2007.

6. S. Cova, M. Ghioni, A. Lacaita, C. Samori, and F. Zappa. Avalanche photodiodes and quenching circuits for single-photon detection. *Applied Optics*, 35(12):1956–1976, 1996.
7. S. Cova, A. Longoni, and A. Andreoni. Towards picosecond resolution with single-photon avalanche diodes. *Review of Scientific Instruments*, 52:408–412, 1981.
8. H. Dautet, P. Deschamps, B. Dion, A. D. MacGregor, D. MacSween, R. J. McIntyre, G. Trottier, and P. P. Webb. Avalanche photodiodes and quenching circuits for single-photon detection. *Applied Optics*, 32(21):3894–3900, 1993.
9. C. J. de Grauw, J. M. Vroom, H. T. M. van der Voort, and H. C. Gerritsen. Imaging properties in two-photon excitation microscopy and effects of refractive-index mismatch in thick specimens. *Applied Optics*, 38(28):5995–6003, October 1999.
10. D. Elson, S. Webb, J. Siegel, K. Suhling, D. Davis, J. Lever, D. Phillips, A. Wallace, and P. French. Biomedical applications of fluorescence lifetime imaging. *Optics and Photonics News*, 13:26–32, 2002.
11. H. Eltoukhy, K. Salama, and A. El Gamal. A 0.18 μm CMOS bioluminescence detection lab-on-chip. *IEEE Journal of Solid-State Circuits*, 41(3):651–662, 2006.
12. K. Findlater, R. Henderson, D. Baxter, J. Hurwitz, L. Grant, Y. Cazaux, F. Roy, D. Herault, and Y. Marcellier. SXGA pinned photodiode CMOS image sensor in 0.35 mum technology. In *Digest of Technical Papers. ISSCC. 2003 IEEE International Solid-State Circuits Conference*, San Francisco, CA, pp. 218–489, 2003.
13. H. Finkelstein, M. Hsu, and S. Esener. Sti-bounded single-photon avalanche diode in a deep-submicrometer cmos technology. *IEEE Electron Device Letters*, 27(11):887–889, November 2006.
14. E. Fossum. CMOS image sensors: Electronic camera-on-a-chip. *Electron Devices*, 44(10):1689–1698, October 1997.
15. M. Furumiya, H. Ohkubo, Y. Muramatsu, S. Kurosawa, F. Okamoto, Y. Fu-Jimoto, and Y. Nakashiba. High-sensitivity and no-crosstalk pixel technology for embedded cmos image sensor. *Electron Devices*, 48(10):2221–2227, October 2001.
16. A. Gamal. *Course on Digital Image Sensors, EE-293B*. Stanford University, Stanford University, Menlo Park, CA, 2001.
17. J. Guo and S. Sonkusale. A 65 nm cmos digital phase imager for time-resolved fluorescence imaging. *IEEE Journal of Solid-State Circuits*, 47(7):1731–1742, 2012.
18. J. Guo, J. Zhang, S. Thomas, and S. Sonkusale. Cmos fluorometer for oxygen sensing. *The IEEE Sensors Journal*, 12(7):2506–2507, 2012.
19. J. Horton, R. Mazza, and H. Dym. The scanistor—A solid-state image scanner. *Proceedings of the IEEE*, 52(12):1513–1528, December 1964.
20. A. Hu and V. P. Chodavarapu. CMOS optoelectronic lock-in amplifier with integrated phototransistor array. *IEEE Transactions on Biomedical Circuits and Systems*, 4(5):274–280, 2010.
21. I. Inoue, H. Nozaki, H. Yamashita, T. Yamaguchi, H. Ishiwata, H. Ihara, R. Miyagawa et al. New lv-bpd (low voltage buried photo-diode) for CMOS imager. In *International Electron Devices Meeting, 1999. IEDM Technical Digest*, Washington, DC, pp. 883–886, 1999.
22. I. Inoue, N. Tanaka, H. Yamashita, T. Yamaguchi, H. Ishiwata, and H. Ihara. Low-leakage-current and low-operating-voltage buried photodiode for a CMOS imager. *IEEE Transactions on Electron Devices*, 50(1):43–47, January 2003.
23. K. Johnson. Photodiode signal enhancement effect at avalanche breakdown voltage. In *Solid-State Circuits Conference. Digest of Technical Papers. 1964 IEEE International*, Philadelphia, PA, vol. VII, pp. 64–65, February 1964.
24. K. Johnson. High-speed photodiode signal enhancement at avalanche breakdown voltage. *IEEE Transactions on Electron Devices*, 12(2):55–63, February 1965.
25. M. Kyomasu. A new MOS imager using photodiode as current source. *IEEE Journal of Solid-State Circuits*, 26(8):1116–1122, August 1991.

26. J. R. Lakowicz and K. W. Berndt. Lifetime selective fluorescence imaging using an rf phase sensitive camera. *Review of Scientific Instruments*, 62(7):1727–1734, July 1991.
27. T. Lule, S. Benthien, H. Keller, F. Mutze, P. Rieve, K. Seibel, M. Sommer, and M. Bohm. Sensitivity of CMOS based imagers and scaling perspectives. *IEEE Transactions on Electron Devices*, 47(11):2110–2122, November 2000.
28. N. Manaresi, A. Romani, G. Medoro, L. Altomare, A. Leonardi, M. Tartagni, and R. Guerrieri. A CMOS chip for individual cell manipulation and detection. *Solid-State Circuits*, 38(12):2297–2305, December 2003.
29. R. McIntyre. Recent developments in silicon avalanche photodiodes. *Measurement*, 3(4):146–152, 1985.
30. B. Mheen, M. Kim, Y.-J. Song, and S. Hong. Operation principles of 0.18-μm four-transistor CMOS image pixels with a nonfully depleted pinned photodiode. *IEEE Transactions on Electron Devices*, 53(11):2735–2740, November 2006.
31. S. Moon, H. O. Keles, A. Ozcan, A. Khademhosseini, E. Hggstrom, D. Kuritzkes, and U. Demirci. Integrating microfluidics and lensless imaging for point-of-care testing. *Biosensors and Bioelectronics*, 24(11):3208–3214, 2009.
32. C. Niclass, M. Gersbach, R. Henderson, L. Grant, and E. Charbon. A single photon avalanche diode implemented in 130-nm CMOS technology. *IEEE Journal of Selected Topics in Quantum Electronics*, 13(4):863–869, July–August 2007.
33. C. Niclass, A. Rochas, P.-A. Besse, and E. Charbon. Design and characterization of a CMOS 3-d image sensor based on single photon avalanche diodes. *IEEE Journal of Solid-State Circuits*, 40(9):1847–1854, September 2005.
34. R. Nixon, S. Kemeny, C. Staller, and E. Fossum. 128 × 128 CMOS photodiode-type active pixel sensor with on-chip timing, control, and signal chain electronics. *Proceedings of SPIE*, 2415:117–123, 1995.
35. P. Noble. Self-scanned silicon image detector arrays. *IEEE Transactions on Electron Devices*, 15(4):202–209, April 1968.
36. A. Ozcan and U. Demirci. Ultra wide-field lens-free monitoring of cells on-chip. *Lab Chip*, 8:98–106, 2008.
37. J.-H. Park, S. Kawahito, and Y. Wakamon. A new active pixel structure with a pinned photodiode for wide dynamic range image sensors. *IEICE Electronics Express*, 2:482–487, 2005.
38. J. D. Pitts and M.-A. Mycek. Design and development of a rapid acquisition laser-based fluorometer with simultaneous spectral and temporal resolution. *Review of Scientific Instruments*, 72:3061–3072, 2001.
39. A. Rochas. Single photon avalanche diodes in CMOS technology. PhD thesis, Ecole Polytechnique Federale De Lausanne, EPFL, Lausanne, Switzerland, 2003.
40. A. Rochas, M. Gani, B. Furrer, P. A. Besse, R. S. Popovic, G. Ribordy, and M. Gisin. Single photon detector fabricated in a complementary metal-oxide-semiconductor high-voltage technology. *Review of Scientific Instruments*, 74:3263–3270, 2003.
41. J. Russell, K. Diamond, T. Collins, H. Tiedje, J. Hayward, T. Farrell, M. Patterson, and Q. Fang. Characterization of fluorescence lifetime of photofrin and delta-aminolevulinic acid induced protoporphyrin IX in living cells using single- and two-photon excitation. *IEEE Journal of Selected Topics in Quantum Electronics*, 14(1):158–166, 2008.
42. D. Schwartz, E. Charbon, and K. Shepard. A single-photon avalanche diode array for fluorescence lifetime imaging microscopy. *IEEE Journal of Solid-State Circuits*, 43(11):2546–2557, 2008.
43. S. Seo, T.-W. Su, D. K. Tseng, A. Erlinger, and A. Ozcan. Lensfree holographic imaging for on-chip cytometry and diagnostics. *Lab Chip*, 9:777–787, 2009.
44. D. Stoppa, D. Mosconi, L. Pancheri, and L. Gonzo. Single-photon avalanche diode CMOS sensor for time-resolved fluorescence measurements. *IEEE Sensors Journal*, 9(9):1084–1090, September 2009.

45. M. Tavakoli, L. Turicchia, and R. Sarpeshkar. An ultra-low-power pulse oximeter implemented with an energy-efficient transimpedance amplifier. *IEEE Transactions on Biomedical Circuits and Systems*, 4(1):27–38, 2010.
46. C. C. Wang. A study of CMOS technologies for image sensor applications. PhD thesis, Massachusetts Institute of Technology, Cambridge, MA, 2001.
47. G. Weckler. Operation of p-n junction photodetectors in a photon flux integrating mode. *IEEE Journal of Solid-State Circuits*, 2(3):65–73, September 1967.
48. O. Yadid-Pecht and R. Etienne-Cummings. *CMOS Imagers: From Phototransduction to Image Processing*. Kluwer Academic Publishers, Dordrecht, the Netherlands, 2004.
49. L. Yao, R. Khan, V. P. Chodavarapu, V. Tripathi, and F. Bright. Sensitivity-enhanced CMOS phase luminometry system using xerogel-based sensors. *IEEE Transactions on Biomedical Circuits and Systems*, 3(5):304–311, 2009.
50. H.-J. Yoon, S. Itoh, and S. Kawahito. A CMOS image sensor with in-pixel two-stage charge transfer for fluorescence lifetime imaging. *IEEE Transactions on Electron Devices*, 56(2):214–221, February 2009.
51. R. Yun and V. Joyner. A monolithically integrated phase-sensitive optical sensor for frequency-domain nir spectroscopy. *IEEE Sensors Journal*, 10(7):1234–1242, 2010.
52. D. Tseng, O. Mudanyali, C. Oztoprak, S. O. Isikman, I. Sencan, O. Yaglidere, and A. Ozcan, Lensfree microscopy on a cellphone. *Lab on a Chip*, 10:1787–1792, 2010.

8 Design of CMOS Microsystems for Time-Correlated Single-Photon Counting Fluorescence Lifetime Analysis

Liping Wei and Derek Ho

CONTENTS

8.1 Introduction 170
8.2 Background... 171
    8.2.1 Fluorescence Lifetime 171
    8.2.2 Operation Principle.. 172
8.3 Applications of Fluorescence Lifetime Analysis.................................... 174
    8.3.1 Förster Resonance Energy Transfer... 174
    8.3.2 Fluorescence Lifetime Imaging Microscopy............................... 175
    8.3.3 Fluorescence Lifetime Correlated Spectroscopy......................... 175
8.4 Conventional TCSPC Systems.. 176
    8.4.1 Excitation Sources 177
    8.4.2 Detectors.. 178
    8.4.3 Constant Fraction Discriminator 180
    8.4.4 Timing Electronics 180
        8.4.4.1 TAC/ADC Implementation 181
        8.4.4.2 TDC Implementation 181
8.5 CMOS TCSPC Systems.. 182
    8.5.1 Chip-Level TDC Integration 183
        8.5.1.1 Single-Detector Implementation..................................... 183
        8.5.1.2 Multi-Detector Implementation 183
    8.5.2 Column-Level TDC Integration 183
    8.5.3 Pixel-Level TDC Integration 185
8.6 Conclusion 188
Acknowledgment 189
References... 189

8.1 INTRODUCTION

Time-resolved fluorescence measurement has wide applications in medical diagnostics and molecular imaging. It is a technique in which the decaying emission intensity of a fluorescence molecule, or fluorophore, is recorded with a high temporal resolution. A time constant, also known as the fluorescence lifetime, is subsequently extracted.[1] This lifetime reflects the surrounding chemical composition of the fluorophore and their interaction. For example, Förster resonance energy transfer (FRET),[2] quenching, and molecular rotation[3] are phenomena that fluorescence lifetime measurements can examine effectively.

Time-domain and frequency-domain methods are currently two prominent techniques for extracting the fluorescence decay characteristics. Since the instruments based on the time-domain methods are relatively small, low cost, and reliable,[4] this chapter mainly introduces the time-domain methods. In this class of methods, there are two key variations, time-gated photon counting[5] and time-correlated single-photon counting (TCSPC).[1] The time-gated method measures a fluorescence decay curve either by using several time gates with equal width[6] or by time gate scanning.[7] In the former method, to measure monoexponential fluorescence decays, two time gates with an equal width are typically sufficient, whereas for multiexponential decays, more time gates are required. The number of detected photons within each time gate and the time difference between the adjacent two time gates are used to calculate the lifetime.

It is rather difficult to resolve lifetimes shorter than a nanosecond. For a good temporal resolution, the time gate scanning method has been introduced. A narrow time gate of tens of picoseconds is used to scan over the decaying fluorescence signal. The temporal resolution is dependent on the time gate width. However, since time gating excludes the majority of the detected photons, photon detection efficiency (PDE) is compromised.[8]

At present, most of the time-resolved fluorescence measurements are performed by utilizing TCSPC, which has high photon collection efficiency and high temporal resolution for measuring fluorescence lifetime in the time domain.[9] In commercially available TCSPC instruments, the temporal resolution, which is characterized by the time bin width, is as low as hundreds of femtoseconds. The acquisition times achieved with these instruments can be as short as several milliseconds.[10]

Despite high performance, bulkiness and high cost are two major limitations in the commercially available TCSPC equipment that prevent them from serving portable and point-of-care applications. This calls for the reduction of system complexity, size, and cost.

Complementary metal-oxide semiconductor (CMOS) technology, as benefited from process scaling, provides several advanced capabilities such as high integration density, high-resolution signal processing, and low power consumption, enabling the realization of sensitive, integrated, and low-cost fluorescence analytical platforms. As a result, CMOS TCSPC systems can simultaneously achieve a high performance (i.e., high temporal resolution and throughput[11]) and small size. State-of-the-art CMOS designs can contain millions of pixels within a die area of only several squared millimeters.

This chapter describes the TCSPC technique, presents recent progress of CMOS TCSPC systems, and forecasts the technological trajectory of TCSPC sensors. The rest of this chapter is organized as follows. Section 8.2 describes the concept of fluorescence lifetime and the operation principle of the TCSPC technique. Section 8.3 surveys well-adopted applications of TCSPC. Section 8.4 discusses conventional TCSPC systems, including key architecture, components, and design parameters. Section 8.5 reviews advanced TCSPC systems implemented in the CMOS technology, with different levels of detection and signal processing parallelism. Section 8.6 concludes the chapter.

8.2 BACKGROUND

8.2.1 FLUORESCENCE LIFETIME

Fluorescence is a phenomenon where a fluorophore absorbs excitation light and then emits light of a longer wavelength. The difference in wavelengths between excitation and emission spectral peaks is known as the Stokes shift. As depicted in the Jablonski diagram of Figure 8.1, the valence electron of a fluorophore is originally in the ground-state S_0 and then transits into a higher energy level S_1 or S_2 by absorbing one or two photons, respectively.[12] When electrons are at highly excited states, such as S_2 or S_3, they transit rapidly into the S_1 state. This is because the electrons at high-energy states are less stable. In this process, energy is released in the form of photons that can be observed as fluorescence.

A fluorophore can also be excited by absorbing two photons simultaneously. The absorbed energy must be higher than the energy difference between S_0 and S_1. In practice, excitation is typically provided by a femtosecond laser. Two-photon excitation is a nonlinear process, as the excitation efficiency increases with the second power of the photon flux.[13]

The average time a fluorophore spends in the excited state prior to return to the ground state is known as the lifetime (τ), which is important characteristic of a fluorophore as it contains information of the chemical environment local to the fluorophore. Not all the electrons return to the ground state by emitting photons. There are a number of additional pathways that the electrons can go through, such as internal conversion and quenching. Neglecting radiationless decay processes, the lifetime of the excited state is the natural lifetime τ_n, which is a constant for a given fluorophore and refractive index of the solvent. However, since internal conversion and quenching can also lead to energy dissipation, the fluorescence lifetime is shorter than the natural lifetime.

Other important concepts of fluorescence are quantum yield and bleaching. Quantum yield is the ratio of the number of emitted photons to the number of absorbed photons, which is proportional to the lifetime.[14] Bleaching is an irreversible process in contrast to quenching. Bleaching is based on the destruction of the fluorophore by prolonged exposure to the excitation light. The amount of bleaching depends on the intensity, exposure time, and the energy of the light source. Each fluorophore has a certain amount of excitation and emission cycles before bleaching occurs.

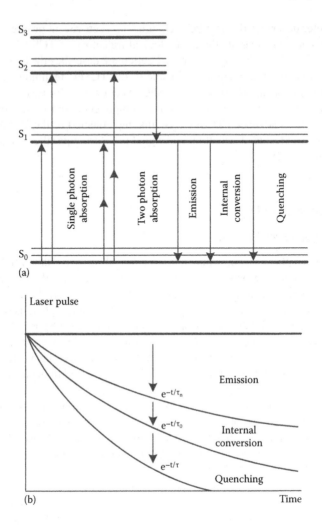

FIGURE 8.1 (a) Jablonski diagram and (b) fluorescence decay curves.

8.2.2 Operation Principle

To measure the fluorescence lifetime efficiently and accurately, the most widely adopted technique is TCSPC, which has high photon collection efficiency and temporal resolution. A typical TCSPC system consists of a pulsed excitation source (a laser or a light-emitting diode [LED]), a single-photon detector, an timing electronics, and a workstation to compute the lifetime, as depicted in Figure 8.2a. The detector is typically a photomultiplier tube (PMT) for a conventional system and a single-photon avalanche detector (SPAD) for a CMOS system. The timing electronics is typically a time-to-digital converter (TDC) or a time-to-amplitude conversion (TAC) combined with an analog-to-digital converter (ADC). When the excitation source illuminates the fluorophore-labeled sample, a synchronization

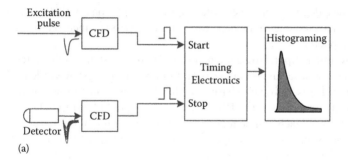

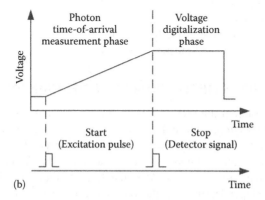

FIGURE 8.2 Time-correlated single-photon counting "start–stop watch" mode operation principle: (a) system schematic and (b) timing diagram.

pulse is outputted by the excitation source driver and is delivered to the *start* input of the timing electronics. After excitation, the fluorophores in the sample emit fluorescence that is detected by the single-photon detector. After detecting a photon, the detector delivers a pulse to the *stop* input of the timing electronics. Both the excitation pulse and detector signal pass through a constant-fraction discriminator (CFD) before reaching the timing electronics. The time interval between the *start* and *stop* pulses is the photon time of arrival. This timing mode is called "start–stop watch" mode, as shown in Figure 8.2b. The fluorescence decay histogram can be obtained after a sufficiently large number of excitation–detection cycles, by binning the photon time of arrival. Since the pulsed excitation source is switched off during fluorescence detection, the background signal is substantially reduced, a key advantage of TCSPC. An optical filter can be utilized to further reject the excitation light.

The single-photon pulse from the detector is a sequence that captures the photon arrival statistics, which is a Poisson process. Some signal periods contain one photon, whereas many signal periods record no photons. The excitation intensity is typically moderated so that the probability of more than one photon arriving within a period is low, which is necessary to prevent signal shape distortion, known as photon pileup. The pileup effect arises when subsequent photons within the same period cannot be detected due to the dead time of the electronics incurred by the detection of the first photon.

The overall temporal resolution of the TCSPC system is characterized by its instrument response function (IRF). Assuming that the system is ideal, that is, the excitation pulse is infinitely sharp and the detectors and timing electronics work with no time uncertainty, the IRF of the system should be infinitely narrow. Time uncertainty from each component results in a broadening of the IRF. Thus, the IRF contains the excitation pulse shape, the transit time spread (TTS) of the detector, and the timing jitter in the timing electronics. The TTS of the detector is the dominant factor of the IRF broadening.[8] Thus, the temporal resolution of the TCSPC system is mainly limited by the TTS of the detector. The single electron response (SER) is the electrical pulse that a detector outputs for each photon detected. Since the TTS is much shorter than the SER, using the TTS to estimate the photon time of arrival rather than SER often provides sufficient accuracy.

8.3 APPLICATIONS OF FLUORESCENCE LIFETIME ANALYSIS

High-temporal-resolution fluorescence lifetime measurement based on TCSPC has become an extremely important tool in a variety of fluorescence spectroscopy and microscopy applications. The lifetime of a fluorophore depends on its neighboring environment, rather than the fluorophore concentration. Therefore, analyte concentration measurements, local environment changes, and molecular interactions can be investigated independently from the unstable fluorophore concentration.[15] For instance, the degradation process of the polymeric nanoparticles can be assessed by the differences of the fluorescence decay profiles.[16] The extracellular Na^+ levels can be determined by measuring reversible fluorescence lifetime changes.[17] Typical TCSPC applications including FRET, fluorescence lifetime imaging microscopy (FLIM), and fluorescence lifetime correlated spectroscopy (FLCS) are discussed subsequently.

8.3.1 Förster Resonance Energy Transfer

FRET is a dipole–dipole coupling process where energy is transferred from an excited donor fluorophore to an acceptor fluorophore nonradiatively. During the FRET process, both the fluorescence intensity and lifetime of the donor decrease, whereas for the acceptor fluorophore, both are increased. The energy transfer can only occur when the distance between the donor and the acceptor is less than approximately 10 nm. The energy transfer efficiency varies with the distance and is usually expressed as an inverse sixth power of the Förster radius. The Förster radius is the distance at which the energy transfer efficiency is 50%.[18]

FRET has become an indispensable technique to explore extracellular and intracellular biological interactions. FRET based on fluorescence lifetime is utilized for selective deoxyribonucleic acid (DNA) or ribonucleic acid detection.[19] The oligonucleotide probes attached to quantum dot (CdSe/ZnS core shell) and organic dye (cyanine-5) are hybridized to target DNA sequences. The quantum dot and the organic dye serve as the FRET donor–acceptor pair. In the absence of the target DNA, only the fluorescence from the donor is detected and the fluorescence lifetime of the donor is measured. However, in the presence of a hybridized DNA, the donor and acceptor come within several nanometers of each other, which is sufficiently close for FRET to occur. Therefore, fluorescence from the acceptor can be detected. The FRET efficiency can be calculated by the fluorescence lifetimes measured in the presence and absence of the acceptor. As the FRET efficiency varies with the distance between the donor and the acceptor, by measuring the different FRET efficiencies, different target DNA sequences can be distinguished.

FRET combined with time-resolved measurement is a powerful technique, but most measurement setups involving microscopes and lenses are bulky and expensive. To mitigate this problem, an active oligonucleotide microarray platform for time-resolved FRET assays is proposed.[20] The use of an active array substrate, fabricated in a standard CMOS technology, provides substantial reduction in system complexity and cost.

8.3.2 FLUORESCENCE LIFETIME IMAGING MICROSCOPY

FLIM is a technique that forms images based on the spatial distribution of fluorescence lifetimes, instead of the fluorescence intensity. This makes the technique insensitive to excitation intensity variations, photon scattering, and probe concentration. FLIM combined with FRET can be utilized to obtain quantitative information about physiological parameters such as Ca^{2+}, pO_2, and pH[21,22] or to study spatial-temporal protein–protein interactions in situ.[23] TCSPC-based FLIM has become a pervasive and essential tool in cell biology and medical diagnosis. This technique has a high temporal resolution and high photon collection efficiency (i.e., a small number of photons are required to accurately obtain the lifetime). With high temporal resolution, lifetime can be obtained accurately, which provides high image quality. FLIM can be classified into scanning technique and wide-field technique. The scanning technique, including confocal and two-photon laser scanning, is based on the conventional TCSPC system. In this system, the excitation laser is scanned over the sample via a small focal plane. The emitted light from this plane propagates through the lenses and reaches the detector. An external marker is used to record the location of the photon, and then timing electronics is used to determine the photon time of arrival and location.[24,25] In the wide-field technique, laser scanning is not required.[26] Attempts to reduce system complexity have been made by using the CMOS technology to implement the detector array and timing electronics. However, complicated optical lenses are still used, preventing these systems from serving portable applications.

8.3.3 FLUORESCENCE LIFETIME CORRELATED SPECTROSCOPY

FLCS is a combination of fluorescence lifetime measurement and fluorescence correlation spectroscopy (FCS). In an FLCS measurement, the fluorescence decay curves of each fluorescent component and their mixture are first obtained. Then the decay curve of each component in the mixture is used to calculate the filter functions. The filter functions are numerically calculated by software. By orthonormalizing the filter functions with respect to the decay curves, the autocorrelation functions (ACFs) of individual components in a fluorophore mixture can be separated. Thus, different fluorescent molecular species in a mixture can be identified or separated according to the ACFs.[27] To obtain FLCS data, a pulsed excitation source and two independent timings are required to simultaneously measure two parameters: the photon time of arrival with respect to the pulse at picosecond resolution and the macroscopic time with respect to the beginning of the experiment at nanosecond resolution. The ability to simultaneously measure these two independent timings makes it possible to separate the ACF of different components.[28] Conventional FCS uses continuous excitation, performs single-channel detection, and requires an external hardware autocorrelator to calculate the ACF of the fluorescence intensity fluctuations. This has the limitation of resolving no more than one fluorescent component simultaneously. FLCS is a powerful tool to improve the quality of the FCS data by removing noise and distortion caused by scattered excitation light, detector thermal noise, and detector after pulsing.[27]

8.4 CONVENTIONAL TCSPC SYSTEMS

Due to the wide applications of the TCSPC, a variety of instrumentations are commercially available. These products are typically based on components, including the pulsed excitation source, the single-photon detector, and the timing electronics. For example, in the Delta Pro system made by Horiba, as shown in Figure 8.3, the detector, excitation source, and sample holder are assembled into the sample chamber, whereas the excitation source driver and timing electronics are each implemented as a stand-alone component. Operations of the individual components are described in the following subsections.

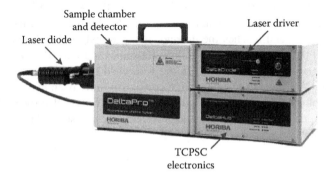

FIGURE 8.3 Commercial time-correlated single-photon counting system from Horiba utilizing stand-alone detector and timing electronics. (DeltaPro Lifetime System, Horiba 29.)

8.4.1 EXCITATION SOURCES

The typical excitation sources for TCSPC experiments are titanium–sapphire laser, fiber laser, laser diodes, and LEDs. The parameters of each type are listed as follows.

As evident from Table 8.1, titanium–sapphire laser provides a large range of wavelengths, femtosecond pulse width, and ultrahigh pulse energy and repetition rate but is very expensive. Fiber lasers deliver femtosecond pulses at several wavelengths and provide high repetition and pulse energy but also are expensive. Laser diodes and LEDs are low-cost excitation sources. They have low power consumption and the repetition rate can be easily controlled by a pulse driver. At present, the dominant excitation sources for TCSPC are laser diodes and LEDs.

Laser diodes are currently available in the wavelengths of 375, 395, 405, 420, and 440 nm and a large number of wavelengths above 450 nm with a full width at half maximum (FWHM) of 20 nm (e.g., 450 ± 10 nm). Pulse width down to 40 ps is readily available, with selected models achieving 25 ps. The pulse width is short enough to record fluorescence lifetime less than 10 ps. The average power ranges from a few hundred μW to a few mW at the 40 MHz repetition rate. With wider pulse width at the same repetition rate, higher power is available.[32] The output power is sufficient to obtain excellent sensitivity in high-efficiency optical systems.

LEDs are more cost-effective than laser diodes but have a wider pulse width, usually in the order of hundreds of picoseconds and even up to 1.3 ns. Wavelength ranges from 245 to 600 nm. The average power is at tens of μW at the 40 MHz repetition rate, which is much lower than laser diodes.[33] Since the susceptibility of damage to the sample must be taken into consideration when choosing appropriate excitation sources, for the samples vulnerable to high-intensity excitations, LEDs are especially suitable.

TABLE 8.1
Main Parameters of Light Sources

Light Source	Wavelength Range (nm)	Pulse Width (ps)	Repetition Rate (Typ. MHz)	Pulse Energy (pJ)	Cost
Titanium–sapphire laser[30]	650–950	<0.006	85	1765	Very high
Fiber laser[31]	515, 780, 1040, 1560	0.09–0.15	100	10–10⁴	High
Diode laser[32]	375, 405, 450, 520	70–130	Up to 40	12.5–250	Medium
	650, 905, 975, 1990	50–120	Up to 80	0.25–50	
Light-emitting diode[33]	245, 285, 325, 575	500–1300	Up to 10	0.04–0.3	Low
	380, 460, 500, 600	600–950	Up to 40	0.8–8	

8.4.2 DETECTORS

A single-photon detector is the key element of the TCSPC system. Upon detecting a photon, an electric pulse is delivered to the timing electronics. As mentioned in Section 8.2.2, the IRF is determined by the TTS of the detector, which limits the temporal resolution.

Currently, photomultiplier tube (PMT), microchannel plate (MCP)-PMT, and SPAD are typical single-photon detector for TCSPC. PMT, the most frequently used detector in conventional TCSPC systems, is among the most sensitive detectors for low-level light detection. A PMT is a vacuum device that contains a photocathode, a number of dynodes, and an anode that delivers the output signal, as shown in Figure 8.4a. When a photon hits the photocathode, a photoelectron is emitted at the photocathode and it is accelerated toward the dynodes by an electric field built up by the operating voltage. When an electron hits a dynode, many additional electrons are released. After a cascade of dynodes, a short output pulse containing millions of electrons is generated at the anode. The timing uncertainty between photon time of arrival and electrical output, on the subnanosecond scale, is typically small enough for TCSPC experiments. However, the throughput of the PMT-based detection system is relatively low due to the lack of parallelism. In addition, PMTs are bulky and expensive and require high operational voltage, making them unattractive to be integrated into a portable system.

The MCP-PMT is a type of PMT that uses MCP. Each plate contains a large number of microscopically small channels with inner wall conductive coating, as shown in Figure 8.4b.[8] After a photoelectron enters a channel, it is reflected between the walls. Under a high voltage applied along the channels, many secondary electrons are generated when hitting the inner wall conductive coating.[34] Typical channel diameters are 3–10 μm, which is so small that electrons are well confined in the channel. Since the distance between the first MCP and the cathode is very small,

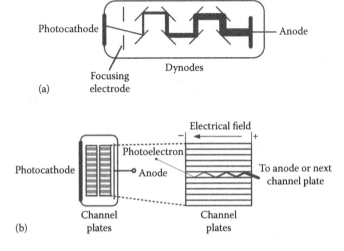

FIGURE 8.4   Structure of a single-photon detector: (a) photomultiplier tube and (b) microchannel plate photomultiplier tube.

the TTS of the MCP-PMT can be as low as 25 ps, which is lower than that of the PMT. Two or three cascaded MCP-PMTs can be used to achieve a high gain of $10^3 \times$ to $10^6 \times$. The SER is in the order of hundreds of picoseconds. Although they are expensive, currently, MCP-PMTs are the fastest commercially available single-photon detectors. Furthermore, by replacing the single anode with multianodes, a position-sensitive MCP-PMT can be realized. In this MCP-PMT, the arrival position of the photon on the photocathode can be determined by detecting the pulses from the individual anode.

SPADs are solid-state photodetectors in which a single photon can trigger an avalanche due to impact ionization. SPADs are avalanche photodiodes (APDs) working in Geiger mode. In Geiger mode, APDs are biased above the breakdown voltage where a single photon may trigger an avalanche of approximately 10^8 carriers, which enables this device to detect low intensity light down to the single-photon level. To prevent the avalanche from destroying the diode and to reset the diode for detecting the next photon, an active or passive quenching circuit is required to set the bias voltage to above the breakdown voltage. Passive quenching is achieved by operating the avalanche diode via a series resistor, whereas active quenching is obtained by an electronic quenching circuit.[35,36] The active area of a SPAD is typically designed to be a circle or an octagon to prevent premature edge breakdown due to high electric fields at sharp edges.[37,38] The significance of the emergence of SPADs is the fact that they can be realized by the CMOS technology, which enables large arrays of detectors. Miniaturization, in turn, improves portability, reduces fabrication costs, and often yields higher overall performance when integrated with high-speed electronics on the same substrate. However, current commercial SPAD arrays are still typically of a single or several pixels.

The general performance characteristics of photon detectors are temporal resolution, SER, PDE, external quantum efficiency (QE), dark count rate (DCR), maximum count rates, dead time, and afterpulsing probability. The temporal resolution of a detector is characterized by the TTS, which is due to the accuracy of the photon time of arrival. For a commercial PMT, SPAD, and MCP-PMT, the TTS is normally less than 180, 40, and 25 ps, respectively. The SER is the output pulse of a detector for a detected photon. The pulse shape and FWHM vary across detector types. The typical FWHMs of an MCP-PMT, PMT, and SPAD are 0.36, 1, and 40 ns, respectively.

PDE and QE are used to characterize the efficiency of photon-counting detectors. PDE is the ratio of the number of detected photons to the number of incident photons. QE is defined as the probability of a photoelectron or an electron–hole pair generation for each incident photon. Since not every photoelectron can lead to a detectable pulse in a PMT and not every electron–hole pair can trigger an avalanche in a SPAD, the PDE is smaller than the QE. In a SPAD, the PDE is determined by three factors: the absorption efficiency of photons, the collection efficiency of generated carriers, and the avalanche-triggering probability. The first two factors correspond to the QE of photodiodes.[39] Both PDE and QE depend on the cathode material properties of a detector and the incident light wavelength. For example, a Hamamatsu PMT with GaAsP cathode has a typical QE of 40% at 600 nm, whereas a Hamamatsu PMT with GaAs cathode has a typical QE of 12% at 800 nm.[40] The PDE of an Excelitas SPAD reaches 65% at 650 nm.[41]

Dark counts are output pulses caused by various noise sources in the detector. The DCR has a unit of counts per second. DCR is dependent on the cathode material, cathode area, and thermal effects. Since the main driven source of dark counts is thermal effect, the DCR is typically higher in detectors that are sensitive in the near-infrared wavelengths than those sensitive in the far-infrared wavelengths. For different types of detector, it varies from tens to hundreds of counts per second.

Dead time is the period of time during which the detector is unable to detect a new photon arrival after the previous photon event. Dead time is typically tens to hundreds of nanoseconds. The longer the dead time, the more unlikely all arrival photons in the same pulse period can be detected. When the first arrival photon is registered but the latter ones are missed, pileup occurs. Therefore, it is necessary to ensure a low probability of more than one photon arrival in a given period. In practice, the average count rate of a detector should be kept below 1%–5% of the excitation repetition rate.[42]

Afterpulsing refers to the generation of false pulses by the detector. This often occurs in a conventional PMT, due to ion feedback or luminescence of the dynode material and the glass of the tube.[8] In a SPAD, afterpulsing is caused by carriers trapped in an impurity after a previous avalanche and then released in about tens to hundreds of nanoseconds. Once such a carrier enters the avalanche region, it may cause an avalanche indistinguishable from that due to real photon detection. The probability of detecting afterpulsing after photon detection is known as the afterpulsing probability. It can be reduced by decreasing the PMT gain or by utilizing active quenching with SPAD.

8.4.3 CONSTANT FRACTION DISCRIMINATOR

The output pulses from a PMT or MCP-PMT have a considerable amplitude jitter due to the random amplification mechanism in the detector. This amplitude jitter leads to a time spread when determining the photon time of arrival. This time spread widens the IRF and introduces timing error in TCSPC systems. To alleviate this problem, a CFD is added between the detector and the timing electronics in conventional TCSPC systems. The CFD operates by adding the delayed version of the single-electron pulse to the original one and triggering at the zero-cross point. The time taken for the signal to cross the zero amplitude level is largely independent of the peak amplitude of the signal; thus, amplitude jitter can be significantly reduced. Modern CFDs typically detect the vertex of each pulse and trigger on that point,[42] which is equal to applying a constant fraction of one.

8.4.4 TIMING ELECTRONICS

The core element of TCSPC system is the timing electronics, which records the photon time of arrival. The operating principle is depicted in Section 8.2.2. Here, the emphasis is on two different kinds of timing systems: TAC/ADC implementation and TDC implementation.

8.4.4.1 TAC/ADC Implementation

In a conventional TAC/ADC system, as shown in Figure 8.2, the synchronization signal from the excitation source is connected to the *start* input of the TAC, and the detector output signal passing through the CFD is connected to the *stop* input. After the TAC receives a *start* signal, it begins to charge a capacitor until the *stop* signal arrives. The TAC output voltage, which is proportional to the time between the excitation source pulse and the first arrival photon, is then transferred to the ADC. If there is no photon detected in the period, the TAC is reset to zero. Based on the final ramp voltage, the ADC provides the digital timing value used to address the fluorescence decay histogram. By using this method, high temporal resolution can be achieved.

Although fast ADCs are often used to minimize dead time, it is inevitable and typically in the order of tens to hundreds of nanoseconds. Since there are many pulse periods in which no photon is detected, the TAC has to be stopped by the reset circuit rather than the photon pulse from the detector. At low excitation repetition rate, this is acceptable. However, at a high excitation repetition rate, the TAC is instructed to reset before the previous reset completes. For example, for a pulsed diode laser with a 50 MHz repetition rate, the TAC must be reset at least within 1/50 MHz = 20 ns. If the dead time is 100 ns, the TAC cannot be reset within a pulse period. In this case, an alternative TAC mode of operation known as the reverse start–stop mode has been introduced. In this mode, the output from the detector and the synchronization signals from the excitation source are connected to the *start* and *stop* inputs of the TAC, respectively. When there is no photon arrival, the TAC is not triggered. Thus, the TAC only needs to work at the rate of the photon detection, but at a much higher rate of the excitation repetition. In the reverse start–stop mode, the TAC output voltage decreases with the increase of photon time of arrival. The signal reversal is subsequently corrected by software. Due to advances in microelectronics, low-noise, high-speed ADCs are readily available. The mature TAC/ADC architecture is still used in advanced TCSPC for its high temporal resolution.

8.4.4.2 TDC Implementation

In most modern TCSPC systems, the TAC/ADC combination is replaced by a single TDC, for their superior time channel width and differential nonlinearity (DNL).[43] DNL describes the nonuniformity of each time bin in a TDC.[44] The basic principle of TDC is measuring the time difference based on the delay times in a chain of logic gates. A start pulse goes through an active delay line built by a large number of buffers, each connected to a D flip-flop. When a *stop* pulse is asserted on the clock inputs of the flip-flops, the buffer outputs are latched into the flip-flops. Based on the flip-flop outputs, the time between the *start* and *stop* pulses can be determined. However, differences in the gate delays cause high DNL. By utilizing ring oscillator–based architecture, the nonuniformity of the buffer delays averages out. To extend the time range of a TDC, two-stage counters can be used to count the periods of the ring oscillator,[8] as shown in Figure 8.5. Within the two-stage architecture, the first stage operates as a coarse counter, and then the time residue not converted in the first stage is converted by the second stage known as a fine counter.

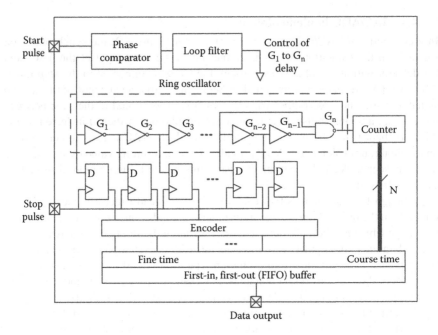

FIGURE 8.5 Architecture of a two-stage complementary metal-oxide semiconductor time-to-digital converter.

Among different kinds of TDC structures, this is the most commonly used, due to the accurate timing information it generates over a continuous range from picoseconds to seconds. It is particularly valuable for single-molecule spectroscopy, where the photon time of arrival with respect to the pulse at picosecond resolution and the macroscopic time with respect to the beginning of the experiment at nanosecond resolution are required simultaneously.

The TDC structures with a large number of fully paralleled channels are especially suitable for TCSPC applications. Currently, TDC chips with up to eight channels of 10 ps resolution are commercially available.[45] Rapid progress in CMOS technology leads to smaller gate delay, which greatly improves the temporal resolution of TDCs.

8.5 CMOS TCSPC SYSTEMS

Commercial TCSPC instruments typically have high temporal resolution and photon collection efficiency, bulkiness, and high cost rendering them unsuitable for portable and point-of-care applications.[46] These shortcomings call for a solution with reduced platform complexity, size, and cost, which can be achieved by implementation using CMOS technology. As benefited from process scaling, CMOS technology provides several advanced capabilities such as high integration density, low-noise signal processing, and low power consumption, which enables sensitive, small-sized, and low-cost TCSPC systems.[47] In particular, high integration density not only refers to the ability to fabricate component dimensions down to the nanoscale but also the ability to implement detector, readout, and signal-processing electronics on the same substrate. In this section, CMOS TCSPC systems are reviewed on three TDC integration levels: chip level, column level, and pixel level.

8.5.1 CHIP-LEVEL TDC INTEGRATION

8.5.1.1 Single-Detector Implementation

Integrating SPAD and TDC on the same chip significantly reduces the TCSPC system size. A compact, high-performance, and low-power consumption CMOS TDC with an internal on-chip SPAD has been presented,[48] as shown in Figure 8.6. The versatile TDC provides 10 ps temporal resolution, 160 ns dynamic range, and less than 1.5% least significant bit (LSB) DNL. This TDC can be operated either as a general-purpose time-interval measurement device when receiving the external *start* and *stop* pulses or in photon-timing mode when employing the on-chip SPAD for single-photon counting and timing measurements. With an overall resolution of 70 ps FWHM, it is suitable for demanding TCSPC applications such as FLIM and FRET. Comparing with state-of-the-art TDC, this TDC consumes very low power. Further, this TDC is optimized by implementing a couple of two-stage interpolators based on a new coarse-fine synchronization circuit and a new single-stage Vernier delay loop fine interpolation.[49] With these improvements, the DNL is better than 0.9% LSBs and the TDC resolution reaches 17 ps.

8.5.1.2 Multi-Detector Implementation

The development of SPAD implemented in CMOS technology makes it possible for TCSPC to be performed with sensitivity and temporal resolution comparable to PMT-based systems. With SPAD array and on-chip TDC, TCSPC systems avoid the bulky PMTs and laser scanning across the sample for FLIM applications.[50]

A SPAD imager specifically designed for FLIM applications consists of an array of 64 × 64 actively quenched and reset SPAD pixels, and an on-chip TDC has been presented.[51] The DCR of the SPAD at room temperature is 1059 Hz with a maximum photon detection probability of 4.7% at 440 nm. The on-chip TDC based on a delay-locked loop (DLL) and calibrated interpolators has a measured temporal resolution of 350 ps (1 LSB), 1.04 LSBs DNL, and 1.37 LSBs integral nonlinearity (INL). This imager is capable of performing both TCSPC and gated-window detection. This system is fabricated on a 4 mm × 4 mm die in a 0.35 μm high-voltage CMOS process. The die micrograph of the prototype chip is shown in Figure 8.7.

8.5.2 COLUMN-LEVEL TDC INTEGRATION

In a conventional TCSPC system, to avoid pileup, the photon-counting rate is typically limited by design. This has the disadvantage of low efficiency. Therefore, processing multiple photon events per excitation period is desirable.[52] Since column-parallel TDC implementation can achieve a high data processing rate, it is proposed in many works.

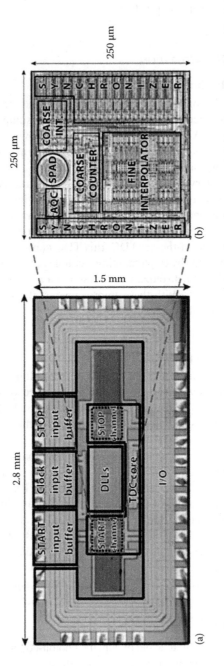

FIGURE 8.6 Microphotograph of a time-correlated single-photon counting: (a) chip-level layout showing two independent *start* and *stop* channels and global electronics for synchronizations and references and (b) *start* channel showing a 20 μm diameter single-photon avalanche detector.

FIGURE 8.7   Die micrograph of an imager prototype chip with a 64 × 64 single-photon avalanche detector array and an on-chip time-to-digital converter.

A miniaturized and high-throughput time-resolved fluorescence lifetime sensor with 32 × 32 SPAD array and 2 × 8 TDC implemented in a 0.13 μm CMOS process has been proposed.[9] As shown in Figure 8.8, a pulse-shortening monostable circuit is added at the output of each SPAD to reduce the pulse width, thereby making multiple overlapping events distinguishable. The array of 8 TDC pairs is used to generate multiple time stamps per excitation period. It is possible to count up to eight photons per excitation period and capable of processing up to 100 Mphoton/s with an excitation repetition rate of 12.5 MHz. The 16-bit TDC in this system has a resolution of about 50 ps, permitting a measurable lifetime range up to about 3 μs.

In addition to usage in fluorescence lifetime measurements, TCSPC microsystems with column-parallel TDCs are also widely used in 3-D imaging and optical range-finding applications. The TCSPC microsystems comprising with a 128 × 128 SPAD array, a bank of 32 TDCs, and a 7.68 Gbps readout system are operated as an optical range finder to reconstruct 3-D scenes with millimetric precisions under extremely low light intensity.[53]

8.5.3   PIXEL-LEVEL TDC INTEGRATION

With the pursuing of new time-resolved image sensors with higher temporal resolution and increased throughput, TCSPC systems need to be designed with a higher level of integration and parallelism.[54,55] By integrating one TDC per pixel, recent works have achieved high data rate processing through parallelism.

To achieve TCSPC systems with high temporal resolution, high frame rate, and small area for time-of-flight or FLIM applications, a 128-channel two-stage TDC utilizing a time difference amplifier (TDA) is implemented in a 0.35 μm CMOS process.[56] The first stage features as a coarse TDC. The time residue from the first stage is amplified by a TDA and subsequently converted by the second-stage TDC.

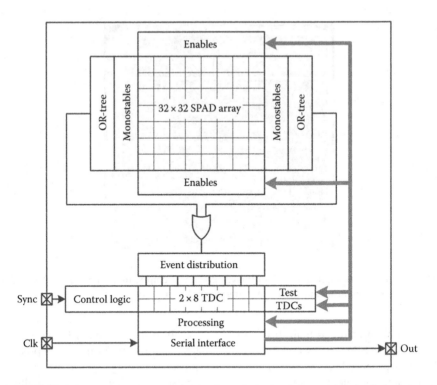

FIGURE 8.8   Top-level block diagram of a 32 × 32 single-photon avalanche detector array and a 2 × 8 time-to-digital converter array.

Each TDC is connected to a SPAD not only to measure DNL and INL using density tests but also to capture line-based 3-D images and FLIM applications. As the gain of the TDA can be adjusted from 8.5 to 20.4, the temporal resolution of the TDC can be tuned from 21.4 to 8.9 ps. The temporal resolution variation due to process–voltage–temperature effects is 5.8% without calibration when the temporal resolution is 12.9 ps. LSB changes due to power supply fluctuation and temperature variation are compensated by calibration.

A fully integrated SPAD array and TDC array for high-speed FLIM are realized in a standard 0.13 μm CMOS process.[11] This imager consists of an array of 64 × 64 SPADs, each with an independent TDC to perform single-photon counting independently. Each pixel has a 48 × 48 μm² area, which contains one octagonal SPAD with a 5 μm diagonal active area and circuitry for SPAD control. The DLL-based TDCs achieve a 62.5 ps resolution with up to 64 ns range. The DNL and INL of the TDC are smaller than 4 LSBs and 8 LSBs, respectively. The IRF is 125 ps, and dead time is 10 ns. By utilizing a data-compression data path to transfer TDC data to off-chip buffers, a data rate of up to 42 Gbps and a FLIM imaging rate of up to 100 frames per second are achieved. The total power consumption of the system is 8.79 W, or 2.15 mW/pixel.

A 32 × 32 smart pixel array with in-pixel SPAD, TDC, memory, and output buffers implemented in a cost-effective 0.35 μm CMOS platform is presented.[57]

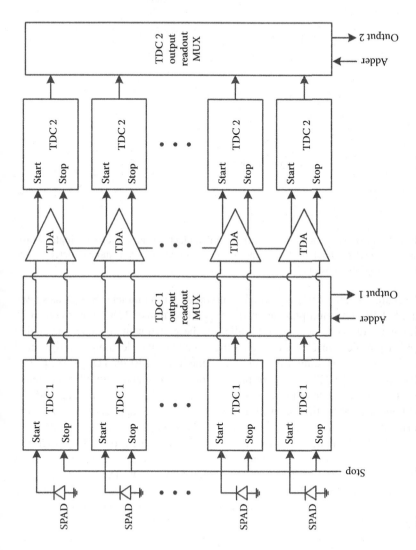

FIGURE 8.9 Chip architecture of a complementary metal-oxide semiconductor time-correlated single-photon counting system with pixel-level two-stage time-to-digital converters.

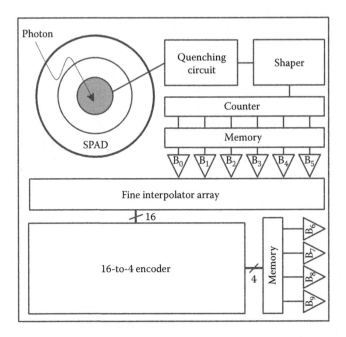

FIGURE 8.10 Complementary metal-oxide semiconductor time-correlated single-photon counting system. Pixel floor plan showing a 30 μm single-photon avalanche detector and 10-bit time-to-digital converter.

Each pixel area is 150 × 150 μm², which contains a 30 μm diameter SPAD, as shown in Figure 8.9. With no multiplexing during either detection or TDC conversion, 1024 fully independent pixels operate in parallel. In the global shutter mode, the analog front-end electronics senses and quenches the avalanche, thus leading to only a small afterpulsing effect. The temporal resolution of the in-pixel 10-bit TDC is 312 ps and the full-scale range is 320 ns. The in-pixel 10-bit memory and output buffers make this smart pixel the viable building block for advanced single-photon imager arrays for 2-D fluorescence lifetime as well as 3-D depth ranging applications.[58] The 3-D reconstruction of a human target under office lighting is shown in Figure 8.10.

A 32 × 32 SPAD array with in-pixel TDC is implemented in a 0.13 μm CMOS technology.[59] The temporal resolution of the TDC is 119 ps, the measured DNL is 0.4 LSBs, and INL is 0.2 LSBs. With a combined SPAD/TDC area of 50 × 50 μm², the design is one of the smallest demonstrated with deep subnanosecond resolution. This high level of performance makes it suitable for many applications, such as 3-D imaging and FLIM.

8.6 CONCLUSION

Fluorescence lifetime measurement based on TCSPC with high temporal resolution and photon collection efficiency has become an extremely important tool in a variety of fluorescence spectroscopy and microscopy applications including FLIM, FRET, and FLCS. With implementation in the CMOS technology, advanced TCSPC systems achieve a high level of integration, low power consumption, high sensory data throughput, small size, and reduced cost. These characteristics enable a new breed of TCSPC microsystems suitable for portable and point-of-care applications.

ACKNOWLEDGMENT

The authors gratefully acknowledge the funding support from the Research Grant Council, under project no. 21200414 and no. 11213515.

REFERENCES

1. O'Connor, D. *Time-Correlated Single Photon Counting*. London, U.K.: Academic Press, 1984.
2. Clegg, R.M. Fluorescence resonance energy transfer. *Current Opinion in Biotechnology* 6 (1) (1995): 103–110.
3. Birch, D.J. and R.E. Imhof. Time-domain fluorescence spectroscopy using time-correlated single-photon counting. In *Topics in Fluorescence Spectroscopy*. (ed.) Joseph R. Lakowicz. Springer, New York, 1999, pp. 1–95.
4. Lakowicz, J.R. *Principles of Fluorescence Spectroscopy*. Springer Science & Business Media, Maryland, 2007.
5. Yoon, H.-J. and S. Kawahito. A CMOS image sensor for fluorescence lifetime imaging. In *2006 IEEE Sensors*, Daegu, South Korea, Vols. 1–3, 2006, pp. 400–403.
6. Sytsma, J., J.M. Vroom, C.J. De Grauw, and H.C. Gerritsen. Time-gated fluorescence lifetime imaging and microvolume spectroscopy using two-photon excitation. *Journal of Microscopy* 191 (1998): 39–51.
7. Pancheri, L., N. Massari, and D. Stoppa. SPAD image sensor with analog counting pixel for time-resolved fluorescence detection. *IEEE Transactions on Electron Devices* 60 (10) (2013): 3442–3449.
8. Becker, W. *Advanced Time-Correlated Single Photon Counting Techniques*. Berlin, Germany: Springer, 2005.
9. Tyndall, D., B.R. Rae, D.D.-U. Li et al. A high-throughput time-resolved mini-silicon photomultiplier with embedded fluorescence lifetime estimation in 0.13 μm CMOS. *IEEE Transactions on Biomedical Circuits and Systems* 6 (6) (2012): 562–570.
10. Becker, W. *The bh TCSPC Handbook*, 5th edn. Becker & Hickl GmbH, Germany, UK, 2012.
11. Field, R.M., S. Realov, and K.L. Shepard. A 100 fps, time-correlated single-photon-counting-based fluorescence-lifetime imager in 130 nm CMOS. *IEEE Journal of Solid-State Circuits* 49 (4) (2014): 867–880.
12. Hof, M., R. Hutterer, and V. Fidler. *Fluorescence Spectroscopy in Biology*, 15 Vols., Vol. 3. Springer, Berlin, Germany, 2005.
13. Masters, B.R. and P.T.C. So. Antecedents of two-photon excitation laser scanning microscopy. *Microscopy Research and Technique* 63 (1) (2004): 3–11.
14. Verity, B. and S.W. Bigger. The dependence of quinine fluorescence quenching on ionic strength. *International Journal of Chemical Kinetics* 28 (12) (1996): 919–923.
15. Becker, W., V. Shcheslavkiy, S. Frere, and I. Slutsky. Spatially resolved recording of transient fluorescence-lifetime effects by line-scanning TCSPC. *Microscopy Research and Technique* 77 (3) (2014): 216–224.
16. Gustafson, T.P., Y.H. Lim, J.A. Flores et al. Holistic assessment of covalently labeled core-shell polymeric nanoparticles with fluorescent contrast agents for theranostic applications. *Langmuir* 30 (2) (2014): 631–641.

17. Schwarze, T., H. Mueller, S. Ast et al. Fluorescence lifetime-based sensing of sodium by an optode. *Chemical Communications* 50 (91) (2014): 14167–14170.
18. Ishikawa-Ankerhold, H.C., R. Ankerhold, and G.P.C. Drummen. Advanced fluorescence microscopy techniques—FRAP, FLIP, FLAP, FRET and FLIM. *Molecules* 17 (4) (2012): 4047–4132.
19. Peng, Y., C. Qiu, S. Jockusch et al. CdSe/ZnS core shell quantum dot-based FRET binary oligonucleotide probes for detection of nucleic acids. *Photochemical and Photobiological Sciences* 11 (6) (2012): 881–884.
20. Schwartz, D.E., P. Gong, and K.L. Shepard. Time-resolved Forster-resonance-energy-transfer DNA assay on an active CMOS microarray. *Biosensors and Bioelectronics* 24 (3) (2008): 383–390.
21. Medintz, I.L., H.T. Uyeda, E.R. Goldman, and H. Mattoussi. Quantum dot bioconjugates for imaging, labelling and sensing. *Nature Materials* 4 (6) (2005): 435–446.
22. Carlin, L.M., R. Evans, H. Milewicz et al. A targeted siRNA screen identifies regulators of Cdc42 activity at the natural killer cell immunological synapse. *Science Signaling* 4 (201) (2011): ra81.
23. Poland, S.P., S. Coelho, N. Krstajic et al. Development of a fast TCSPC FLIM-FRET imaging system. *Multiphoton Microscopy in the Biomedical Sciences XIII* 8588 (2013): 85880X-1–85880X-8.
24. Wahl, M., T. Roehlicke, H.-J. Rahn et al. Integrated multichannel photon timing instrument with very short dead time and high throughput. *Review of Scientific Instruments* 84 (4) (2013): 043102.
25. Wahl, M., H.-J. Rahn, T. Roehlicke et al. Scalable time-correlated photon counting system with multiple independent input channels. *Review of Scientific Instruments* 79 (12) (2008): 123113.
26. Yoon, H.-J., S. Itoh, and S. Kawahito. A CMOS image sensor with in-pixel two-stage charge transfer for fluorescence lifetime imaging. *IEEE Transactions on Electron Devices* 56 (2) (2009): 214–221.
27. Kapusta, P., R. Machan, A. Benda, and M. Hof. Fluorescence lifetime correlation spectroscopy (FLCS): Concepts, applications and outlook. *International Journal of Molecular Sciences* 13 (10) (2012): 12890–12910.
28. Kapusta, P., M. Wahl, A. Benda, M. Hof, and J. Enderlein. Fluorescence lifetime correlation spectroscopy. *Journal of Fluorescence* 17 (1) (2007): 43–48.
29. DeltaPro Lifetime System, Horiba, Japan, 2013. http://www.horiba.com/fileadmin/uploads/Scientific/Documents/Fluorescence/DeltaPro_Spec_Sheet.pdf. Accessed January 2013.
30. Ti:Sapphire Lasers, Thorlabs, New York, 2015. http://www.thorlabs.hk/newgrouppage9.cfm?objectgroup_id=3163. Accessed August 2015.
31. Femtosecond Fiber Lasers, Thorlabs, New York, 2015. http://www.thorlabs.hk/newgrouppage9.cfm?objectgroup_id=5924. Accessed August 2015.
32. Picosecond Pulsed Diode Laser, PicoQuant, Berlin, Germany, 2015. https://www.picoquant.com/products/category/picosecond-pulsed-sources/ldh-series-picosecond-pulsed-diode-laser-heads#custom1. Accessed March 2015.
33. Sub-Nanosecond Pulsed LEDs, PicoQuant, Berlin, Germany, 2014. https://www.picoquant.com/products/category/picosecond-pulsed-sources/pls-series-sub-nanosecond-pulsed-leds#specification. Accessed August 2014.
34. Kume, H., K. Koyama, K. Nakatsugawa, S. Suzuki, and D. Fatlowitz. Ultrafast microchannel plate photomultipliers. *Applied Optics* 27 (6) (1988): 1170–1178.
35. Cova, S., A. Longoni, and G. Ripamonti. Active-quenching and gating circuits for single-photon avalanche-diodes (SPADs). *IEEE Transactions on Nuclear Science* 29 (1) (1982): 599–601.
36. Cova, S., M. Ghioni, A. Lacaita, C. Samori, and F. Zappa. Avalanche photodiodes and quenching circuits for single-photon detection. *Applied Optics* 35 (12) (1996): 1956–1976.

37. Palubiak, D.P. and M.J. Deen. CMOS SPADs: Design issues and research challenges for detectors, circuits, and arrays. *IEEE Journal of Selected Topics in Quantum Electronics* 20 (6) (2014): 6000718.
38. Dandin, M., A. Akturk, B. Nouri, N. Goldsman, and P. Abshire. Characterization of single-photon avalanche diodes in a 0.5 μm standard CMOS process—Part 1: Perimeter breakdown suppression. *IEEE Sensors Journal* 10 (11) (2010): 1682–1690.
39. Savuskan, V., I. Brouk, M. Javitt, and Y. Nemirovsky. An estimation of single photon avalanche diode (SPAD) photon detection efficiency (PDE) nonuniformity. *IEEE Sensors Journal* 13 (5) (2013): 1637–1640.
40. Photosensor Modules H7422 Series, Hamamatsu, Tokyo, Japan, 2015. http://www.hamamatsu.com/resources/pdf/etd/m-h7422e.pdf. Accessed on 2015.
41. Single Photon Counting Module, SPCM-AQR Series, Excelitas, New York, 2015. http://www.excelitas.com/Downloads/DTS_SPCM-AQRH.pdf. Accessed on 2015.
42. Michael Wahl, Time-Correlated Single Photon Counting, PicoQuant, Berlin, Germany, 2014.
43. Bohmer, M., F. Pampaloni, M. Wahl et al. Time-resolved confocal scanning device for ultrasensitive fluorescence detection. *Review of Scientific Instruments* 72 (11) (2001): 4145–4152.
44. Kalisz, J. Review of methods for time interval measurements with picosecond resolution. *Metrologia* 41 (1) (2004): 17–32.
45. TDC-GPX Multifunctional High-End Time-to-Digital Converter, ACAM. Berlin, Germany. http://www.acam.de/products/time-to-digital-converters/tdc-gpx/. Accessed on 2015.
46. Singh, R.R., D. Ho, A. Nilchi et al. A CMOS/thin-film fluorescence contact imaging microsystem for DNA analysis. *IEEE Transactions on Circuits and Systems I—Regular Papers* 57 (5) (2010): 1029–1038.
47. Guo, N., K. Cheung, H.T. Wong, and D. Ho. CMOS time-resolved, contact, and multispectral fluorescence imaging for DNA molecular diagnostics. *Sensors* 14 (11) (2014): 20602–20619.
48. Markovic, B., D. Tamborini, F. Villa et al. 10 ps Resolution, 160 ns full scale range and less than 1.5% differential non-linearity time-to-digital converter module for high performance timing measurements. *Review of Scientific Instruments* 83 (7) (2012): 074703.
49. Markovic, B., S. Tisa, F.A. Villa, A. Tosi, and F. Zappa. A high-linearity, 17 ps precision time-to-digital converter based on a single-stage Vernier delay loop fine interpolation. *IEEE Transactions on Circuits and Systems I—Regular Papers* 60 (3) (2013): 557–569.
50. Schwartz, D.E., E. Charbon, and K.L. Shepard. A single-photon avalanche diode imager for fluorescence lifetime applications. In *2007 Symposium on VLSI Circuits, Digest of Technical Papers*. Tokyo, Japan: Japan Society of Applied Physics, 2007.
51. Schwartz, D.E., E. Charbon, and K.L. Shepard. A single-photon avalanche diode array for fluorescence lifetime imaging microscopy. *IEEE Journal of Solid-State Circuits* 43 (11) (2008): 2546–2557.
52. Tyndall, D., B. Rae, D. Li et al. A 100 Mphoton/s time-resolved mini-silicon photomultiplier with on-chip fluorescence lifetime estimation in 0.13 μm CMOS imaging technology. In *ISSCC 2012*, San Francisco, CA, 2012.
53. Niclass, C., C. Favi, T. Kluter, M. Gersbach, and E. Charbon. A 128 × 128 single-photon image sensor with column-level 10-bit time-to-digital converter array. *IEEE Journal of Solid-State Circuits* 43 (12) (2008): 2977–2989.
54. Veerappan, C., J. Richardson, R. Walker et al. A 160×128 single-photon image sensor with on-pixel 55 ps 10b time-to-digital converter. In *ISSCC 2011*, San Francisco, CA, 2011.

55. Gersbach, M., Y. Maruyama, R. Trimananda et al. A time-resolved, low-noise single-photon image sensor fabricated in deep-submicron CMOS technology. *IEEE Journal of Solid-State Circuits* 47 (6) (2012): 1394–1407.

56. Mandai, S. and E. Charbon. A 128-channel, 8.9-ps LSB, column-parallel two-stage TDC based on time difference amplification for time-resolved imaging. *IEEE Transactions on Nuclear Science* 59 (5) (2012): 2463–2470.

57. Villa, F., B. Markovic, S. Bellisai et al. SPAD smart pixel for time-of-flight and time-correlated single-photon counting measurements. *IEEE Photonics Journal* 4 (3) (2012): 795–804.

58. Villa, F., R. Lussana, D. Bronzi et al. CMOS imager with 1024 SPADs and TDCs for single-photon timing and 3-D time-of-flight. *IEEE Journal of Selected Topics in Quantum Electronics* 20 (6) (2014): 364–373.

59. Gersbach, M., Y. Maruyama, E. Labonne et al. A parallel 32 × 32 time-to-digital converter array fabricated in a 130 nm imaging CMOS technology. In *2009 Proceedings of ESSCIRC*, Athens, Greece, 2009, pp. 197–200.

Section III

Biomedical Applications

9 Fiber Delivery of Femtosecond Pulses for Multiphoton Endoscopy

Shuo Tang and Jiali Yu

CONTENTS

9.1 Introduction 195
    9.1.1 Principle of MPM 196
    9.1.2 Literature Review on MPM Endoscopy 198
9.2 Optical Fiber for Delivering Femtosecond Pulses 199
    9.2.1 Single-Mode Fiber 200
    9.2.2 Double-Clad Fiber 200
    9.2.3 Photonic Crystal Fiber 201
    9.2.4 CPCF 202
9.3 MPM System of Fiber Delivery of Femtosecond Pulses 204
    9.3.1 Setup of MPM Imaging System 204
    9.3.2 MPM System Dispersion Management 205
9.4 Image Demonstration with CPCF Fiber-Based MPM System 208
    9.4.1 MPM Imaging of the Skin.. 208
    9.4.2 MPM Imaging of the Cornea... 209
    9.4.3 MPM Imaging of the Tooth 210
9.5 Conclusions 211
References 211

9.1 INTRODUCTION

Optical imaging uses the interaction between light and tissue to probe tissue morphology and functions. Compared to other medical imaging techniques, optical imaging is noninvasive, has a very high resolution at the cellular level, and has biochemically specific contrast from light absorption, scattering, and fluorescence. It provides critical information about the morphology of cells and extracellular matrix and the anatomy and physiology of tissues. Optical imaging has found many applications in medicine and is very useful in the early-stage diagnosis of diseases and monitoring the treatment outcomes.

A variety of optical imaging techniques have been developed and applied medically. Photoacoustic imaging is a relatively new technique that uses a laser to excite photoacoustic waves and ultrasound transducers to detect the waves [1]. It can image tissues for up to a few centimeters of penetration depth; optical coherence tomography uses light interference to detect backscattered light from tissue interfaces [2]. It can image the cross-sectional view of layered tissue structures from a few millimeters deep; confocal microscopy is a laser scanning microscopy for imaging cells with high resolution in 3D [3]. By labeling cells and organelles with exogenous fluorescent dyes or genetically modified fluorescent proteins, confocal microscopy can detect subcellular organelles and differentiate cancer cells from normal cells with high specificity; multiphoton microscopy (MPM) is a nonlinear microscopy that uses a femtosecond laser to excite intrinsic autofluorescence signals from tissues [4,5]. It has similar resolution as confocal microscopy. However, MPM does not require fluorescent staining, which is especially useful for *in vivo* imaging where staining is not applicable.

9.1.1 PRINCIPLE OF MPM

MPM has become a widely used image modality for turbid tissues since the past 30 years [4,5]. Neuroscientists have used MPM to monitor the structure of neural compartments and to observe calcium dynamic networks in live animals [6,7]. Cancer researchers have used MPM for *in vivo* imaging of cells and extracellular matrix for detecting cancer and studying metastasis and angiogenesis [8,9]. Thus, MPM is an indispensable tool for high-resolution structural and functional imaging of tissues.

MPM excites nonlinear contrast signals including two-photon-excited fluorescence (TPEF) and second harmonic generation (SHG). The MPM contrasts are of high biochemical specificity. For instance, TPEF can be excited from intrinsic tissue constituents such as nicotinamide adenine dinucleotide (NADH), elastin, flavins, and melanin [10,11]. Fluorescence from NADH can be used as an indicator for cellular metabolic state because the coenzyme NADH is involved in redox reaction. Substantial NADH is found in cell cytoplasm, and neoplastic metabolism is associated with changes in the concentration of NADH. SHG is detected from extracellular matrix such as collagen and muscle myosin in various tissues [12]. Collagen is abundant in connective tissues, and SHG imaging of collagen can show the organization and orientation of collagen fibers and their interaction with cells. It is reported that the modification of collagen in tissues is associated with various pathological processes such as cancer.

In response to light excitation, tissue generates induced polarization, which can be written as [13]

$$P = \varepsilon_0(\chi^{(1)}E + \chi^{(2)}EE^* + \chi^{(3)}EE^*E + \cdots) \tag{9.1}$$

where
   E is the applied electric field of light
   ε_0 is the permittivity of free space
   $\chi^{(1)}$ is the linear susceptibility
   $\chi^{(n)}$ is the nth order nonlinear susceptibility (for $n > 1$)

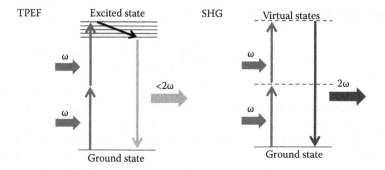

FIGURE 9.1   Energy diagrams of two-photon-excited fluorescence and second harmonic generation.

Linear effects, such as normal absorption and scattering, relate to the linear susceptibility $\chi^{(1)}$. Nonlinear effects depend upon the higher-order susceptibilities $\chi^{(n)}$. Specifically, $\chi^{(2)}$ contributes to SHG, and $\chi^{(3)}$ contributes to TPEF and third harmonic generation. Therefore, TPEF and SHG have different fundamental origins compared with single-photon microscopy such as confocal microscopy.

Figure 9.1 shows the energy diagram of TPEF and SHG, respectively. In TPEF, two photons with low energy are absorbed simultaneously to excite a molecule from ground state to excited state. In comparison, a single high-energy photon is absorbed to excite the molecule in single-photon excitation. When the molecule returns to the ground state, energy is released as a fluorescence photon. The fluorescence photon carries slightly less energy than the sum of the two excitation photons due to energy loss in the transitions. In the SHG process, instead of being absorbed, two excitation photons are simultaneously scattered and an SHG photon at exactly twice the energy of the excitation photon is created.

In TPEF, the number of photons absorbed through a two-photon process per fluorophore per pulse is given by [14]

$$n_2 \approx \frac{P_{ave}^2 \delta}{\tau_p f_p^2} \left(\frac{NA^2}{2hc\lambda} \right)^2 \tag{9.2}$$

where
  δ is the two-photon absorption cross section
  P_{ave} is the average laser intensity
  τ_p is the pulse width
  f_p is the pulse repetition rate
  NA is the numerical aperture (NA) of objective lens
  λ is the wavelength
  h is Planck's constant
  c is the speed of light

For SHG, the emission strength is given by

$$n_{SHG} \propto \left(\chi^{(2)} \right)^2 \frac{P_{ave}^2}{\left(\tau_p f_p^2 \right)} \tag{9.3}$$

As we can see, both TPEF and SHG intensities scale quadratically with the excitation laser power but inversely with the pulse width of the laser.

The two-photon absorption cross section and $\chi^{(2)}$ coefficient are extremely small. Due to the extremely low efficiency of the nonlinear processes, high flux of excitation photons is necessary in order to generate sufficient MPM signals. Focusing the excitation light in the spatial and temporal domains is required to achieve high photon flux. In other words, the light should be focused by an objective lens with high NA and an ultrashort pulsed laser should be used to achieve high peak intensity.

MPM has advantages over single-photon microscopy (i.e., confocal microscopy) in several aspects. First, both TPEF and SHG use low-energy photons in the near-infrared (NIR) wavelength range for excitation. As a result, the penetration depth is increased due to less scattering in turbid tissue at NIR. Phototoxicity is also greatly reduced due to the lack of endogenous absorbers in tissue at NIR. Second, the intensity-squared dependence of TPEF and SHG enables localized excitation in the subfemtoliter focal volume that offers high resolution in 3D. Typically, the lateral resolution is ~0.5 μm with a depth resolution of ~1.5 μm. Third, the inherently localized excitation at the focal volume reduces the photodamage in the light path and thus maintains the viability of living tissues and cells. Finally, TPEF and SHG provide the biochemical specificity from intrinsic signals of cells and extracellular matrix. Imaging the intrinsic signals from tissues without labeling is important for early diagnosis and treatment of cancer in patients.

9.1.2 LITERATURE REVIEW ON MPM ENDOSCOPY

Currently, most MPM systems are based on tabletop microscopes, which have been used on excised tissues and small animals. However, the microscope systems are bulky and use mostly free-space components, which greatly limit their applications for *in vivo* imaging in the clinic. Fiber-based multiphoton endoscopy uses optical fiber for light delivery and signal collection, which makes it more portable and robust. MPM endoscopy has the potential to be used in the clinic on patients for early cancer diagnosis and long-term disease monitoring [15,16]. Thus, there is growing interest in developing fiber-based multiphoton endoscopes for clinical applications due to their flexibility and miniaturization [17,18].

A fiber-optic two-photon microscopy was reported by D. Bird and M. Gu. using a single-mode fiber (SMF) [19]. After light delivery by a 1 m length SMF, the pulses with original duration of 80 fs were stretched to approximately 3 ps at the output of the fiber because of group velocity dispersion (GVD) and strong effective nonlinearity from the fiber [19]. W. Göbel et al. developed a miniaturized two-photon microscope based on a coherent fiber bundle with 30,000 individual cores. Even though a pair of grating was employed for dispersion precompensation, the two-photon excitation remained suboptimal due to the severe nonlinear broadening even at a relatively low laser power of 5 mW [20]. A portable two-photon microendoscope was proposed by B. A. Flusberg et al. using a commercial photonic bandgap fiber, which had zero GVD at around 795 nm. Pulses of 150 fs pulse width and 791 nm center wavelength suffered slight temporal broadening after transmitting in the 1.5 m length photonic bandgap fiber, perhaps due to higher-order dispersion [21]. Similarly, Göbel et al. demonstrated nearly distortion-free delivery of 170 fs pulses at the zero-GVD wavelength (812 nm). However, at above or below the zero-GVD wavelength, the output pulse is severely broadened, for example, to 2 ps at 795 nm [22]. M. T. Myaing et al. achieved a fiber-optic scanning endoscope based on a double-clad fiber (DCF) for the delivery of illumination light and collection of fluorescence. They reported that the 1.37 m length DCF introduced ~59,000 fs^2 dispersion under low excitation power (<10 mW), and it required a grating-based stretcher for dispersion compensation [23]. The application of double-clad photonic crystal fiber (PCF) in the multiphoton field was investigated by several groups such as L. Fu et al. and S. Tang et al. [24,25]. Pulses with 170 fs duration were broadened to 2.5 ps after a double-clad PCF delivery with 2 m length [25].

From earlier, although each type of the reported fibers has its own advantages, none of them could directly deliver sub-30 fs pulses with high fidelity and high power. Even with grating/prism stretchers to precompensate the fiber dispersion, the reported systems mentioned earlier could only deliver relatively long pulses of 80~200 fs duration. Ultrashort pulses with duration less than 80 fs are desirable to increase the signal strength in MPM because the MPM signal intensity scales inversely with the pulse duration.

A key challenge in MPM endoscopy is how to deliver ultrashort pulses from the excitation laser to the imaging site efficiently through a fiber. Next, the pros and cons of various types of optical fibers will be compared in Section 9.2. A fiber-based MPM system based on chirped photonic crystal fiber (CPCF) will be introduced in Section 9.3. The critical issues such as dispersion, pulse broadening, and dispersion precompensation will also be discussed. Representative images obtained by the fiber-based MPM system will be presented in Section 9.4.

9.2 OPTICAL FIBER FOR DELIVERING FEMTOSECOND PULSES

Optical fiber is a flexible waveguide that can guide and deliver light over a long distance. Delivering the excitation and emission light with optical fiber is a key part of multiphoton endoscopy that features portability, flexibility, and miniaturization of the system. However, it is challenging to deliver ultrashort pulses reliably through optical fiber because femtosecond pulses suffer from severe temporal and spectral distortion due to dispersion and nonlinear effects such as self-phase modulation (SPM) in optical fiber [26]. Dispersion is the optical property where light of different wavelengths propagates at different velocities in material. Femtosecond pulses from titanium–sapphire lasers have a pulse width of 10–100 fs, which can be broadened to picosecond pulses due to dispersion after propagating through lenses or optical fiber. Thus, dispersion needs to be compensated by dispersion compensation components such as gratings, prisms, or chirped mirrors. SPM is a fiber nonlinear effect where the phase change depends on the light intensity. SPM is proportional to the cube of the intensity and inversely proportional to the square of the fiber diameter. Therefore, increasing the fiber diameter could significantly reduce the effect of SPM. In the following, various kinds of fibers that are currently used in MPM are compared.

9.2.1 SINGLE-MODE FIBER

SMF is widely used in many areas including fiber-optic communications. Figure 9.2 illustrates the cross section of SMF that contains a core and a cladding. The core has a higher refractive index than the cladding in order to meet the total internal reflection condition. The typical core size is ~5–10 μm depending on the wavelength and its NA is ~0.1–0.3. Due to the small core size, the power density can be high, which causes excessive power-dependent nonlinear effect. For instance, for 100 fs pulses at center wavelength 800 nm with an average power of 400 mW, nonlinearity dominates the propagation and quickly distorts the spectrum even after 1.2 cm fiber delivery [27]. Meanwhile, the zero-dispersion wavelength of standard SMF is near 1300 nm wavelength. This zero-dispersion wavelength is far away from the typical operating wavelength of 800 nm in MPM. Thus, a large amount of GVD occurs in an SMF at 800 nm wavelength, which produces severe pulse broadening. In SMF, the nonlinear effects restrict the maximum delivery power and the GVD severely broadens the pulse width, which reduces the efficiency of MPM signals.

9.2.2 DOUBLE-CLAD FIBER

Figure 9.3 shows the cross section of a DCF, which contains a core, an inner cladding, and an outer cladding. The refractive index varies and descends from

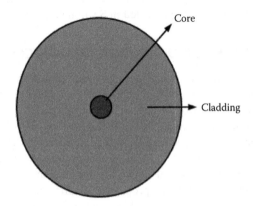

FIGURE 9.2   Single-mode fiber.

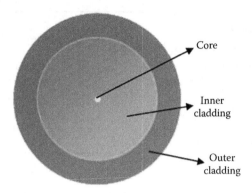

FIGURE 9.3   Double-clad fiber.

the core to the inner cladding to the outer cladding. The core functions as an SMF and the inner cladding as a multimode fiber. Femtosecond pulses can be delivered through the core and MPM signals can be collected by the core and the inner cladding. The collection efficiency is enhanced due to higher NA from the inner cladding. Therefore, the advantage of DCF is that delivery of illumination light and collection of fluorescence can be achieved by the same fiber. The limitation, similar to the SMF, is pulse broadening and nonlinear effect from its small core.

9.2.3 Photonic Crystal Fiber

PCF is a microstructured fiber that can overcome the limitations of a conventional solid fiber and revolutionize the field of optical fiber [28,29]. PCFs are characterized by a periodic array of wavelength-scale air holes that extend over the entire length of the fiber. Various types of PCF have been designed for different properties. Figure 9.4 shows the cross section of a hollow-core PCF where the core is hollow. The guiding mechanism of a PCF is different from solid fiber,

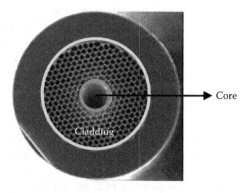

FIGURE 9.4   Photonic crystal fiber.

TABLE 9.1

Specifications of HC-800-01

Core diameter	9.5 ± 1 μm
Diameter of the holey region	40 μm
Zero-GVD wavelength	812 nm
GVD (k_2) at 790 nm	45,000 fs^2/m
Attenuation (790~870 nm)	<0.35 dB/m
Numerical aperture	0.28

which requires the core to have higher refractive index than cladding. In PCF, the periodic air-hole structure in the cladding creates a photonic bandgap (similar to the electronic bandgap in a periodic lattice) that can confine light inside the air core. PCFs with hollow core are specifically named as hollow-core fibers (HCF). Since light propagates inside the air core, HCF can circumvent the nonlinear effects imposed by available materials and reduce the material dispersion as well as managing the waveguide dispersion [30].

The specifications of a commercial HCF, specifically HC-800-01 (Thorlabs) [31], are shown in Table 9.1. The core diameter is 9.5 μm and transmission band is ~70 nm. The loss is relatively low at ~0.35 dB/m due to the relative mature technology. The zero-GVD wavelength is 812 nm. However, it still has significant GVD at 790 nm of approximately $D = -150$ ps/nm/km, corresponding to $k_2 = 45,000$ fs^2/m. It also has a significant amount of third-order dispersion (k_3) due to the large slope of its dD/dλ curve.

9.2.4 CPCF

CPCF is a special type of HCF, whose geometrical dimension of the cladding changes in radial direction [31]. Figure 9.5 shows the cross section of a CPCF of 50 μm core size. It consists of a five-layered cladding with increasing hole size from the inner layer to the outer layer. The size of the unit cells keeps increasing in a certain ratio between any two adjacent layers from the inner side to the outer side. The radial chirp in cell size controls the dispersion property of the fiber, analogous to the principle of chirped mirrors. Light of different wavelengths is reflected at different layers of the chirped cladding in the CPCF, which introduces group delay variations with wavelength and tailors the dispersion profile effectively. Its GVD parameter k_2 appears on the order of a few fs^2/m to 100 fs^2/m over the whole transmission band (720–820 nm). This unique fiber design enables almost total elimination of GVD. The 50 μm core size of the CPCF is sufficiently large to relax the power limitation for nonlinear effects. All these advantages come with the cost of increased propagation and attenuation. Distinct from commercial HCF, CPCF is based on quasi-guiding that results in increased guiding loss.

The specifications of the customized CPCF are summarized in Table 9.2. The core diameter is 50 μm. It has a relatively broad transmission bandwidth of ~130 nm

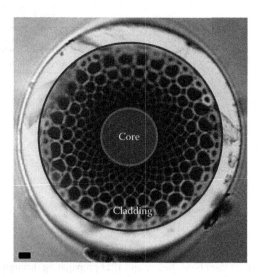

FIGURE 9.5   Chirped photonic crystal fiber.

around a 790 nm center wavelength. It has a very low GVD k_2 of ~150 fs^2/m in this wavelength range. Its dispersion slope is small, indicating relatively low third-order dispersion. Compared to normal HCF with uniform cell size, CPCF has a much higher loss at 5.1 dB/m and its NA is 0.027, which is quite low, due to the technical challenges in fabricating such fiber, which can be improved in the future.

In TPEF, different fluorophores have different excitation peak wavelengths. Tunable lasers have been used to selectively excite TPEF signals with different excitation wavelengths. Meanwhile, the signal intensity of MPM depends inversely on the pulse width. Shorter pulses can excite MPM signals more efficiently than broader pulses. However, shorter pulses also have broader bandwidth. For example, ultrashort femtosecond pulses of 10 fs pulse width have a spectral bandwidth of ~100 nm. For tunable lasers and ultrashort femtosecond lasers with broad bandwidth, the transmission band of the fiber needs to be sufficiently large in order to cover the full laser wavelength bandwidth.

Figure 9.6 shows the spectrum of the output of an ultrashort laser and the spectrum after delivery through a CPCF and an HC-800-01, respectively. The titanium–sapphire laser (Femtolasers, Fusion PRO 400) has a full width at half

TABLE 9.2
Specifications of the Custom Chirped Photonic Crystal Fiber

Core diameter	50 μm
Diameter of the holey region	200 μm
GVD (k_2) near 790 nm	~150 fs^2/m
Attenuation	5.1 ± 2 dB/m
Numerical aperture	0.027 ± 0.003

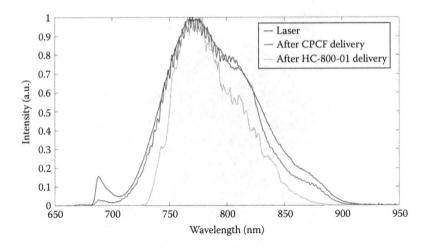

FIGURE 9.6   Spectra of laser beam before and after propagating through a chirped photonic crystal fiber and a HC-800-1.

maximum (FWHM) of 90 nm from 740 to 830 nm. After transmitting through the CPCF, the FWHM bandwidth is slightly reduced to 85 nm. The bandwidth is more significantly reduced to 64 nm after transmitting through the HC-800-01. For the HC-800-01, the wavelengths shorter than 730 nm are forbidden to propagate due to strict bandgap. The CPCF does not have a cutoff wavelength in the range of 700–900 nm, which enables a wider bandwidth. Furthermore, CPCF has an extended bandwidth especially in the short wavelength range compared to HCF. This feature can increase the two-photon fluorescence signals for some important endogenous fluorophores (e.g., NADH, elastin), which have two-photon absorption peaks between 700 and 730 nm.

9.3   MPM SYSTEM OF FIBER DELIVERY OF FEMTOSECOND PULSES

Using optical fiber for delivering femtosecond pulses for MPM imaging can increase the portability and flexibility of the system. In MPM endoscopy, the imaging head in the distal end needs to be coupled to the laser source through a flexible fiber. Various types of fibers have been investigated. From the discussions in Section 9.2, the HCF and CPCF have relatively low dispersion that can minimize the pulse broadening for propagating femtosecond pulses. CPCF is especially suitable for ultrashort pulses with broad bandwidth. In the following texts, we will show experimental results of using the CPCF for pulse delivery in an MPM system.

9.3.1   Setup of MPM Imaging System

The experimental setup for the MPM imaging system using the CPCF is shown in Figure 9.7. Light generated from the titanium–sapphire laser (Femtolasers, Fusion

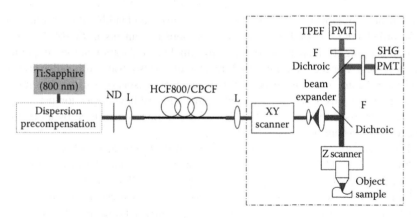

FIGURE 9.7   Schematic of the multiphoton microscopy system.

PRO 400) first passes through a dispersion precompensation unit. The center wavelength of the laser is around 800 nm with a spectral bandwidth of 90 nm. The pulse width is ~10 fs, and the repetition rate is 80 MHz. A continuously variable neutral density (ND) filter controls the laser power entering the fiber. The beam is then coupled into a piece of the CPCF or the commercial HC-800-01. A pair of objective lenses (Olympus, Plan N 4X NA = 0.1 and MPlanFL 10X NA = 0.3) is used as the coupling and collimation lens. The attenuator is adjusted to ensure identical output power level after light is delivered by the CPCF or the HCF. The beam after passing through the fiber enters an imaging subsystem to examine the MPM imaging quality. For comparison, a free-space coupling scheme without any fiber is also tested, where the laser beam after the ND filter is directed to the XY scanner directly.

The imaging subsystem is shown in the dashed box in Figure 9.7. The beam is raster scanned by two galvanometer mirrors (Cambridge Technology, 6215H). After beam expansion by two lenses, the beam overfills the back aperture of an objective. The objective (Olympus, LUMPLANFL N) has 40× magnification and 0.8 NA. It eventually focuses the beam onto the sample. The objective lens is mounted on a depth scanner (Piezosystem Jena, nanoMIPOS 400). The TPEF and SHG signals are collected in the backward direction by the same objective lens. The received signals are separated from the excitation light by passing through a short-pass dichroic mirror (Semrock, FF670-SDi01). TPEF and SHG signals are further separated from each other by a second dichroic mirror (Chroma, 450DCXRU) and then are respectively detected by two photomultiplier tubes (Hamamatsu, H6780 and H5783P).

9.3.2   MPM SYSTEM DISPERSION MANAGEMENT

Since the MPM signal intensity scales inversely with the pulse width of the excitation light, shortening the pulse width at the sample location can enhance the MPM intensity, especially when the power on sample is limited by photodamage or photobleaching. However, a short pulse with broad bandwidth suffers from severe pulse broadening due to dispersion from the optical elements in the MPM system. Thus, managing the dispersion to obtain minimal pulse duration at the sample location is critical. For the titanium–sapphire laser of ~10 fs pulse width, pulse broadening is significant, and thus, dispersion management is especially critical. Figure 9.8 shows how the pulses get broadened by dispersion and how dispersion precompensation compresses the pulses back. The pulse duration is measured by an intensity autocorrelator (Femtochrome, FR-103MN).

The pulse after the laser output is shown in Figure 9.8a, which shows a pulse width of ~14 fs. The ultrashort pulse is then launched into the MPM system by three pulse delivery schemes, free-space optics, CPCF, and HCF, respectively. In Figure 9.8b through g, the pulse durations are measured at the sample plane before (left column) and after (right column) implementing dispersion precompensation for the three pulse delivery cases. First, the pulse width of the free-space system without using any fiber is measured. As shown in Figure 9.8b and c, the pulse width is stretched to ~290 fs by the objective lens and other optics before applying the dispersion precompensation, and it is compressed back to ~27 fs after optimizing a prism pair for dispersion precompensation. Second, one piece of CPCF with ~1 m length and the corresponding coupling and collimation lenses are added to deliver the excitation pulses. As shown in Figure 9.8d, the pulse duration of the CPCF delivery system is broadened to ~300 fs after all the optical components and the CPCF. Figure 9.8e demonstrates that the pulse is compressed back to ~23 fs after optimizing the prism-based precompensation unit. Third, the CPCF is replaced by a piece of commercial HCF (Thorlabs, HC-800-01) with ~0.5 m length and the pulse width is measured again. A shorter HCF is selected because of its severe pulse broadening. Figure 9.8f indicates that the pulse is severely stretched to 2200 fs after all the lenses and the HCFs. This stretching is attributed to the broad laser bandwidth, the relatively large GVD far away from the zero-dispersion wavelength, and significant third-order dispersion of the HCF in the laser bandwidth window. Even after precompensation, the observed pulse width is ~770 fs as shown in Figure 9.8g, which is likely caused by the strong third-order dispersion of HC-800-01. Comparing the pulse width after propagating through the free-space-, CPCF-, and HCF-based systems, it is found that the CPCF can achieve similar pulse width as the free-space system, which indicates that the CPCF introduces very minimum dispersion. The HCF still significantly broadens the pulse due to its limited transmission bandwidth and relatively large third-order dispersion.

This result demonstrates that sub-30 fs pulses can be successfully delivered to the sample location after ~1 m CPCF delivery. It also confirms that the dedicated CPCF design enables the ultrashort pulse delivery with extremely low GVD and third-order dispersion. Furthermore, the comparable pulse width obtained from the CPCF-delivered system and the free-space system at sample location shows that the CPCF minimizes the fiber effects for the ultrashort pulse delivery in the MPM system. As a result, the dispersion management from the CPCF itself potentially simplifies the higher-order dispersion control of fiber-based MPM endoscopes and enables a significant increase of MPM signals by delivering ultrashort pulses.

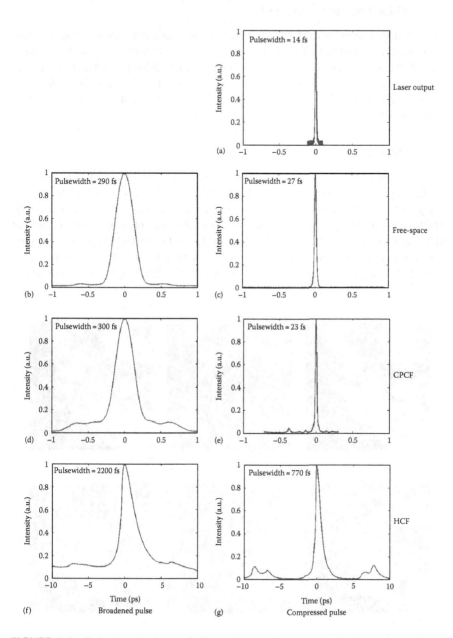

FIGURE 9.8   Pulse broadening and dispersion precompensation in free-space, chirped photonic crystal fiber, and hollow-core fiber delivery systems. (a) At laser output. (b) and (c), With free-space delivery. (d) and (e), With CPCF deliver. (f) and (g), With HCF delivery.

9.4 IMAGE DEMONSTRATION WITH CPCF FIBER-BASED MPM SYSTEM

The MPM system using the CPCF for femtosecond pulse delivery is tested on imaging biological samples, including the skin, cornea, and tooth. Compared with free-space MPM system, the images obtained by the CPCF-delivered pulses have similar imaging quality, showing that CPCF is a good candidate for delivering ultrashort pulses in MPM endoscopy. In the following texts, the images obtained by the CPCF-delivered MPM system are demonstrated.

9.4.1 MPM IMAGING OF THE SKIN

Figure 9.9 shows the representative MPM images of the dermis layer of a skin sample. Figure 9.9b and c shows the separate TPEF and SHG channels in gray scale, respectively. In the dermis, collagen fibers provide SHG contrast and elastin fibers provide TPEF contrast. The grayscale images show the absolute intensity of the TPEF and SHG channels, where the collagen fibers in the SHG channel show higher signal intensity than the elastin fibers in the TPEF channel. Figure 9.9a shows the combined TPEF and SHG image in pseudocolor where the TPEF signal is color-coded in

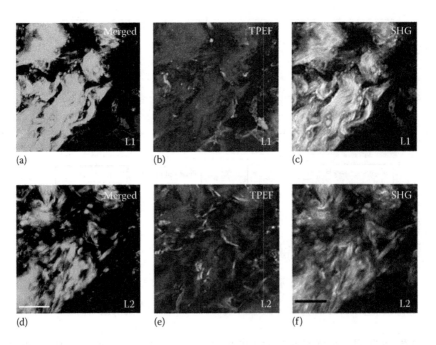

(a)   (b)   (c)

(d)   (e)   (f)

FIGURE 9.9   Images of the dermis layer of the human skin. Images on the left side are a combination of two-photon-excited fluorescence/second harmonic generation, while images on the right side are obtained from respective single channel. The optical power on sample is 5 mW. Pixel dwell time is 10 μs. The scale bar is 50 μm. (a)–(c), Image at location 1. (d)–(f) Image at location 2.

red and SHG in green. Figure 9.9d through f shows the MPM images at a different location. Based on anatomical features of the skin structure, thick collagen bundles and thin elastin fibers are clearly distinguished. The power delivered by the CPCF on the fresh skin sample is about 5 mW, and the pixel dwell time is 10 μs.

9.4.2 MPM IMAGING OF THE CORNEA

Figure 9.10 shows the TPEF and SHG images of a fish cornea acquired at different depths from the anterior to the posterior side of the cornea. The TPEF signal is color-coded in red and SHG in green. Good signal-to-noise ratio is obtained even though the power delivered through the CPCF on the sample is ~4 mW. In Figure 9.10a, the cells correspond to the epithelium layer. In Figure 9.10b, the TPEF signals, from the autofluorescence in cells, in combination with SHG signals, from collagen fibers,

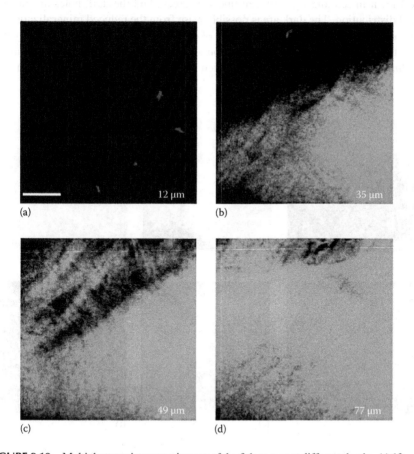

(a)   (b)

(c)   (d)

FIGURE 9.10   Multiphoton microscopy images of the fish cornea at different depths: (a) 12 μm, (b) 35 μm, (c) 49 μm, and (d) 77 μm. Each of these numbers denotes the depth of the current image from the cornea surface. Power delivered on the sample is 4 mw. Pixel dwell time is 60 μs. The scale bar is 50 μm.

suggest the boundary between the epithelium and the Bowman's layer. In Figure 9.10c and d, the SHG signal shows the orientation of collagen fibers, which alters at different depths in the corneal stroma. The endothelium is not observed in the image stack due to the low illumination power and attenuation of light by the tissue.

9.4.3 MPM IMAGING OF THE TOOTH

Figure 9.11 shows a stack of MPM images of the human tooth acquired at different depths from 4 to 84 μm. In Figure 9.11a and b, the interconnected enamel rods are clearly observed. The TPEF is color-coded in red and SHG in green, and the color shows yellow where the TPEF and SHG signals overlap. In the enamel layer, there is significant overlap of the TPEF signal with the SHG channel, which may be caused by the TPEF signal whose bandwidth extends into the collection band of the SHG selection filter. Figure 9.11c shows the dentin–enamel junction. The enamel rods are shown in red and the collagen fibers in green, and the dark holes are in scattered distribution. The dark areas possibly arise from the reduced mineralization. In Figure 9.11d through f, the fine structure of dentin tubules is clearly revealed by tiny dark tubules in the bright SHG background. Noticeably, the size of dentin tubules gradually increases at greater depth, which matches with the structure of tooth.

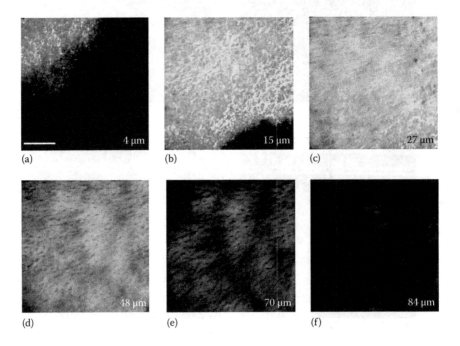

(a)　　　　　　(b)　　　　　　(c)

(d)　　　　　　(e)　　　　　　(f)

FIGURE 9.11 Multiphoton microscopy images of the human tooth at different depths: (a) 4 μm, (b) 15 μm, (c) 27 μm, (d) 48 μm, (e) 70 μm, and (f) 84 μm. Each of these numbers denotes the depth of the current image from the tooth surface. The power delivered on the sample is 5 mw. Pixel dwell time is 10 μs. The scale bar is 50 μm.

9.5 CONCLUSIONS

To develop multiphoton endoscopy, which can be used in the clinic, optical fibers are needed to deliver the femtosecond excitation pulses from the laser source to the sample location. Various types of optical fibers are compared for their dispersion and pulse broadening effects. Solid-core fibers like the SMF and DCF suffer from sever pulse broadening due to dispersion. Hollow-core PCFs have the advantages of significantly reduced dispersion and pulse broadening. Our results show that CPCF can reliably deliver sub-30 fs pulses and obtain high-quality multiphoton images. Multiphoton endoscopy with flexible optical fiber for guiding light makes the system portable and compact and thus suitable for clinical applications.

REFERENCES

1. L. H. V. Wang and S. Hu, Photoacoustic tomography: In vivo imaging from organelles to organs, *Science*, 335, 1458–1462, 2012.
2. D. Huang, E. A. Swanson, C. P. Lin, J. S. Schuman, W. G. Stinson, W. Chang, M. R. Hee et al., Optical coherence tomography, *Science*, 254, 1178–1181, 1991.
3. J. A. Conchello and J. W. Lichtman, Optical sectioning microscopy, *Nature Methods*, 2, 920–931, 2005.
4. W. Denk, J. H. Strickler, and W. W. Webb, Two-photon laser scanning fluorescence microscopy, *Science*, 248, 73–76, 1990.
5. W. R. Zipfel, R. M. Williams, and W. W. Webb, Nonlinear magic: Multiphoton microscopy in the biosciences, *Nature Biotechnology*, 21, 1369–1377, 2003.
6. C. Stosiek, O. Garaschuk, K. Holthoff, and A. Konnerth, In vivo two-photon calcium imaging of neuronal networks, *Proceedings of the National Academy of Sciences of the United States of America*, 100, 7319–7324, 2003.
7. D. A. Dombeck, K. A. Kasischke, H. D. Vishwasrao, M. Ingelsson, B. T. Hyman, and W. W. Webb, Uniform polarity microtubule assemblies imaged in native brain tissue by second-harmonic generation microscopy, *Proceedings of the National Academy of Sciences of the United States of America*, 100, 7081–7086, 2003.
8. E. B. Brown, R. B. Campbell, Y. Tsuzuki, L. Xu, P. Carmeliet, D. Fukumura, and R. K. Jain, In vivo measurement of gene expression, angiogenesis and physiological function in tumors using multiphoton laser scanning microscopy, *Nature Medicine*, 7, 864–868, 2001.
9. M. C. Skala, K. M. Riching, A. Gendron-Fitzpatrick, J. Eickhoff, K. W. Eliceiri, J. G. White, and N. Ramanujam, In vivo multiphoton microscopy of NADH and FAD redox states, fluorescence lifetimes, and cellular morphology in precancerous epithelia, *Proceedings of the National Academy of Sciences of the United States of America*, 104, 19494–19499, 2007.
10. A. Zoumi, A. Yeh, and B. J. Tromberg, Imaging cells and extracellular matrix in vivo by using second-harmonic generation and two-photon excited fluorescence, *Proceedings of the National Academy of Sciences of the United States of America*, 99, 11014–11019, 2002.
11. W. R. Zipfel, R. M. Williams, R. Christie, A. Y. Nikitin, B. T. Hyman, and W. W. Webb, Live tissue intrinsic emission microscopy using multiphoton-excited native fluorescence and second harmonic generation, *Proceedings of the National Academy of Sciences of the United States of America*, 100, 7075–7080, 2003.
12. P. J. Campagnola, A. C. Millard, M. Terasaki, P. E. Hoppe, C. J. Malone, and W. A. Mohler, Three-dimensional high-resolution second-harmonic generation imaging of endogenous structural proteins in biological tissues, *Biophysical Journal*, 82, 493–508, 2002.

13. P. J. Campagnola and L. M. Loew, Second-harmonic imaging microscopy for visualizing biomolecular arrays in cells, tissues and organisms, *Nature Biotechnology*, 21, 1356–1360, 2003.

14. P. T. So, C. Y. Dong, B. R. Masters, and K. M. Berland, Two-photon excitation fluorescence microscopy, *Annual Review of Biomedical Engineering*, 2, 399–429, 2000.

15. F. Helmchen, M. S. Fee, D. W. Tank, and W. Denk, A miniature head-mounted two-photon microscope: High-resolution brain imaging in freely moving animals, *Neuron*, 31, 903–912, 2001.

16. C. L. Hoy, N. J. Durr, P. Chen, W. Piyawattanametha, H. Ra, O. Solgaard, and A. Ben-Yakar, Miniaturized probe for femtosecond laser microsurgery and two-photon imaging, *Optics Express*, 16, 9996, 2008.

17. B. A. Flusberg, E. D. Cocker, W. Piyawattanametha, J. C. Jung, E. L. Cheung, and M. J. Schnitzer, Fiber-optic fluorescence imaging, *Nature Methods*, 2, 941–950, 2005.

18. L. Fu and M. Gu, Fibre-optic nonlinear optical microscopy and endoscopy, *Journal of Microscopy*, 226, 195–206, 2007.

19. D. Bird and M. Gu, Fibre-optic two-photon scanning fluorescence microscopy, *Journal of Microscopy*, 208, 35–48, 2002.

20. W. Göbel, J. N. Kerr, A. Nimmerjahn, and F. Helmchen, Miniaturized two-photon microscope based on a flexible coherent fiber bundle and a gradient-index lens objective, *Optics Letters*, 29, 2521–2523, 2004.

21. B. A. Flusberg, J. C. Jung, E. D. Cocker, E. P. Anderson, and M. J. Schnitzer, In vivo brain imaging using a portable 3.9 gram two-photon fluorescence microendoscope, *Optics Letters*, 30, 2272–2274, 2005.

22. W. Göbel, A. Nimmerjahn, and F. Helmchen, Distortion-free delivery of nanojoule femtosecond pulses from a Ti: Sapphire laser through a hollow-core photonic crystal fiber, *Optics Letters*, 29, 1285–1287, 2004.

23. M. T. Myaing, D. J. MacDonald, and X. Li, Fiber-optic scanning two-photon fluorescence endoscope, *Optics Letters*, 31, 1076–1078, 2006.

24. L. Fu, A. Jain, H. Xie, C. Cranfield, and M. Gu, Nonlinear optical endoscopy based on a double-clad photonic crystal fiber and a MEMS mirror, *Optics Express*, 14, 1027–1032, 2006.

25. S. Tang, W. G. Jung, D. McCormick, T. Q. Xie, J. P. Su, Y. C. Ahn, B. J. Tromberg, and Z. P. Chen, Design and implementation of fiber-based multiphoton endoscopy with microelectromechanical systems scanning, *Journal of Biomedical Optics*, 14, article #034005, 2009.

26. G. P. Agrawal, *Nonlinear Fiber Optics*, Springer, Berlin, Germany, 2000.

27. L. Fu, Fibre-optic nonlinear optical microscopy and endoscopy, PhD Thesis, Faculty of Engineering and Industrial Sciences, Swinburne University of Technology, Melbourne, Victoria, Australia, 2007.

28. P. Russell, Photonic crystal fibers, *Science*, 299, 358–362, 2003.

29. J. C. Knight, Photonic crystal fibres, *Nature*, 424, 847–851, 2003.

30. F. Benabid, Hollow-core photonic bandgap fibre: New light guidance for new science and technology, *Philosophical Transactions of the Royal Society A: Mathematical, Physical and Engineering Sciences*, 364, 3439–3462, 2006.

31. J. L. Yu, H. Zeng, H. Lui, J. S. Skibina, G. Steinmeyer, and S. Tang, Characterization and application of chirped photonic crystal fiber in multiphoton imaging, *Optics Express*, 22, 10366–10379, 2014.

10 In Vivo Flow Cytometer
A Powerful Tool to Study Cancer Metastasis

Xun-Bin Wei, Dan Wei, Ping Yang, Zhen-Yu Niu, and Huamao Ye

CONTENTS

10.1 Introduction 213
    10.1.1 Metastasis and Circulating Tumor Cells...213
    10.1.2 Significance of CTC Detection..214
    10.1.3 Limitations of Conventional *In Vitro* Detection Methods...............215
    10.1.4 IVFC and Its History...215
10.2 Basic Principle of IVFC 216
10.3 Instrument Setup..217
    10.3.1 Setup of Fluorescence-Based IVFC 217
    10.3.2 Setup of PAFC 219
10.4 Data Analysis...220
    10.4.1 IVFC Signal Processing 220
    10.4.2 PAFC Signal Processing..222
10.5 Application of IVFC in Cancer Metastasis Research............................223
    10.5.1 Study of Hepatocellular Carcinoma with Fluorescence-Based IVFC...223
    10.5.2 Study of Melanoma with PAFC..224
References..225

10.1 INTRODUCTION

10.1.1 METASTASIS AND CIRCULATING TUMOR CELLS

Cancer is one of the major threats to human health all over the globe. It is the leading cause of death in developed countries and the second leading cause of death in developing countries [1,2]. Cancer prognosis has been greatly improved recently with the development of earlier diagnosis methods and better therapies. Nevertheless, metastasis and recurrence still hinder the pursuit of long-term survival for cancer patients.

Metastasis refers to the process in which tumor cells spread to other tissues or organs. The illustration of cancer metastasis process is depicted in Figure 10.1. Tumor cells have access to the circulatory system as healthy cells do. This accessibility enables malignant cells to detach from the tumor to enter the blood

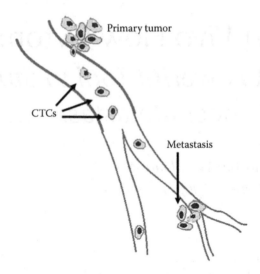

FIGURE 10.1   The process of cancer metastasis. Tumor cells detaching from the primary tumor invade into the blood circulation to form circulating tumor cells (CTCs). Most of CTCs undergo apoptosis or are cleared by immune system. Only a small fraction of CTCs can survive and arrest at the vessel wall in distant tissues. These arrested cells might generate secondary tumors, which are called metastases.

circulation [3–6]. Circulating tumor cells (CTCs) are defined as tumor cells origi-nating from either primary sites or metastases and circulating freely in the periph-eral blood of patients and are extremely rare in the healthy people. The blood circulation plays an important role in determining where CTCs travel and stay. CTCs usually get trapped in the first set of capillaries they encounter downstream from where they enter into the blood. Frequently, these capillaries are in the lung and liver. After being trapped, the CTCs will reattach themselves in the distant site and penetrate the basement membrane of the blood vessel and then establish themselves in the distant tissue. A small fraction of these cells will grow to form metastatic tumor and cause recurrence. Thus, it makes cancer difficult to be cured and causes most cancer death. Thus, understanding the pathogenesis of metastasis is essential in cancer research.

10.1.2   SIGNIFICANCE OF CTC DETECTION

CTCs have long been considered a reflection of tumor aggressiveness [7–10]. Tumor-induced angiogenesis occurs along with tumor invasion. This gives rise to the possibility that highly invasive primary tumors may unleash CTCs into periph-eral circulation before any bona fide metastases are established. CTC detection has become a new and promising method to monitor patients with metastatic disease [11]. A number of CTC detection platforms have been developed and verified in various clinical settings, which strongly suggests that CTC detection has enormous poten-tial in malignancy diagnosis, prognosis, and efficacy estimation of the anticancer therapy [12]. According to previous studies, CTC count would be used as a potential biomarker for cancer diagnosis and prognostic prediction [7,8], for example, in liver cancer [13], lung cancer [14,15], breast cancer [8,9] and prostate cancer [16].

10.1.3 LIMITATIONS OF CONVENTIONAL *IN VITRO* DETECTION METHODS

Conventional CTC detection methods are based on the analysis of extracted blood samples. All these *in vitro* methods can be categorized into two types: immunological assays using monoclonal antibodies directed against histogenic proteins and polymerase chain reaction (PCR)-based molecular assays exploiting tissue-specific transcripts. In immunological approaches, monoclonal antibodies for protein markers of cancer cells are used. The most widely used approaches of immunological assays are CellSearch™ system [8,17], fluorescence-activated cell sorting analysis by conventional flow cytometry [18,19], and the use of high-throughput CTC chip [20,21]. PCR assays are used in CTC detection since 1990 [22], which have higher sensitivity [23].

The *in vitro* detection methods mentioned earlier have the following limitations. First, CTCs are rare cells in the circulating system. At the early stage of cancer metastasis, there might be only a few CTCs in 1 mL blood sample, which is easily to be missed by conventional *in vitro* flow cytometry. Second, the experimental animals have limited total blood volume. Drawing blood samples frequently from the subject will affect it significantly, which makes it unfeasible for continuous, real-time, and long-term detection of CTCs. Third, invasive extraction of cells from a living organism may lead to cell property changes and prevents cell analysis in their native environment. Thus, the conventional methods mentioned earlier are limited due to their invasiveness, low sensitivity caused by limited blood sample volumes, and difficulty to record the dynamics of CTCs.

10.1.4 IVFC AND ITS HISTORY

In order to overcome the aforementioned limitations, *in vivo* flow cytometer (IVFC) has been developed to monitor CTCs in live animals. The first IVFC was reported by Novak from Lin's group in 2004 [24]. They showed that IVFC could be used as a powerful tool to study cell dynamics in the blood circulation. Afterward, Lin's group presented the depletion kinetics of injected prostate cancer cells in mice and rats and demonstrated that IVFC might become a novel tool to monitor CTCs dynamically [25].

Zharov and his coworkers [26,27] have, for the first time, realized the photoacoustic flow cytometry (PAFC). His group mainly focused on the application of PAFC in the study of circulating metastatic melanoma cells, carbon nanotube kinetics in the blood circulation and lymph vascular system.

In 2007, Low's group developed a multiphoton intravital flow cytometry [18]. Qu's group reported label-free IVFC based on two-photon autofluorescence imaging in 2012 [28]. Our group used IVFC for the research of liver cancer metastasis [19,29]. Recently, our group designed and utilized a photoacoustic flow cytometer to study circulating melanoma cells [30].

IVFC can noninvasively monitor various types of circulating cells with high sensitivity. In the study of cancer metastasis, the acquired real-time information of CTC dynamics may provide novel insights into tumor progresses, metastasis processes, and responses to anticancer treatments.

10.2 BASIC PRINCIPLE OF IVFC

The basic principle of IVFC is confocal excitation and detection of circulating cells in the circulatory system with the concept of conventional flow cytometer [24]. The blood circulation system and lymphatic vascular system of live animals act as natural sheath flow system of the conventional flow cytometer. Generally, a laser beam is modulated into a slit-shaped beam through a set of optical components. Then, the laser spot will be placed over a blood vessel of interest with the guidance of live imaging. When the fluorescently labeled cells flow across the radiated volume of the laser beam, a burst of fluorescence will be generated and recorded by the system. As for photoacoustic flow cytometer, a train of ultrasound is generated while the cell goes across the laser slit [27]. An ultrasound transducer is utilized to collect the generated photoacoustic signal. Figure 10.2 illustrates the basic principles of the two modalities.

In a typical experiment, the animal is anesthetized and placed on the sample stage with its ear adhered to the microscope slide by glycerine, as shown in Figure 10.3. The slit-shaped laser beam is positioned across the blood vessel with the guidance of IVFC's image-guided navigation module. The long dimension of the laser beam covers the full width of the vessel at the mouse ear in order to make sure that no cells will be missed. A burst of fluorescence will be generated when a fluorescently labeled cell flows through the detection region. The emitted fluorescence can be detected by optoelectronic module of the device with appropriate filtering. For photoacoustic modality, the generated ultrasound can be collected by an ultrasound transducer.

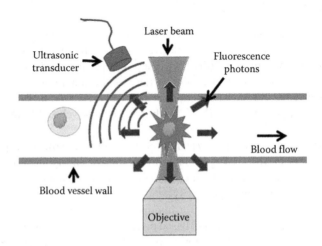

FIGURE 10.2   Basic principles of both fluorescence-based *in vivo* flow cytometry and photoacoustic flow cytometry.

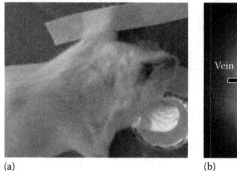
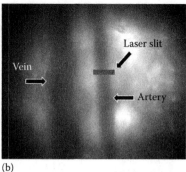

(a)   (b)

FIGURE 10.3   (a) An anesthetized mouse on the sample stage. (b) The image of microvessels acquired by the image-guided navigation module of *in vivo* flow cytometer (IVFC) with a 40× objective. The smaller vessel is an artery and the bigger one is a vein. The green slit is a laser slit of IVFC, which is positioned across the artery.

After that, the signal is converted into digital data and recorded by the computer. The data trace is processed by the homemade software with its statistical results being extracted for further analysis.

10.3   INSTRUMENT SETUP

10.3.1   SETUP OF FLUORESCENCE-BASED IVFC

Fluorescence-based IVFC includes three major modules: optical module, electronic module, and software module. Figure 10.4 depicts a classic setup of a two-color two-channel IVFC [29].

Optical module plays an important role in the laser beam modulation, fluorescence collection, and image-guided navigation. Accordingly, this module is further divided into three submodules: image-guided navigation submodule, laser excitation submodule, and fluorescence collection submodule.

The image-guided navigation submodule provides live imaging of the mouse ear, which is used to guide the laser spot to the specific blood vessel of interest. This submodule comprises of an LED light source, an objective lens (objective), a beam splitter (BS3), a filter (F1), an achromatic lens (AL2), and a CCD camera. The LED light source is placed above the sample stage. It typically has a central wavelength of 530 nm, which can provide good contrast between blood vessels and surrounding tissues. This phenomenon is due to the high absorption of the light at this wavelength by the blood. The green light that transmits through the mouse ear is collected by the objective lens and then directed toward the CCD camera by BS3 (Figure 10.4). Before entering the CCD camera, the filter F1 filters out unwanted light whose wavelength does not range between 520 and 550 nm. The achromatic lens AL2 is used to refocus the transmitted light. Thus, the image captured by the CCD camera depicts the structure information of the mouse ear.

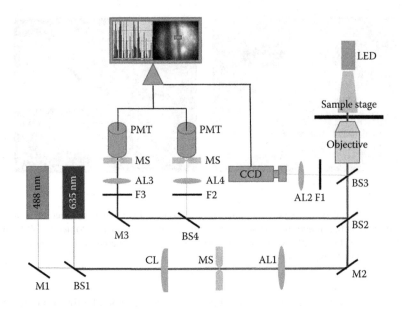

FIGURE 10.4   Schematic of a two-color two-channel *in vivo* flow cytometer.

The CCD is connected to an image acquisition card of a personal computer (PC). The real-time images are displayed on the screen and act as the guidance for laser spot navigation.

The laser excitation submodule consists of two diode lasers (488 and 635 nm), two mirrors (M1 and M2), three beam splitters (BS1, BS2, and BS3), a cylindrical lens (CL), a mechanical slit (MS), an achromatic lens (AL1), a couple of pinholes, an objective lens (objective), and a sample stage. The dimension of the diode laser is smaller compared to solid-state laser that makes the whole system more compact. Two laser beams emitted by the diode lasers are integrated into a single beam by the dichroic beam splitter BS1. BS1 is a low-pass beam splitter that transmits the 488 nm laser beam while reflecting 635 nm laser beam. Then the integrated laser beam is condensed in one dimension by the CLs. The shape of the beam looks like a vertical slit at the focal plane of the CLs. An MS is positioned at this focal plane in order to make the slit sharper. Then the slit-shaped beam is directed to the back focal plane of the objective lens by the achromatic lens AL1. Finally, a horizontal slit-shaped beam is formed at the front focal plane of the objective lens. The approximate length and width of the slit-shaped laser beam here are 30 and 5 μm, respectively.

The fluorescence detection submodule collects the fluorescence photons and converts them into electronic signals. This submodule mainly consists of an objective lens (objective), three beam splitters (BS2, BS3, and BS4), a mirror (M3), two filters (F2 and F3), two achromatic lens (AL3 and AL4), two MSs, and two photomultiplier tubes (PMT). The fluorescence emitted from the fluorescently labeled cells is collected by the objective lens. Then this beam passes through BS3 and reaches BS2. BS2 guides the fluorescence light toward the BS4 that is a high-pass beam splitter. BS4 reflects green fluorescence toward F2 and transmits the red fluorescence to M3.

The passband of filter F2 is from 500 to 520 nm, while the passband of filter F3 is between 650 and 680 nm. These filters can filter out the light from LED and laser sources. The achromatic lenses, AL3 and AL4, refocus the fluorescence beam to their focal planes, respectively. The fluorescence beam at the focal plane appears like a vertical slit as well. Two MSs are placed right here in order to eliminate the fluorescence that is not emitted from the focal plane of the objective lens. The function of the MSs in the IVFC system is similar to that of the pinholes in a confocal microscope. The design of the whole optical system enables confocal excitation and detection that improves the signal-to-noise ratio.

The electronic module converts the optical signal into digital signal. This system consists of two multiplier tubes, a current amplifier and an analog-to-digital (A/D) converter. The PMTs convert the fluorescence photons into the current that can be interfaced and processed by electronic devices. Since the amplitude of the current generated by PMTs is too low as the input for the A/D converter, a current amplifier is added between them. After being amplified, the signal is digitized and recorded in the PC.

The software module is responsible for device control, real-time signal display, and data analysis. More details concerning data analysis will be introduced in the following sections.

10.3.2 SETUP OF PAFC

In PAFC, optophysical time-dependent photoacoustic phenomena are generated through the interaction between the pulsed laser and the target cells. Blood or lymph vessels with circulating target cells are illuminated by an LED light and imaged by a CCD camera. A laser beam is focused on the vessel of interest. The photoacoustic (PA) signals generated from the targets are detected by an ultrasound transducer attached to the living sample [30].

The setup of PAFC includes three parts as shown in Figure 10.5: illumination part, excitation part, and detection part. The design is introduced briefly as follows.

The illumination part is responsible for image-based navigation of the laser spot. The typical detection positions in the animal are either the mouse ear or the mesenteric vasculature [31]. In order to locate the appropriate blood vessels for detection, LED light (green line in Figure 10.5, 524 nm) transmits the sample while the structure information of the illuminated area is imaged by a CCD camera. With the real-time imaging of sample vessels, the ideal position for cell detection can be located.

The excitation part modulates the laser beam through a set of optical components. In PAFC experiments, PA signals from cells can be covered by background PA signals from the skin and blood [32,33]. Thus, we select pulsed laser source at the wavelength of 1064 nm in our setup to minimize the background absorption and its PA signals. The CLs reshape the circular laser spot into a slit-shaped laser spot.

In the PA signal detection part, an ultrasonic transducer is utilized to transform the ultrasound signals to electrical signals. The electrical signals are amplified by preamplifier. Then the data are digitized and recorded in the PC through a data-acquisition card.

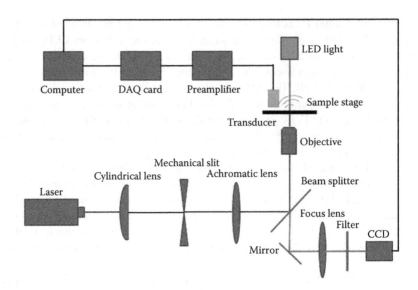

FIGURE 10.5 Framework of photoacoustic flow cytometer includes three parts: (1) illumination part (LED light, objective, beam splitter, mirror, focus lens, filter, CCD), (2) excitation part (pulsed laser, cylindrical lens, mechanical slit, achromatic lens [focus lens], beam splitter, objective), (3) detection part (ultrasonic transducer, preamplifier, data-acquisition card, computer).

10.4   DATA ANALYSIS

10.4.1   IVFC SIGNAL PROCESSING

The fluorescence signals yielded by fluorescently labeled cells are digitized and recorded in the PC. Then the data will be processed with a software coded in-house using MATLAB®. Figure 10.6 depicts a typical data trace acquired at the artery of a mouse ear using fluorescence-based IVFC. From the data trace, we can observe two types of signal components: one is called the baseline, which has relatively lower intensity and flat shape; the other one is called the peak, which has a pulse shape with high intensity. In CTC detection, baseline level is generally a reflection of background noise caused by autofluorescence. The peak corresponds to single fluorescently labeled cell that is excited when it flows through the laser radiated volume. There are two major features of the peak signal: peak height and width, as defined in Figure 10.7.

Figure 10.8 shows a common workflow for IVFC data processing and analysis. The noise in the recorded raw data generally degrades the accuracy and efficiency of subsequent procedure. Thus, the very first goal is denoising. The noise in IVFC data varies due to the difference in the working environments. However, power line noise, thermal noise, and static electricity pulse noise are the three types of common noise.

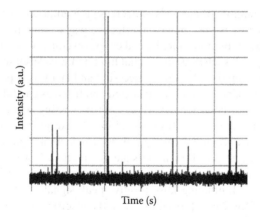

Time (s)

FIGURE 10.6 Typical data trace of fluorescence-based *in vivo* flow cytometry. The horizontal axis indicates the time, while the vertical axis stands for the fluorescence intensity.

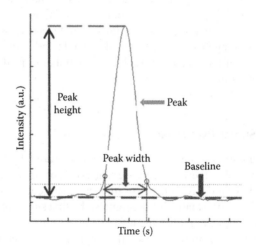

Time (s)

FIGURE 10.7 Definition of *in vivo* flow cytometer signal components: peak and baseline. Peak is a pulse signal having two parameters: the height, which is the maximum intensity of the pulse, and the width, which is its time duration. Baseline corresponds to the rest of the data trace. The baseline usually appears flat with low intensity.

FIGURE 10.8 The workflow of *in vivo* flow cytometry data processing.

There are a great number of methods to decrease the influence of the noise. In most cases, moving average is the simple and effective way to smooth the data trace and reduce the noise level. In more complicated cases, several advanced denoising methods have been applied, such as digital signal filter and wavelet-based denoising algorithm. The second step in IVFC data processing and analysis is peak identification, in which the pulse-shaped signals are detected and listed as candidates for real peak signals. This step is crucial for the whole data processing procedure. Many peak detection algorithms are based on threshold gating like global threshold gating or adaptive local threshold gating. Advanced peak detection algorithms are utilized to improve the accuracy, such as finite state automation. Not all the pulse-shaped signals are truly positive peak signal. Some static electric noise signals also appear in the traces. Thus, eliminating false-positive peak signals with additional criteria is the third step of the entire procedure. Take the electric noise signal for example; its intensity is usually higher than the gating threshold. Therefore, after the second step, these signals are regarded as peak candidates. However, the peak width feature of static electric noise signals is very different from the real positive peak signals. The peak width of positive peak signals is much wider than that of static electric noise signals. With this additional information as criteria, the false peak signals are eliminated. At last, the remaining peak candidates are listed as positive peak signals with their statistics being calculated. The peaks' height and width, time stamp, and the statistical results of the data trace, such as the number of peaks per minute, are recorded in the output files.

10.4.2 PAFC SIGNAL PROCESSING

The raw PAFC data often contains some noise signals caused by electronic devices, such as background noise, thermal noise, electric noise, or static electric pulse noise. Those noise signals would have a negative impact on subsequent signal analysis. In order to obtain further information of the signals, the first step is to reduce the noise. A number of approaches can reduce the noise.

Wavelet denoising method is one of the approaches to reduce the noise. In wavelet denoising method, the raw data are imported into homemade software coded by MATLAB. First, the signal is divided into a number of different layers according to the frequency. The number of layers and wavelet function are selected as parameters. Second, denoising threshold values of the wavelet transform are calculated. Finally, these threshold values are set as parameters in denoising process to reduce the noise from the raw data. After these steps, the desirable signals with less noise are acquired. Figure 10.9 shows the PA signal before and after denoising.

By comparison, the photoacoustic signal after denoising is smoother than the original signal. The amplitude of the noise in the data is reduced significantly after denoising, which is helpful for identifying and distinguishing photoacoustic signals of target cells, such as melanoma CTCs. To obtain more information of the signals from target cells, the photoacoustic signal after denoising can be analyzed in both time domain and frequency domain.

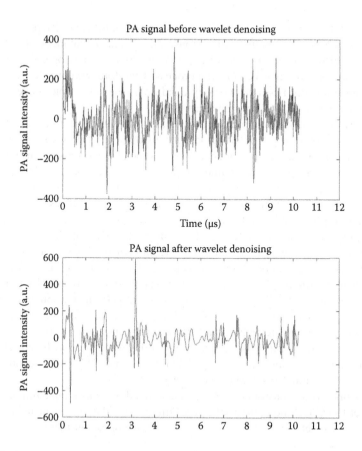

FIGURE 10.9   The raw photoacoustic signal before and after wavelet denoising.

10.5   APPLICATION OF IVFC IN CANCER METASTASIS RESEARCH

10.5.1   STUDY OF HEPATOCELLULAR CARCINOMA WITH FLUORESCENCE-BASED IVFC

Our group has focused on the study of liver cancer metastasis using IVFC [19,29]. Previously, we injected liver cancer cells via the tail vein in order to mimic CTCs. With IVFC, we revealed the depletion kinetics of liver cancer cells with different metastasis potential [19,29,34]. This CTC-mimicking model with intravenous injection has been used by many groups before. However, it does not well mimic tumor metastasis. Despite that it provides the depletion kinetics of CTCs, such a large number of injected CTCs actually do not occur in pathologic conditions [35,36]. Thus, we have applied an orthotopic hepatocellular carcinoma green fluorescence protein (GFP-HCC) model to solve this problem [19]. It is the first study of CTC dynamics under clinically relevant oncology condition using IVFC. It reflects the process of CTCs targeting to host tissues as well as the process of CTC forming from the primary tumor. Interestingly, we have found out significant differences in CTC dynamics between orthotopic tumor mouse model and subcutaneous (s.c.) model. Our study also confirms that local environment is essential for CTC-dependent metastasis, especially in CTC-forming process. Therefore, we have shown by IVFC that under clinically relevant oncology conditions, the orthotopic model is better than s.c. model to study cancer metastasis.

We have also used our model to investigate a disputed question: whether liver resection promoted or restricted hematogenous metastasis in advanced liver cancer. We have provided direct evidence to assess the effectiveness of surgical resection for cancer metastasis and found out that both the number of CTCs and early metastases decrease significantly after tumor resection. Tumor progression and distant metastases development are greatly slowed down after resection. We have also found out that the counts of CTCs are correlated with tumor growth in the orthotopic tumor model, including the number and size of distant metastases. Furthermore, we have demonstrated that the number of CTCs drops to undetectable level after the primary tumor is removed. The count of CTCs in this model could not maintain in circulation without the supply of cells from solid tumor. The detected number of CTC is just the homeostasis of CTC formation and depletion (including apoptosis, immune system killing, or metastasizing to target organs). Furthermore, we have revealed that after the primary tumor is removed, all the mice with CTC recurrence have metastases, while all the mice without CTC recurrence have no noticeable metastases. This implies that CTCs may become a biomarker to show tumor residual or recurrence.

Our work has demonstrated that, when combined with orthotopic tumor models, the novel IVFC technique offers the capability to elucidate mechanisms that drive hematogenous metastasis and to monitor the efficacy of cancer therapy.

10.5.2 Study of Melanoma with PAFC

To validate the melanoma detection capacity of PAFC, s.c. melanoma model was established [30]. Melanoma-bearing mice are constructed by s.c. inoculation of B16F10 tumor cells. Then PAFC is used to monitor the dynamics of circulating melanoma cells without labeling (the endogenous melanin serves as the light absorber to generate the photoacoustic signal). The number of CTCs is counted on a weekly basis after inoculation of the B16F10 melanoma cells, while the data length is 10 min for each mouse. Within 10 min, there are about 1~3 CTCs at the end of the first week after inoculation. At the end of the second week after inoculation, this number increases to 2~8 CTCs. More CTCs (6~17 CTCs) are detected with the progress of melanoma at the end of the third week after inoculation. Finally, 16~29 CTCs are detected at the end of the fourth week after inoculation. These results have demonstrated that PAFC is capable of realizing label-free detection of circulating melanoma cells.

In conclusion, IVFC provides a novel and powerful tool to study CTC dynamics *in vivo*. It is potentially helpful to solve some disputed biomedical problems and useful to guide clinical therapies.

REFERENCES

1. Mathers, C., D.M. Fat, and J.T. Boerma, The global burden of disease: 2004 update 2008. World Health Organization, Geneva, Switzerland.
2. Jemal, A. et al., Global cancer statistics. *CA: A Cancer Journal of Clinicians*, 2011. 61(2): 69–90.
3. Brooks, P.C. et al., Integrin αvβ3 antagonists promote tumor regression by inducing apoptosis of angiogenic blood vessels. *Cell*, 1994. 79(7): 1157–1164.
4. Bedner, E. et al., High affinity binding of fluorescein isothiocyanate to eosinophils detected by laser scanning cytometry: A potential source of error in analysis of blood samples utilizing fluorescein—Conjugated reagents in flow cytometry. *Cytometry*, 1999. 36(1): 77–82.
5. Blankenberg, F.G. et al., Imaging cyclophosphamide-induced intramedullary apoptosis in rats using 99mTc-radiolabeled annexin V. *Journal of Nuclear Medicine*, 2001. 42(2): 309–316.
6. Aotake, T. et al., Changes of angiogenesis and tumor cell apoptosis during colorectal carcinogenesis. *Clinical Cancer Research*, 1999. 5(1): 135–142.
7. Sun, Y.F. et al., Circulating tumor cells: Advances in detection methods, biological issues, and clinical relevance. *Journal of Cancer Research and Clinical Oncology*, 2011. 137(8): 1151–1173.
8. Cristofanilli, M. et al., Circulating tumor cells, disease progression, and survival in metastatic breast cancer. *New England Journal of Medicine*, 2004. 351(8): 781–791.
9. Cohen, S.J. et al., Relationship of circulating tumor cells to tumor response, progression-free survival, and overall survival in patients with metastatic colorectal cancer. *Journal of Clinical Oncology*, 2008. 26(19): 3213–3221.
10. Cristofanilli, M. et al., Circulating tumor cells: A novel prognostic factor for newly diagnosed metastatic breast cancer. *Journal of Clinical Oncology*, 2005. 23(7): 1420–1430.
11. Andreopoulou, E. and M. Cristofanilli, Circulating tumor cells as prognostic marker in metastatic breast cancer. *Expert Review of Anticancer Therapy*, 2010. 10(2): 171–177.
12. Pantel, K., R.H. Brakenhoff, and B. Brandt, Detection, clinical relevance and specific biological properties of disseminating tumour cells. *Nature Reviews Cancer*, 2008. 8(5): 329–340.
13. Vona, G. et al., Impact of cytomorphological detection of circulating tumor cells in patients with liver cancer. *Hepatology*, 2004. 39(3): 792–797.
14. Krebs, M.G. et al., Evaluation and prognostic significance of circulating tumor cells in patients with non–small-cell lung cancer. *Journal of Clinical Oncology*, 2011. 29(12): 1556–1563.
15. Hofman, V. et al., Preoperative circulating tumor cell detection using the isolation by size of epithelial tumor cell method for patients with lung cancer is a new prognostic biomarker. *Clinical Cancer Research*, 2011. 17(4): 827–835.
16. Coumans, F. et al., All circulating EpCAM+ CK+ CD45– objects predict overall survival in castration-resistant prostate cancer. *Annals of Oncology*, 2010. 21(9): 1851–1857.
17. Riethdorf, S. et al., Detection of circulating tumor cells in peripheral blood of patients with metastatic breast cancer: A validation study of the CellSearch system. *Clinical Cancer Research*, 2007. 13(3): 920–928.
18. He, W. et al., *In vivo* quantitation of rare circulating tumor cells by multiphoton intravital flow cytometry. *Proceedings of National Academy of Sciences of the United States of America*, 2007. 104(28): 11760–11765.
19. Fan, Z.C. et al., Real-time monitoring of rare circulating hepatocellular carcinoma cells in an orthotopic model by *in vivo* flow cytometry assesses resection on metastasis. *Cancer Research*, 2012. 72(10): 2683–2691.
20. Uhr, J.W., Cancer diagnostics: One-stop shop. *Nature*, 2007. 450(7173): 1168–1169.

21. Nagrath, S. et al., Isolation of rare circulating tumour cells in cancer patients by microchip technology. *Nature*, 2007. 450(7173): 1235–1239.

22. Smith, B. et al., Detection of melanoma cells in peripheral blood by means of reverse transcriptase and polymerase chain reaction. *The Lancet*, 1991. 338(8777): 1227–1229.

23. Brakenhoff, R.H. et al., Sensitive detection of squamous cells in bone marrow and blood of head and neck cancer patients by E48 reverse transcriptase-polymerase chain reaction. *Clinical Cancer Research*, 1999. 5(4): 725–732.

24. Novak, J. et al., *In vivo* flow cytometer for real-time detection and quantification of circulating cells. *Optics Letters*, 2004. 29(1): 77–79.

25. Georgakoudi, I. et al., *In vivo* flow cytometry: A new method for enumerating circulating cancer cells. *Cancer Research*, 2004. 64(15): 5044–5047.

26. Galanzha, E.I., J.W. Kim, and V.P. Zharov, Nanotechnology-based molecular photoacoustic and photothermal flow cytometry platform for in-vivo detection and killing of circulating cancer stem cells. *Journal of Biophotonics*, 2009. 2(12): 725–735.

27. Zharov, V.P. et al., *In vivo* photoacoustic flow cytometry for monitoring of circulating single cancer cells and contrast agents. *Optics Letters*, 2006. 31(24): 3623–3625.

28. Zeng, Y. et al., Label-free *in vivo* flow cytometry in zebrafish using two-photon autofluorescence imaging. *Optics Letters*, 2012. 37(13): 2490–2492.

29. Li, Y. et al., Circulation times of prostate cancer and hepatocellular carcinoma cells by *in vivo* flow cytometry. *Cytometry A*, 2011. 79(10): 848–854.

30. Liu, R. et al. *In vivo*, label-free, and noninvasive detection of melanoma metastasis by photoacoustic flow cytometry. In *SPIE BiOS*. 2014. International Society for Optics and Photonics, *Proc. SPIE*. 8944, Biophotonics and Immune Responses IX, 89440Q (1–7).

31. Tuchin, V.V., A. Tárnok, and V.P. Zharov, *In vivo* flow cytometry: A horizon of opportunities. *Cytometry Part A*, 2011. 79(10): 737–745.

32. Nedosekin, D.A. et al., Ultra-fast photoacoustic flow cytometry with a 0.5 MHz pulse repetition rate nanosecond laser. *Optics Express*, 2010. 18(8): 8605–8620.

33. Poellinger, A. et al., Near-Infrared laser computed tomography of the breast. *Academic Radiology*, 2008. 15(12): 1545.

34. Li, Y. et al., Circulation times of hepatocellular carcinoma cells by *in vivo* flow cytometry. *Chinese Optics Letters*, 2010. 8(10): 953–956.

35. Chang, Y.S. et al., Mosaic blood vessels in tumors: Frequency of cancer cells in contact with flowing blood. *Proceedings of the National Academy of Sciences of the United States of America*, 2000. 97(26): 14608–14613.

36. Méhes, G. et al., Circulating breast cancer cells are frequently apoptotic. *The American Journal of Pathology*, 2001. 159(1): 17–20.

11 Transition Metal Luminophores for Cell Imaging

*Christopher S. Burke, Aisling Byrne,
and Tia E. Keyes*

CONTENTS

11.1 Introduction 228
11.2 Luminescent Coordination Compounds..229
11.3 Main Classes of Transition Metal Luminophores Used in Imaging:
Their Synthetic Design and Modification...229
11.4 Uptake of Metal Complexes for Cell Imaging...231
    11.4.1 Mechanisms of Cell Uptake 231
    11.4.2 Improving Cell Permeability of Metal Complex Luminophores 231
    11.4.3 Lipophilicity 232
    11.4.4 Charge..232
    11.4.5 Targeting Metal Luminophores within the Cell by Conjugation......234
        11.4.5.1 Peptides...234
        11.4.5.2 Sugar Conjugates 237
        11.4.5.3 Sterols 238
        11.4.5.4 Covalent Reactive-Linker Moieties: Rhenium
              "Prolabels".. 238
        11.4.5.5 Polymer Conjugates 238
11.5 Sensing Applications of Transition Metal Luminophores in Cellular
Imaging...239
    11.5.1 Fluorescent Lifetime Imaging 240
11.6 Molecular Oxygen Sensing...241
11.7 Membrane Sensing 243
    11.7.1 DNA...243
11.8 Conclusions and Future Directions...244
Acknowledgments..246
References...246

11.1   INTRODUCTION

The demand for luminescent probes for sensing and imaging in live cells and tissues is growing due to the increasing diversity and sophistication of luminescence-based imaging technologies available. Luminescence-based imaging methods range from conventional intensity-based fluorescence imaging such as confocal and wide-field microscopies through superresolution and multiphoton methods to luminescence-lifetime-based imaging. Traditional organic fluorophores, although universally used in fluorescence imaging, suffer from a number of limitations that metal complex luminophores can potentially overcome. The drawbacks include poor photochemical stability (this is particularly important for long imaging processes, e.g., in tracing dynamic change in cells); small Stokes shifts, that is, closely overlapping absorbance and emission spectra (this can lead to self-quenching and distortion of the emission spectra and intensity, particularly if the probe tends to accumulate in a given region of the cell, causing local dye concentrations to well exceed the administered concentration), and poor aqueous solubility, which can demand the use of organic solvents, which disrupt the cell membrane. Finally, organic fluorophores tend to be short lived, with fluorescence lifetimes in the range of 1–10 ns. This tends to render them relatively insensitive to their environment, particularly to diffusing an analyte. And, in particular because they are fluorescent (rather than phosphorescent), they tend to be insensitive to oxygen.

Conversely, luminescent inorganic probes, principally in the context of this review, coordination compounds of the platinum group metals, typically exhibit strong Stokes shift, long-lived, and environmentally sensitive and polarized emission. With appropriate ligand modification, they can show excellent photostability and tuneable emission wavelengths and through the use of appropriate counterions can be rendered water soluble. Although the use of luminescent platinum group metal complexes such as the prototype ruthenium(II) trisbipyridyl complex had been mooted for application in the cell imaging domain, before the beginning of 2006, there were very few examples in the literature of their application in this regard.[1,2] A significant barrier is likely the poor cell membrane permeability of such complexes in their native form.[2] Conversely, there has been considerable research activity focused on the interactions of complexes such as Ru(II) complexes with biomolecules, particularly with DNA, and this work has been an important platform from which metal–cell studies have evolved. Sparked, perhaps, by a combination of greater understanding of cell uptake and conjugation of metal complexes along with much wider availability of microscopy, over the past 8 years, there has been a veritable explosion of interest in the application of metal complex probes to cell imaging and sensing.

Herein, we overview the key types of transition metal luminophore that have been explored so far, their properties, and the approaches taken to render them cell permeable and for organelle targeting. We examine selected representative examples of the application of metal complex probes in cell imaging and discuss briefly their limitations that might be addressed in the coming generations. It is important to note that another class of metal complex, based on lanthanide luminophores, has been developed in parallel to transition metal probes. Many of the strategies for targeting these materials within tissues are analogous to those used in transition metal probes, and indeed the lanthanides are well advanced in this regard due to the broader application of metals of this class as magnetic resonance contrast agents. The lanthanides have numerous advantages as cellular imaging probes but also have a number of challenges including the need for sensitized excitation of their emission. The coverage of these materials is beyond the scope of this chapter, but the reader is directed to the following reviews.[3,4]

11.2   LUMINESCENT COORDINATION COMPOUNDS

The profound differences in photophysical behavior of metal complexes compared to organic fluorophores expounded previously arise primarily because the emissive states typically originate from triplet charge transfer transitions in metal complexes influenced by the heavy nuclear mass of the metal. This blurs formal assignment of spin state, yielding Stokes-shifted and often long-lived excited states, but without the exceedingly long-lived and weak phosphorescence observed in organic molecules. A detailed discussion on the photophysics of coordination compounds of, for example, ruthenium and iridium, is beyond the scope of this short chapter, so the reader is directed to the following authoritative literature.[5]

11.3   MAIN CLASSES OF TRANSITION METAL LUMINOPHORES USED IN IMAGING: THEIR SYNTHETIC DESIGN AND MODIFICATION

Ru(II), Ir(III), Rh(III), Re(I), and Pt(II) are probably the most explored luminescent complexes in the broader literature, which is the reason for their translation to imaging. While it is important that the inorganic luminophore exhibits the appropriate photophysical characteristics, there are additional structural demands and associated synthetic challenges that must be considered for a metal complex to be suited to cell imaging. Among these are appropriate lipophilicity, limited cytotoxicity, and photostability, and these considerations will be discussed further later.[6]

The Ru(II) cellular probes explored to date have been based on polypyridyl coordination compounds, yielding structures related to the prototype $[Ru(bpy)_3]^{2+}$ complex of the form $[Ru(N^{\wedge}N)_2(N'^{\wedge}N')]^{2+}$ (Figure 11.1), where $N'^{\wedge}N'$ is a polypyridyl 2,2'-bypyridine (bpy) or 1,10-phenanthroline (phen) derivative, due to the well-established synthetic route via the bis-chelated ruthenium dichloride $[RuCl_2(N^{\wedge}N)_2]$.[7,8] The ruthenium polypyridyl complexes typically have an emission spectral range of 600–700 nm and exhibit good thermal stability and photostability. In the case of Ir(III) complexes, the most widely studied probes for cell imaging, have the general formula $[Ir(C^{\wedge}N)_2(N'^{\wedge}N')]^{+}$ containing cyclometalated aryl-pyridine ligands such as 2-phenylpyridine (ppy), where reaction from commercial precursor yields a dimer intermediate.[9] Unlike Ru(II) and its analog, Os(II), complexes whose emissive states tend largely to originate from triplet metal-to-ligand charge transfer (MLCT) states, the Ir(III) complexes tend to emit from excited states with mixed MLCT and triplet[3] $(\pi{-}\pi^*)$ ligand state character. This renders their emission wavelength more tuneable than ruthenium(II) polypyridyl complexes, and they tend to have greater quantum yields and emission lifetimes.[10] However, cyclometalated iridium complexes tend to

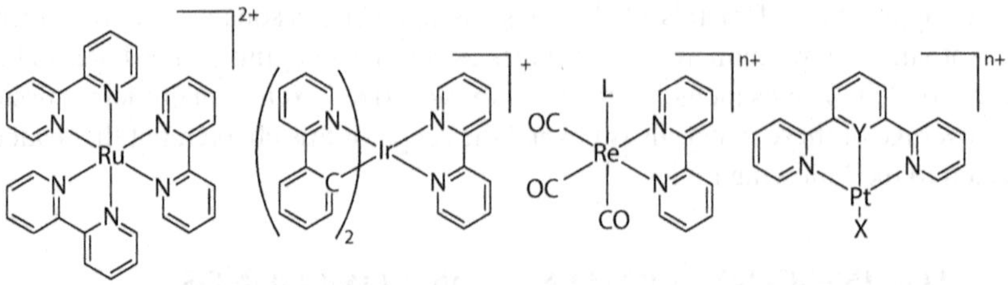

FIGURE 11.1 General forms of the main classes of transition metal luminophores used in cell imaging. Left to right: the Ru(II) archetype, $[Ru(bpy)_3]^{2+}$; the Ir(III) cyclometallated core unit, $[Ir(ppy)_2(bpy)]^+$; the Re(I) class, *fac*-$[Re(CO)_3(L)(bpy)]^{n+}$, L is typically Cl, Br (n = 0), or Py (n = 1); and a Pt(II) case, $[Pt(N^\wedge Y^\wedge N)(X)]^{n+}$, usually Y is C and X is Cl (n = 0) or Y is N and X is Cl (n = 1).

have rather poor visible absorption, requiring blue excitation. They often show relatively low optical cross sections at wavelength regions commonly used for microscopy. Although, the uptake of Ir(III) cyclometallates is favored by their lipophilic balance conferred by phenyl moieties and the reduction of the net formal charge from +3 to +1 at the iridium core relative to their N_6 coordination analogs. They tend to exhibit higher cytotoxicity than ruthenium(II) complexes, possibly relating to their greater lipophilicity and photosensitization of $^1O_2$ generation in situ.

The lesser reported Rh(III) probes closely resemble those of the cyclometalated complexes of Ir(III), yielding complexes of the form $[Rh(C^\wedge N)_2(N^\wedge N)]^+$, but Rh(III) complexes tend only to emit strongly at low temperature, limiting their application in cell imaging. Re(I) complexes are also quite prominent transition metal luminophores and typically are reported as variants of the easily accessible *fac*-$[Re(CO)_3(N^\wedge N)(L)]^+$ complex where L is usually a halogen or a pyridyl derivative.[11] However, similar to the cyclometalated Ir(III) probes, they tend to be blue absorbers and they can also undergo photochemically induced CO loss. Re(I) probes have also been reported to be incorporated with bis-quinoline ligands, particularly in the early work on Re(I) cell probes.[12] Again, the halogen ligand is quite labile, which can lead to direct Re coordination to biological substrates via donor atoms such as N, O, and S. Displacement with pyridyl units inhibits this interaction while also allowing outer-sphere functionalization through substitution on the pyridyl rings, usually without affecting the photophysical transitions that typically localize on the $N^\wedge N$ chelating ligand. Pt(II) complexes have also been applied to cell imaging, and these complexes are d^8 metals that tend to exist in four-coordinate square planar geometries. In general, Pt(II) constructs conform to one of the two series: the first that bears tridentate ligands, $[Pt(N^\wedge Y^\wedge N)(X)]^{n+}$ (Y is N or C and X is typically a halogen), or the second that incorporates bidentate ligands, $[Pt(N^\wedge N)(C^\wedge N)]^+$.

Overall, because of the CT nature of the optical transitions and emissive states in metal complexes, they can be relatively readily tuned by altering the σ-donor and/or π-accepting ability of the coordinated ligands. The modular construction of metal complexes presents a key advantage in synthetic tunability of their optical properties over organic probes where excited states tend to be localized π–π^* transitions that are not so readily synthetically tuned.

11.4 UPTAKE OF METAL COMPLEXES FOR CELL IMAGING

Confocal laser scanning microscopy is probably the most widely used imaging method in biology. Its drawback is that it requires continuous laser scanning of the sample that can lead to photobleaching, self-quenching, and damage to the cell membrane.[13] Transition metal complexes have frequently shown improved photostability compared to conventional organic fluorophores that can allow for continuous monitoring of biological samples. Ru(II),[14,15] Ir(III),[16–19] and Re (I)[20–22] have been quite widely explored as general imaging probes in live cell imaging. As described, a major hurdle to their broad application in imaging has been their uptake by living cells.

11.4.1 MECHANISMS OF CELL UPTAKE

A key historical limitation in the application of metal probes to cell and tissue imaging has been their poor uptake across the cell membrane. Over the past 30 years, our understanding of the mechanisms of cell uptake has grown considerably, and this knowledge has enabled the application of much broader types of imaging probes. There are several routes to cargo transport across the cell membrane, and the mechanism can affect the rate of uptake and the intracellular fate of the complex.[23–25] Cell uptake can occur via energy-dependent mechanisms; these include endocytosis, ATP-powered pumps, and cell surface proteins. In uptake by endocytosis, a vesicle derived from the plasma membrane surrounds and engulfs the incoming molecule that has come in contact with the cell. Endocytosis can be encouraged for more specific applications by targeting cell-surface receptors and transporters present on the membrane surface.[4,26] And it is highly advantageous in the design of therapeutic agents for targeting specific diseased states such as apoptosis, cell-proliferation, and diabetes.[27,28]

Energy-independent methods include passive diffusion and transport through ion channels.[29] Passive diffusion is the preferred mechanism for drug delivery as it has the broadest applicability and is nonselective. The charge of a complex and its lipophilicity strongly influence whether a complex will passively diffuse across the cell membrane. As the net charge inside the cell is negative, positively charged molecules are more likely to cross the lipid bilayer driven by the membrane potential.

The temperature dependence of uptake is usually a key indicator of the mechanism behind the transport. Active uptake switches off at 4°C as all metabolic pathways are switched off. Therefore, all active endocytosis is inhibited at this temperature, whereas passive diffusion can continue.

11.4.2 IMPROVING CELL PERMEABILITY OF METAL COMPLEX LUMINOPHORES

Transition metal complexes typically do not diffuse freely across the lipid bilayer of live cells due to their polarity and charge. Organic solvent such as DMSO or ethanol or detergents (typically <1%) such as Triton X in low concentrations (e.g., 1%–5% v/v) in the dye incubation medium have been used to promote permeability of metal complexes across the cell membrane. The solvent or detergent affects the packing of the cell membrane, making the lipid bilayer more permeable. However, such methods are not ideal because not only do they damage the cell but also the complexes cannot be said to be truly cell permeable, which is important if they are applied outside the basic cell culture conditions.[30] Methods that have successfully rendered metal complexes as water soluble and cell permeable include conjugation of the complex with cell-penetrating peptide (CPP) sequences,[31-33] or polymers such as poly(ethylene glycol) (PEG),[34,35] as well as structurally altering the complex using hydrophobic ligands such as 2,3-bis(2-pyridyl)pyrazine, 1,10-phenanthroline[36] or lipophilic pendants such as estradiol.[37] A balance of lipophilicity, charge, and solubility is key to designing a successful cell-permeable imaging probe.

11.4.3 LIPOPHILICITY

Lipophilicity is a key physicochemical property that is predicative of molecular permeability across biological membranes. It determines absorption, distribution, and toxicity of a complex within a cell.[38,39] The most widely used denotation of lipophilicity of a compound is its log P value that describes the partition equilibrium of an unionized solute between water and an immiscible organic solvent such as octane. The larger its log P, the greater is the molecule's lipophilicity. There have been many reports on the importance of lipophilicity in transition metal uptake in live cells and their toxicity.[36,40-43] It has also been shown that increased lipophilicity is linked to increased cytotoxicity of metal complexes. For example, Lv et al. demonstrated that by altering the alkyl side groups in a series of Ru(II)-N-heterocyclic carbenes, the lipophilicity was transformed, affecting the toxicity of the complexes when introduced into live cells.[44] Similarly, in ruthenium polypyridyl complexes, replacement of 2,2-bipyridyl ligands with 1,10-diphenyl-4,7-phenanthroline ligands leads to a significant increase in lipophilicity, with retention of comparable photophysical properties.[65] Lipophilicity has also been shown to be important in targeting complexes, to organelles such as the mitochondria.[45] Keene et al. examined a series of Ru(II) complexes that differed in the number of methylene groups in the binding ligands, altering the lipophilicity but not the charge of the complex that influenced the localization of these materials at the mitochondria.[46] Confocal microscopy showed uptake of the complexes in L1210 cells after 4 h incubation and localization at the mitochondria, which was confirmed by costaining with MitoTracker. The toxicity of the complexes varied, depending on the number of methylene groups, with IC_{50} values ranging from 5 to 200 µM, demonstrating the importance of lipophilicity in designing metal complexes.

11.4.4 CHARGE

Healthy cells have an internal negative charge, and passive transport of imaging probes can be promoted if they are cationic. For example, a family of Ir(III) cyclometalated complexes with the general formula [Ir(ppy)2(N^N)]+, where N^N is a chelating ligand, are monocationic and are taken up by the cell due to membrane potential.[47] The charge of the complex has also been shown to be related to its toxicity and localization within the cell.[48,49] For example, Dickerson et al. presented two analogous ruthenium complexes that differed in charge and log P values (Figure 11.2). Uptake studies in A549 human cells showed that Compound 1, with a

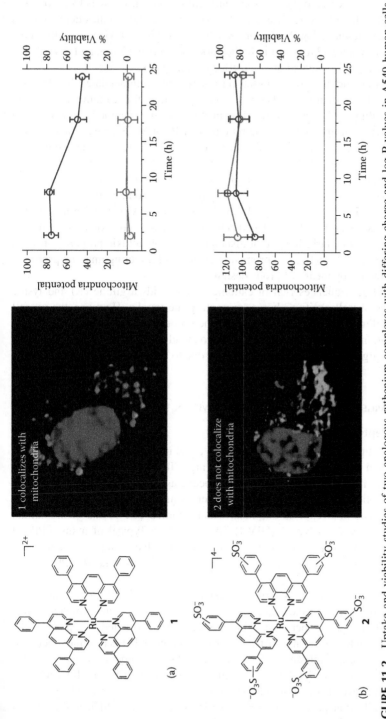

FIGURE 11.2 Uptake and viability studies of two analogous ruthenium complexes with differing charge and log P values in A549 human cells. Confocal microscopy showed Compound 1 entered the mitochondria and lysomes and was highly toxic toward the cells (a), while Compound 2 remained in the cytosol having no toxic effects on the cells (b). (Reprinted with permission from Dickerson, M., Sun, Y., Howerton, B., and Glazer, E.C., Modifying charge and hydrophilicity of simple Ru(II) polypyridyl complexes radically alters biological activities: Old complexes, surprising new tricks, *Inorg. Chem.*, 53, 10370–10377. Copyright 2014 American Chemical Society.)

charge value of +2, entered the lysosome and mitochondria, whereas Compound 2, with a charge value of −4, remained in the cytosol. Viability studies demonstrated that Compound 1, with a log P value of 1.8 ± 0.02, was highly toxic toward the cells, inducing toxicity after 72 h exposure in the dark. In contrast, Compound 2 with a log P value of −2.2 ± 0.12 induced no toxic effects on the cells.

Gupta and coworkers have reported on a series of cyclometalated Ir(III) complexes for cellular imaging.[50] They demonstrated that the cationic nature and hydrogen bonding interactions of Ir(III) contributed to the specific localization and emission intensity in the endoplasmic reticulum (ER) of MCF7 breast cancer cells. Using confocal microscopy, they demonstrated that cells treated with complex 1 and 3, which contained −OH groups in position 4′ and at 2′ and 3′ of the appended phenolic ring, localized within the cytoplasm and exhibited low fluorescence in the nuclei. In contrast, complex 2, with an −OH group at position 2 of the phenolic ring, showed intense fluorescence in the ER. Under irradiation with a 405 nm laser, the complex caused extensive photoinduced cell death, whereas no toxicity toward cells were observed in the dark. In spite of displaying high phototoxic properties toward cells, they have shown Ir(III) as a promising example of directing imaging probes to specific organelles by means of probe design.

Using peptide sequences rich in cationic amino acids arginine (R) and lysine (K) increases overall positive charge, promoting the uptake. However, the overall charge can influence the final destination of the probe within the cell. For example, lipophilic cations typically localize to the mitochondria. However, an excess of the molecular charge can hinder access to the cell due to hydrophobicity.[51,52]

11.4.5 TARGETING METAL LUMINOPHORES WITHIN THE CELL BY CONJUGATION

11.4.5.1 Peptides

CPPs are short peptide sequences, <40 amino acids in length, that can be exploited to facilitate and promote uptake of molecular cargo. To date, the primary application of peptides has been in transportation of therapeutic agents across the cell membrane in clinical applications, but they are being increasingly applied to transport imaging probes.[53] The first reported and most widely studied CPP is that of the human immunodeficiency virus type-1 (HIV-1) TAT peptide.[54] Typical of many CPPs, it contains a high lysine/arginine content whose cationic charge improves membrane penetrability. TAT has inspired the structures behind many related CPPs subsequently reported.[27,55]

Our group have successfully targeted Ru(II), Os(II), and Ir(III) complexes to the cytoplasm using polyarginine sequences.[14,18,112] For the ruthenium complex, it was demonstrated that the number of arginine residues present affected the uptake of the complex wherein R = 5 was not sufficient for uptake, while R = 8 promoted penetration through the cell membrane.[56] Uptake was found to be temperature dependent for the arginine conjugates. Conversely, a rhenium complex described by Raszeja et al. conjugated to neurotensin, an amino acid sequence from a neuropeptide as a CPP, was successfully transported across the membrane via passive diffusion in several cancer cell lines and found to be located within the cytoplasm.[57]

As the understanding of peptide roles in cell penetration has grown, they are increasingly being used to target specific cell organelles by exploiting the peptide sequences found in proteins of organelle membranes. Such targeting opens the possibility to study and monitor subcellular events and to deliver therapeutics to specific sites. Recently, this work has been extended to the targeting of metal complexes, and to date, CPPs have successfully been shown to be capable of directing and accumulating luminescent Ru(II) polypyridyl complexes selectively within selected organelles.

There has been extensive interest in the interactions of metal-based luminophores with DNA, in particular, from the perspective of their application in chemotherapy and photodynamic therapy.[58,59] Historically, much of the focus has been on the solution-based study of DNA–metal complex association and DNA damage. However, realizing the potential of metal complex and DNA photochemistry requires the complex to be directed through the highly restricted nuclear envelope. Nuclear localizing signalling (NLS) peptides hold great potential in this regard. They are typically derived from transcription factors and are particularly synthetically accessible as they are typically less than 12 residues in length. NLS peptides have the ability to cross the cell membrane and nuclear envelope and have been demonstrated to be capable of carrying molecular cargo such as plasmid DNA from the cytoplasm to the nucleus.[60] NLS sequences that have been successfully derived from transcription factors include NF-κB, TCF1-α, TFIIC-β, and SV40,[61,62] of which NF-κB and SV40 have shown to enter the nucleus selectively.[63–65] For example, Ragin et al. have demonstrated a series of fluorescently (fluorescein) labeled NLS sequence peptides from transcription factors that from live uptake studies in MCF-7 breast carcinoma cells showed uptake across all NLS sequences.[66] The NF-κβ peptide sequence is localized to the cytoplasm and nuclear regions of the cells, whereas TFIIE-β and Oct-6 are localized to the cytoplasm only, possibly due to their highly positive charge. At 4°C, the uptake of NLS sequences NF-κβ, TFIIE-β, and HATF-3 was greatly reduced, whereas Oct-6, TCFI-α, SV40, and *Caenorhabditis elegans* SDC3 were unaffected. NLS sequences, in particular the NF-κB sequence, has shown to be successful in directing metal complexes to the nucleus of cells.[65] NF-κβ very effectively and selectively transported ruthenium complexes to the nucleus of CHO cells (Figure 11.3).[65] Interestingly, it was found that by increasing the lipophilicity of the metal complex by replacing bipyridyl counter ligands with lipophilic diphenylphenanthroline ligands, the complex could be directed to the nucleolus.

Mitochondria are responsible for the cell's energy generation by ATP synthesis, supplying oxygen to the cell and regulating apoptosis. Mitochondrial dysfunction is implicated both in disease and toxicity.[67–70] Therefore, mitochondria-targeting peptide sequences are potentially important in targeting both therapeutic and imaging agents to treat and monitor dysfunctional mitochondria. For example, reactive oxygen species (ROS) is implicated in the pathogenesis of a range of important disease states[71] and is a by-product of O_2 production.[72–74] Mitochondria-penetrating peptides (MPPs) typically consist of 4–8 positively charged hydrophobic residues.[75,76] For example, Kelley et al. carried out a detailed systematic study on MPP sequences to evaluate the design rules around CPP sequences of value for mitochondria localization and identified a balance of charge and lipophilicity that enabled mitochondrial

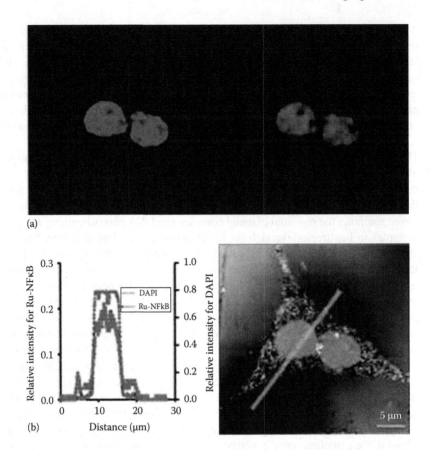

FIGURE 11.3 (a) Nuclear localization of $[Ru(bpy)_2pic-\beta Ala-NF\kappa B]^{6+}$ and nuclear stain DAPI in CHO cells. (b) Merged image of complex and DAPI with backscatter demonstrate colocalization, confirmed by the corresponding cross section and distribution plot of both complexes in the nucleus. (From Boulikas, T., Putative nuclear localization signals (NLS) in protein Transcription factor, *J. Cell. Biochem.*, 55, 32–58, 1994. Reproduced by permission of The Royal Society of Chemistry.)

targeting.[77,78] They demonstrated that a sequence of eight amino acids was most successful in localizing to the mitochondria compared to chain lengths with fewer residues. Their optimal sequence also included D-arginine residues (r) to promote cellular stability, FrFKFrFK. Using this sequence, our group demonstrated that a dimeric ruthenium polypyridyl complex could be effectively and selectively targeted to the mitochondria of mammalian cells, as shown in Figure 11.4.

Protein-binding peptides have also been used to target metal complexes to specific proteins within cells. Peptide ligands containing the Arg-Gly-Asp (RGD) tripeptide sequence has been particularly explored in cell imaging because of the prevalence of integrin $\alpha v\beta 3$ at the surface of tumor cells.[79] And this sequence has been applied to the study of ruthenium polypyridyl complexes with platelet integrin.[80] In a similar vein, the peptide somatostatin has been conjugated to a ruthenium polypyridyl

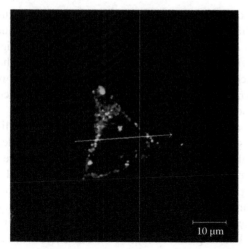

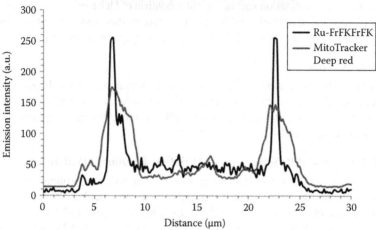

FIGURE 11.4 Confocal image of $[Ru(bpy)_2phen-Ar)2-FrFkFrFk]^{7+}$ in a HeLa cell in green, with MitoTracker Red as a colocalizing agent in red, and their colocalizing in yellow. The corresponding cross-sectional plot demonstrates the collocation of the two complexes within the mitochondria.

complex that showed receptor-mediated endocytotic uptake that showed also cellular selectivity attributed to the targeting of the somatostatin receptors that are upregulated in tumor cells. The focus of this report was photodynamic therapy rather than use as an imaging agent as this material showed strong photocytotoxicity, but low cytotoxicity in the absence of light.[81]

11.4.5.2 Sugar Conjugates

Carbohydrate-modified luminescent metal complexes were originally developed as lectin probes for recognition.[82] However, they have been increasingly used in cell imaging due to their enhanced biocompatibility and capacity to drive complexes into cells. Carbohydrate-modified complexes can serve as new imaging reagents as well as glucose-uptake indicators in live cells. A series of rhenium(I) polypyridine complexes bearing an α-D-glucose, for example, have been designed by Lo et al.[83] Upon incubation with HeLa cells, complex 3 accumulates in the mitochondria. It is thought that glucose transporters, GLUTs, play an important role in the cellular uptake of these complexes as they become overexpressed in cancer cells, catabolizing glucose at a much higher rate than healthy cells.

Lo et al. have also recently demonstrated the effects of lipophilicity on cell uptake and toxicity of a series of phosphorescent iridium(III) complexes conjugated to D-glucose and D-galactose.[84] The log P values of complexes were characterized and lie in the range from 1.4 to 2.59. Complexes with an extra hydrophobic phenyl ring at the cyclometalating ligands exhibited higher lipophilicity than those without. It was noted that the polar sugar entity slightly reduces the log P values compared to the sugar-free complexes. They also demonstrate a difference in cell uptake between the two sugar residues. Focusing on the complex that has a log P value of 2.44, the authors demonstrated its localization within the mitochondria of HeLa cells, which was attributed to the combined cationic and lipophilic nature of this complex and is consistent with the dual requirements noted for mitochondrial targeting using peptides.

11.4.5.3 Sterols

Sterols have also been demonstrated to promote metal complex cell permeability. For example, estradiol conjugation to a series of ruthenium complexes containing DPP functionalities was found to promote cellular uptake to the cytoplasm but with little organelle uptake.[85]

11.4.5.4 Covalent Reactive-Linker Moieties: Rhenium "Prolabels"

An interesting approach to targeting probes to organelles or proteins in live cells is to incorporate functional ligands that will covalently attach to its target, such as a thiol-reactive chloromethyl group Ar-CH$_2$Cl, in situ. These linkers have, to date, been used with a rhenium core because of their synthetic versatility where the diimine and axial ligands can be functionalized with polar and hydrophobic moieties to improve the photophysical properties, lipophilicity, and toxicity.[86] For example, Re(I) complexes with long alkyl chains have improved lipophilicity and serve as good imaging probes and have shown to have DNA-binding abilities.[87] Pope et al. have examined a fac-3 chloromethylpyrine rhenium tricarbonyl bipyridyl complex as a biological imaging probe.[88] Confocal imaging of the complex showed uptake after 1.5 h in MCF-7 cells, which are localized within the mitochondria due to the lipophilic thiol-reactive nature of the complex. This was confirmed using TMRE as a colocalization dye, shown in the cross-sectional image in Figure 11.5.

11.4.5.5 Polymer Conjugates

Macromolecules are widely used to promote lipid membrane permeability, and they can be utilized to drive cargo into cells and tissues.[89,90] The most prevalent polymer in this regard is PEG. PEG has been both covalently and noncovalently linked to transition metal complexes, which increases their hydrophilicity and improves their solubility, water solubility, and cell permeability.[91–93] Furthermore, PEGs are

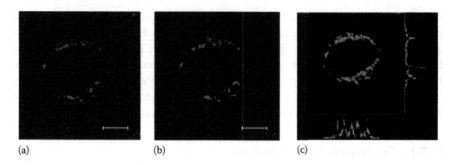

(a)   (b)   (c)

FIGURE 11.5 Live cell imaging of a rhenium prolabel complex (a), TMRE mitochondrial stain (b) and their colocalization cross section (c) in CF-7 cells after 1.5 h exposure. (From Samanta, D., Kratz, K., Zhang, X., and Emrick, T., A Synthesis of PEG- and phosphorylcholine-substituted pyridines to afford water-soluble ruthenium benzylidene metathesis catalysts, *Macromolecules*, 41, 530–532, 2008. Reproduced by permission of The Royal Society of Chemistry.)

an attractive means of cell targeting as they generally exhibit low cytotoxicity. The reduced toxicity is attributed to the reduced nonselective interaction of the probe with DNA, proteins, and organelles in the presence of PEG.[94] Lo et al. have done extensive work with PEGylated metal complexes and have examined their effects on cell uptake and imaging. Recently, they presented a water-soluble cyclometalated Ir(III) PEG complex.[95] Uptake studies using confocal microscopy showed the complexes successfully entered into HeLa cells and were localized in the mitochondrial region of the cell. The complex showed no toxicity toward the cells in the dark but, interestingly, exhibited high phototoxicity toward the cells, indicating significant potential as a therapeutic photosensitizer.

11.5 SENSING APPLICATIONS OF TRANSITION METAL LUMINOPHORES IN CELLULAR IMAGING

To date, the vast majority of transition metal luminophores applied to imaging have been applied to conventional confocal luminescence imaging. This is the most prevalent imaging method, and as the field of transition metal cell-based probes is relatively young, most work has focused on demonstrating uptake, localization, and cytotoxicity of metal complexes by confocal microscopy.

As described, many d^6 transition metal luminophores exhibit long-lived and environmentally sensitive emission, which makes them suitable for sensing. However, to release their full potential in imaging/sensing, a concentration-independent means of assessing luminescence changes to emission properties is needed. One approach that can be exploited with confocal imaging is to incorporate self-referencing into the probe. However, this is relatively difficult on a molecular scale, but has been applied widely to particles, for example, on silicate particles, where an analyte-dependent and analyte-independent probe are coincorporated into the structure.[96,97] However, particles are more difficult to transport and target predictably than molecules. A related approach is to combine the luminophore with a quencher that eliminates emission except when the quencher is released from the compound by cleavage with the analyte of interest. This approach was recently reported for in-cell detection of HOCl by a ferrocene ruthenium dyad. The ruthenium emission is initially weak due to quenching by the ferrocene in the dyad. Cleavage of the peptide bond bridging the two units in macrophage cells, which was attributed to HOCl, eliminates quenching and so is accompanied by an increase in luminescence from regions of high HOCl concentration within the cell.[98]

11.5.1 FLUORESCENT LIFETIME IMAGING

An alternative approach is to use fluorescence lifetime imaging (FLIM), more correctly in the context of metal complex luminophore sometimes referred to as phosphorescence lifetime imaging (PLIM) that maps the spatially resolved average luminescence lifetime of a probe across a sample. Since the lifetime of a lumino-phore is typically independent of its concentration, FLIM is the most reliable way of assessing the probe environment without the need for ratiometric signal. The application of FLIM to live cells was first reported in 1992,[99] and although it is not yet widely adopted in cell biology laboratories, its use is growing. FLIM systems are typically used in tandem setups implemented in conventional confocal and mul-tiphoton microscopes.[100,101] The fluorescence lifetime can be measured using either time or frequency domain measurements. The most common detection method is probably time-correlated single-photon counting (TCSPC) systems. The advantage of this method over time domain and gated systems is that it counts all the photons that reach the detector, giving much higher efficiency.[102] However, acquisition times can be long, particularly for long-lived probes, and this can limit the dynamic range of TCSPC systems. Detailed technical description of the FLIM method is beyond the scope of this chapter, but the following reference is recommended.[103]

In contrast to organic dyes whose lifetimes are relatively short, typically between 1 and 5 ns, the long luminescent lifetime of transition metals such as Ru, Ir, and Re, typically on the scale of hundreds of nanoseconds, is well outside the range of the autofluorescence of biomaterials. Therefore, shorter lifetimes can be eliminated from the decay based on their luminescent lifetimes using a time-gate system to 'gate' what lifetimes are collected from the sample. In addition, metal complexes frequently show better photostability than organic luminophores, which is a particular advantage in lifetime imaging where, in order to collect a representative lifetime image, several images need to be collected to reduce noise and correct for background fluorescence and autofluorescence from the cell. This can lead to long acquisition times and pho-tobleaching, leading to artifacts in the lifetime data if the probe is photounstable.[104]

Their long luminescent lifetimes and triplet states make many transition metal probes suited to pO_2 measurement in cells, judicious synthetic modification can ren-der their lifetimes shorter than O_2 diffusion time while providing sufficient lifetime for response to other processes, for example, through time-resolved luminescence anisotropy, where protein or membrane binding can be used to study cell membrane structures and viscosity based on fluorescent lifetime changes.[104-107] Conversely, as described earlier, a drawback is that the long luminescent lifetimes of transi-tion metal luminophores require a large dynamic range of lifetime microscopy.

This is particularly a challenge in TCSPC imaging approaches as the cycle time for collection of each photon is increased for such long-lived probes, slowing imaging compared to organic fluorophores. Phase fluorimetry and pulsed fluorescence methods are perhaps easier to implement for such probes. Nonetheless, the excellent environmental sensitivity and good photostability of such probes generally offset the drawback of long acquisition times.

11.6 MOLECULAR OXYGEN SENSING

Intracellular dioxygen (icO_2) is a key sensing target and one to which luminescent metal complex probes are particularly well suited. Oxygen is a crucial marker of cell viability; for example, it is known that cancerous cells often foster hypoxic environments.[108] The ability to measure the oxygen concentration in discrete cellular locales is of key importance in understanding molecular mechanisms of redox-driven cellular activity. In an oxygenated environment, excited states of metal complexes such as Ru(II) are quenched as dioxygen in its triplet state and ground state deactivates the complex triplet excited state through energy transfer, in turn generating singlet oxygen.[109] The photophysical process is accompanied by a decrease in emission lifetime and intensity of the probe. Excluding porphyrin constructs (generally Pt or Pd based), the archetypal metal complexes of Ru(II) or Ir(III) discussed earlier are particularly suited to such determinations as they typically possess long luminescent triplet excited states that enables remarkable oxygen sensitivity.[110]

One of the earliest examples of [icO_2] sensing by a d^6 metal complex luminophore using FLIM was reported by Mycek et al. who demonstrated the oxygen-dependent response of [Ru(bpy)$_3$]$^{2+}$ in human bronchial epithelial cell lines. The complex was taken up by this cell line in sufficient levels for FLIM though requiring 4–6 h incubation and high detector gain. Nonetheless, O_2 concentration modulated in the cytoplasm by bubbling N_2 into the contacting medium was imaged by measuring Ru complex luminescent decay time.[111]

Recently, icO_2 was estimated in cytoplasm and membrane regions of SP2 myeloma cells by polyarginine-conjugated ruthenium(II) polypyridyl complex using FLIM. Peptide conjugation, as described earlier, improved the solubility and cytotoxicity of the ruthenium center. In particular, it dramatically improved cell uptake that was complete within less than 20 min with distribution of the complex throughout the cytoplasm.[56] In a more recent report, a dynamic response to changing icO_2 using FLIM was demonstrated for a dinuclear Ru(II) probe that is localized to the mitochondria of HeLa cells via the directing effects of a conjugated CPP, complex [(Ru(bpy)$_2$-phen-Ar)$_2$-MPP]$^{7+}$ (2), where MPP is the mitochondrial-penetrating peptide sequence FrFKFrFK.[112] In aerated PBS solution, the complex exhibited a lifetime of 458 ± 7 ns that increased to 948 ± 6 ns when fully de-aerated, and a calibration curve for the lifetime of the complex toward molecular O_2 was established across this range. Confocal imaging confirmed the peptide conjugate was successfully located within the mitochondria in HeLa cells (Figure 11.4) and FLIM showed the lifetime of the complex in the mitochondria to be 525 ± 10 ns. Upon introduction of Antimycin A, a mitochondrial uncoupler, to the cell medium to induce oxidative stress, the lifetime of the complex within the mitochondria decreased to 423 ns after 10 min exposure and further to a plateau of 228 ns after 100 min. Figure 11.6 shows the PLIM images and lifetime distribution of the probe over 100 min. A control experiment with no Antimycin A showed no changes in luminescent lifetime over the time course. As $[O_2]$ cannot exceed saturation, the separate quenching pathway was attributed to ROS production induced by the inhibition of mitochondrial function. It was concluded that this Ru(II) probe was responding to O_2 concentrations and reporting on ROS generation, monitored and measured using PLIM.

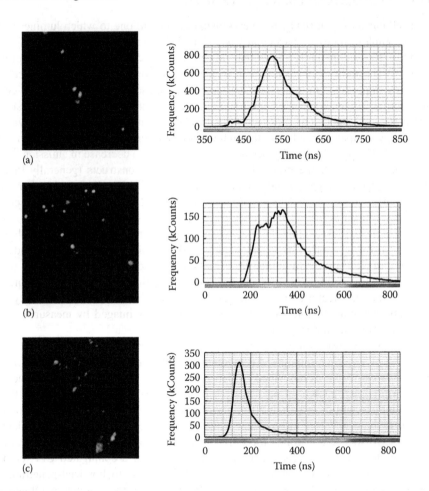

FIGURE 11.6 Luminescent lifetime images and distributions of complex 2 in HeLa cells in response to the mitochondrial uncoupler Antimycin A (200 μM/mL). Lifetime of complex 2 was measured before addition (a), after 10 min exposure (b), and after a further 100 min (c). (n = 2). (Reprinted with permission from Blackmore, L., Moriarty, R., Dolan, C., Adamson, K., Forster, R.J., Devocelle, M., and Keyes, T.E., Peptide directed transmembrane transport and nuclear localization of ru(II) polypyridyl complexes in mammalian cells, *Chem. Commun.*, 49, 2658. Copyright 2013 American Chemical Society; Reprinted with permission from reference 112. The American Chemical Society.)

Importantly, the calibration curve was established for the complex at 37°C as ruthenium polypyridyl complexes, in particular, exhibit temperature-dependent photophysics, and this is an important consideration in cell imaging as it is frequently conducted at temperatures elevated above room temperature.[112]

11.7 MEMBRANE SENSING

The ability to discriminate membrane structures in cells and in tissues is important in imaging, and a cell-permeable peptide conjugated Ru(II) complex coordinated to the dppz ligand has been applied in this regard to live cells. Ru(II) complexes of the dppz ligand have been widely investigated as a light switch in DNA binding, that is, its luminescence switches on in an aqueous environment upon DNA intercalation when water binding to the phenazine ligand is prevented. This effect was exploited in membrane imaging where luminescence from the octaarginine conjugated complex was observed only at the membrane structures. Resonance Raman spectral imaging, enabled by the large Stokes shift of the complex, revealed that the complex was cell permeable and distributed through the cytoplasm.[113]

11.7.1 DNA

The emergence of platinum-based chemotherapeutics such as the famous cisplatin complex, cis-$[Pt(NH_3)_2Cl_2]$,[114] has paved the way for the evolution of metallodrugs to potentially include a host of other transition metals.[115] The mechanism of their antimetastatic action is DNA binding that disrupts the structural integrity of the duplex, leading to strand cleavage and ultimately cell death by apoptosis. Although research in this domain is primarily concerned with DNA destruction for therapeutic effect, a burgeoning approach employs metal complexes both as therapy and as a diagnostic of nucleic acid sequence and structure, in many cases using luminescence as a primary modality that can be translated to DNA imaging in cells and tissues.

An early work by Dwyer and coworkers on metal *tris* chelates of phenanthroline ligands and Sigman on cupric binders has identified the bacteriostatic activity of coordinatively saturated complexes that hinted at their ability to interact with DNA.[116,117] Later, the work by Yamagishi and Barton and coworkers sought to exploit the chirality of d[6] metal *tris* phen chelates as a probe of nucleotide structure given that DNA itself is inherently chiral due to its helicity.[118,119] The lead compound in this endeavor was promptly established as $[Ru(phen)_3]^{2+}$, a luminescent, kinetically inert complex that demonstrated DNA binding by a debated mechanism likely to arise from both groove binding and semi-intercalation, although electrostatic interaction has also been mooted.[120] However, $[Ru(phen)_3]^{2+}$ did not demonstrate any capability as a conclusive diagnostic of nucleic acid sequence or structure; its spectroscopic response upon binding was moderate. Importantly, the binding constant was relatively weak compared to conventional DNA stains such as ethidium bromide. Barton et al. reported on a probe of DNA conformation using more sterically cumbersome diphenylphenanthroline (dpp) complexes such as $[Ru(dpp)_3]^{2+}$,[121] but it was not until the work from the same group using $[Ru(bpy)_2(dppz)]^{2+}$, a complex first described by Amoyal et al.,[122] that a metal complex fit the criteria as a probe of DNA for use in cell imaging was reported.[123] The [Ru(bpy)$_2$(dppz)]$^{2+}$ complex exhibits a strong binding constant, and crucially, it was shown to be a "light switch for DNA," displaying virtually no luminescence in aqueous solution but exhibiting moderately intense, long-lived luminescence upon intercalation between the base pairs of the duplex. The effect is attributed to the protection of the phenazine nitrogens on the dppz ligand from hydrogen bonding once the planar aromatic ligand is incorporated into the π-stack of the DNA helix. While this prototypical Ru-dppz complex is an excellent general stain of DNA, it does not demonstrate any structural specificity with respect to sequence or conformation. Consequently, many reports have since emerged seeking to alter the planarity of the intercalator or the steric properties of the ancillary ligands to attempt to confer DNA sequence or isomer specificity in binding, and the use of such light-switch complexes has been extended to other metals such as Os(II), Rh(III), and Re(I). A prominent example in this regard was the selective binding of Rh(III) metalloinsertors for mismatch sites in DNA using complexes such as [Rh(bpy)$_2$(chrysi)]$^{3+}$ (chrysi = chrysene-5,6-quinone diimine).[124] The work by Hannon et al. on luminescent dinuclear Ru(II) molecular cylinders was also shown to demonstrate sequence specificity.[125] Another notable class of DNA-binding compounds is the Ru(II) polyazaaromatic series that comprise at least two tap or hat ligands and possess a redox potential sufficiently positive to photooxidize DNA at guanine residues, yielding a covalent adduct. Primarily investigated by the Kirsch-De Mesmaeker group and collaborators, these complexes may yet find application in gene-silencing therapy.[126] While there have been remarkable advances in understanding and optimization of DNA–metal chelate binding, there have been relatively few reports, to date, of such complexes applied to targeting in cellulo DNA. This is due primarily to the difficulty in uptake and localization of the probes at the DNA target, as outlined in previous sections herein. Where nuclear localization has been achieved, specific imaging of the subnuclear structures of DNA has remained elusive. The outstanding example where this has been achieved is described by Gill et al. using the dinuclear Ru(II) probe [(phen)$_2$Ru(tpphz)Ru(phen)$_2$]$^{4+}$ (tpphz = tetrapyrido[3,2-a:2′,3′-c:3″,2″-h:2‴,3‴-j] phenazine), which was taken up by a nonendocytotic mechanism, they observed that it acted as an intracellular stain of DNA structure in MCF-7 human breast cancer cells.[127] Nuclear localization was confirmed by costaining with nuclear target DAPI and STYO 9, and remarkably, confocal luminescence imaging captures the four phases of chromosome aggregation in the nucleus during cell division. In a more recent publication, the group used the same probe to describe the two-photon PLIM imaging of the process in HeLa cells, obtaining enhanced levels of sensitivity and autofluorescence free imaging (Figure 11.7).[128]

11.8   CONCLUSIONS AND FUTURE DIRECTIONS

Luminescent transition metal complexes are finding increasing application across cell imaging and sensing. This is driven by their very attractive photophysical properties and fuelled in particular by a growing understanding of the strategies available to promote their permeability and targeting within cells.

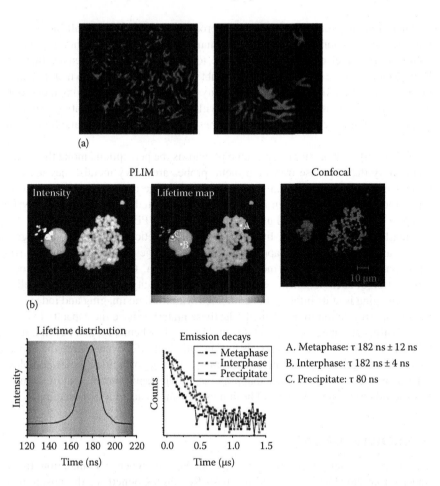

FIGURE 11.7 (a) Metaphase spreads of HeLa chromosomes stained with 2 (left) or 1 (right) (100 mm, 30 min) and imaged by confocal microscopy, (b) 2P-PLIM (left) and confocal (right) imaging of HeLa metaphase spreads labelled with 1. (Reproduced from reference 128 with permission from John Wiley & Sons.)

The extensive literature on interactions of metal complexes with biopolymers, particularly those of Ru(II) and Ir(III) polypyridyl complexes with DNA, has shown that such complexes hold extraordinary potential in the field of theranostics. Translation of these interactions to whole cells and tissues is inevitable with the advancements in the understanding of metal complex uptake and targeting. To date, the majority of complexes used in imaging have coordination compounds of iridium and ruthenium. The future is also likely to see increased development of stable luminescent compounds from other metals such as copper and gold. Au(I) d^{10} complexes have thus far been utilized far more rarely than the transition metals mentioned earlier but have also shown potential in biophotonics application in various both mono- and polynuclear complexes.[19,20]

There remains some barriers to their broad application across cell biology. The intensity of emission is not as great as many commercial fluorophores; however, there are a number of strategies available to increase emission intensity that have been explored in the literature, and no doubt these will be applied to metal complex probes. In addition, the improving sensitivity of the detection in imaging methodologies makes it increasingly easy to work with lower emission intensities, especially in the case of metal complexes, where probes exhibit far-red to NIR and Stokes shift emissions.

A more difficult barrier to overcome perhaps is the perception among the biology community that, because transition metal probes are heavy metals, they are cytotoxic. Certainly this can be an issue and an advantage in therapeutic applications, but cytotoxicity is not a given as recently demonstrated for an Os(II) polypyridyl complex and can be mitigated using strategies such as PEGylation.

Finally, there is extensive literature on the synthetic development of Pt group metals as luminophores for application across a range of sensing domains, from emission-based detection of metals, for example, Ca, Cu, Fe, to microviscocity. Exploration of such complexes and their sensing capacity in the domain of cells or tissue imaging is at its infancy, but combined with lifetime imaging and indeed with newer superresolution imaging methods, these materials have the capacity to have a very significant impact on imaging in cell biology, biochemistry, and beyond. It will take careful design of such materials to translate their effectiveness into the cell and tissues for targeting, quantitative assessment. The opportunities metal complexes offer as probes for dynamically mapping the cellular environment and in theranostics should overcome any residual hesitancy in using metals for cell imaging.

ACKNOWLEDGMENTS

The authors gratefully acknowledge the support of Science Foundation Ireland under Grant No. 14/IA/2488 and the Irish Research Council for the postgraduate scholarship.

REFERENCES

1. Malak H, Dobrucki JW, Malak MM, and Swartz HM. Oxygen sensing in a single cell with ruthenium complexes and fluorescence time-resolve microscopy, *Biophys. J.* **74**, A189 (1998).
2. Dobrucki JW. Interaction of oxygen-sensitive luminescent probes Ru(phen)2+3 and Ru(bipy)2+3 with animal and plant cells in vitro: Mechanism of phototoxicity and conditions for non-invasive oxygen measurements, *J. Photochem. Photobiol. B.* **65**, 136–144 (2001).
3. Heffern MC, Matosziuk LM, and Meade TJ. Lanthanide probes for bioresponsive imaging, *Chem. Rev.* **8**, 4496–4539 (2014).
4. Montgomery CP, Murray BS, New EJ, Pal R, and Parker D. Cell-penetrating metal complex optical probes: Targeted and responsive systems based on lanthanide luminescence, *Acc. Chem. Res.* **42**, 925–937 (2009).
5. Balzani V and Campagna S. Photochemistry and photophysics of coordination compounds. *Topics Curr. Chem.* **280**, 1–36 (2007).

6. Madani F, Lindberg S, Langel U, Futaki S, and Graslund A. Mechanisms of cellular uptake of cell-penetrating peptides. *J Biophys.* 2011, 1–10 (2011).
7. Sullivan BP, Salmon DJ, and Meyer TJ. Mixed phosphine 2,2′-bipyridine complexes of ruthenium, *Inorg. Chem.* **17**, 3334–3341 (1978).
8. Spiccia L, Deacon GB, and Kepert CM. Synthetic routes to homoleptic and heteroleptic ruthenium(II) complexes incorporating bidentate imine ligands. *Coord. Chem. Rev.* **248**, 1329–1341 (2004).
9. Lamansky S, Djurovich P, Murphy D, Abdel-Razzaq F, Kwong R, Tsyba I, Bortz M, Mui B, Bau R, and Thompson ME. Synthesis and characterization of phosphorescent cyclometalated iridium complexes. *Inorg. Chem.* **40**, 1704–1711 (2001).
10. Sprouse S, King KA, Spellane PJ, and Watts RJ. *J. Am. Chem. Soc.* **106**, 6647–6653 (1984).
11. Balasingham RG, Coogan MP, and Thorp-Greenwood FL. Complexes in context: Attempting to control the cellular uptake and localisation of rhenium fac-tricarbonyl polypyridyl complexes. *Dalton Trans.* **40**, 11663–11674 (2011).
12. Stephenson KA et al. Bridging the gap between in vitro and in vivo imaging: Isostructural Re and 99mTc complexes for correlating fluorescence and radioimaging studies. *J. Am. Chem. Soc.* **126**, 8598–8599 (2004).
13. Nwaneshiudu A, Kuschal C, Sakamoto FH, Anderson RR, Schwarzenberger K, and Young RC. Introduction to confocal microscopy. *J. Invest. Dermatol.* **132**, 1–5 (2012).
14. Neugebauer U, Cosgrave L, Pellegrin Y, Devocelle M, Forster RJ, and Keyes TE. Mulitmodal, muliparameter cell imaging using ruthenium polypyridyl peptide conjugates. *Proc. SPIE.* **8427**, 84270C (2012).
15. Xu W, Zuo J, Wang L, Ji L, and Chao H. Dinuclear ruthenium(II) polypyridyl complexes as single and two-photon luminescence cellular imaging probes. *Chem. Commun.* **50**, 2123 (2014).
16. Xiong L, Zhao Q, Chen H, Wu Y, Dong Z, Zhou Z, and Li F. Phosphorescence imaging of homocysteine and cysteine in living cells based on a cationic iridium(III) complex. *Inorg. Chem.* **49**, 6402–6408 (2010).
17. Zhao Q, Yu M, Shi L, Liu S, Li C, Shi M, Zhou Z, Huang C, and Li F. Cationic iridium(III) complexes with tunable emission color as phosphorescent dyes for live cell imaging. *Organometallics* **29**, 1085–1091 (2010).
18. Dolan C, Moriarty RD, Lestini E, Devocelle M, Forster RJ, and Keyes TE. Cell uptake and cytotoxicity of a novel cyclometalated iridium(III) complex and its octaarginine peptide conjugate. *J. Inorg. Biochem.* **119**, 65–74 (2013).
19. Mandal S, Poria DK, Ghosh R, Ray PS, and Gupta P. Development of a cyclometalated iridium complex with specific intramolecular hydrogen-bonding that acts as a fluorescent marker for the endoplasmic reticulum and causes photoinduced cell death. *Dalton Trans.* **43**, 17463–17474 (2014).
20. Balasingham RG, Coogan MP, and Thorp-Greenwood FL. Complexes in context: Attempting to control the cellular uptake and localisation of rhenium fac-tricarbonyl polypyridyl complexes. *Dalton Trans.* **40**, 11663 (2011).
21. Choi AW, Liu H, and Lo KK. Rhenium(I) polypyridine dibenzocyclooctyne complexes as phosphorescent bioorthogonal probes: Synthesis, characterization, emissive behavior, and biolabeling properties. *J. Inorg. Biochem.* **148**, 2–10 (2015).
22. Lo KK and Hui W. Design of rhenium(I) polypyridine biotin complexes as a new class of luminescent probes for avidin. *Inorg. Chem.* **44**, 1992–2002 (2005).
23. Puckett CA, Barton JK. Mechanism of cellular uptake of a ruthenium polypyridyl complex. *Biochemistry* **47** (45), 11711–11716 (2008).
24. Barton JK. DNA-mediated signaling. *Biophys J.* **104**(2 Suppl. 1), 2a (2013).
25. Madani F, Lindberg S, Langel U, Futaki S, and Graslund A. Mechanisms of cellular uptake of cell-penetrating peptides. *J Biophys.* **2011**, 1–10 (2011).

26. Doherty GJ and McMahon HT. Mechanisms of endocytosis. *Ann. Rev. Biochem.* **78**, 857–902 (2009).

27. Heitz F, Morris MC, and Divita G. Twenty years of cell-penetrating peptides: From molecular mechanisms to therapeutics. *Br. J. Pharmacol.* **157**, 195–206 (2009).

28. Deshayes S, Morris MC, Divita G, and Heitz F. Cell-penetrating peptides: Tools for intracellular delivery of therapeutics. *Cell Mol. Life Sci.* **62**, 1839–1849 (2005).

29. Puckett CA, Ernst RJ, and Barton JK. Exploring the cellular accumulation of metal complexes. *Dalton Trans.* **39**, 1159–1170 (2010).

30. Dobrucki JW. Interaction of oxygen-sensitive luminescent probes ru(phen) and 3 21 ru(bipy) with animal and plant cells in vitro 3 mechanism of phototoxicity and conditions for non-invasive oxygen measurements. *J. Photochem. Photobiol. B Biol.* **65**, 136–144 (2001).

31. Peacock AFA, Habtemariam A, Fernández R, Walland V, Fabbiani FPA, Parsons S, Aird RE, Jodrell DI, and Sadler PJ. Tuning the reactivity of osmium(II) and ruthenium(II) arene complexes under physiological conditions. *J. Am. Chem. Soc.* **128**, 1739–1748 (2006).

32. Fuchs SM and Rainesa RT. Internalization of cationic peptides: The road less (or more?) travelled. *Cell Mol. Life Sci.* **63**, 1819–1822 (2006).

33. Jones SW et al. Characterisation of cell-penetrating peptide-mediated peptide delivery. *Br. J. Pharmacol.* **145**, 1093–1102 (2005).

34. Li SP, Lau CT, Louie M, Lam Y, Cheng SH, and Lo KK. Mitochondria-targeting cyclometalated iridium(III)–PEG complexes with tunable photodynamic activity. *Biomaterials* **34**, 7519–7532 (2013).

35. Marin V, Holder E, Hoogenbooma R, and Schubert US. Functional ruthenium(II)- and iridium(III)-containing polymers for potential electro-optical applications. *Chem. Soc. Rev.* **36**, 618–635 (2007).

36. Komatsu H et al. Ruthenium complexes with hydrophobic ligands that are key factors for the optical imaging of physiological hypoxia. *Chem. Eur. J.* **19**, 1971–1977 (2013).

37. Lo KK, Lee TK, Lau JS, Poon W, and Cheng S. Luminescent biological probes derived from ruthenium(II) estradiol polypyridine complexes. *Inorg. Chem.* **47**, 200–208 (2008).

38. Arnott JA and Planey SL. The influence of lipophilicity in drug discovery and design. *Expert Opin. Drug Discov.* **7**, 863–875 (2012).

39. Tang TS, Yip AM, Zhang KY, Liu H, Wu PL, Li KF, Cheah KW, and Lo KK. Bioorthogonal labeling, bioimaging, and Photocytotoxicity Studies of phosphorescent ruthenium(II) Polypyridine Dibenzocyclooctyne complexes. *Chem. Eur. J.* **21**, 1–13 (2015).

40. Law WH, Lee LC, Louie M, Liu H, Ang TW, and Lo KK. Phosphorescent cellular probes and uptake indicators derived from cyclometalated iridium(III) bipyridine complexes appended with a glucose or galactose entity. *Inorg. Chem.* **52**, 13029–13041 (2013).

41. Lv G, Guo L, Qiu L, Yang H, Wang T, Liu H, and Lin J. Lipophilicity-dependent ruthenium N-heterocyclic carbene complexes as potential anticancer agents. *Dalton Trans.* **44**, 7324–7331 (2015).

42. Mazuryk O, Magiera K, Rys B, Suzenet F, Kieda C, and Brindell M. Multifaceted interplay between lipophilicity, protein interaction and luminescence parameters of non-intercalative ruthenium(II) polypyridyl complexes controlling cellular imaging and cytotoxic properties. *J. Biol. Inorg. Chem.* **19**, 1305–1316 (2014).

43. Pisani MJ, Weber DK, Heimann K, Collins JG, and Keene FR. Selective mitochondrial accumulation of cytotoxic dinuclear polypyridyl ruthenium(II) complexes. *Metallomics* **2**, 393–396 (2010).

44. Lv G, Guo L, Qiu L, Yang H, Wang T, Liu H, and Lin J. Lipophilicity-dependent ruthenium N-heterocyclic carbene complexes as potential anticancer agents. *Dalton Trans.* **44**, 7324–31 (2015).

45. Tan C et al. Synthesis, structures, cellular uptake and apoptosis-inducing properties of highly cytotoxic ruthenium-norharman complexes. *Dalton Trans.* **40**, 8611–8621 (2011).

46. Pisani MJ, Weber DK, Heimann K, Collins JG, and Keene FR. Selective mitochondrial accumulation of cytotoxic dinuclear polypyridyl ruthenium(II) complexes. *Metallomics* **2**, 393–396 (2010).

47. Thorp-Greenwood FL, Balasingham RG, and Coogan MP. Organometallic complexes of transition metals in luminescent cell imaging applications. *J. Org. Chem.* **714**, 12–21 (2012).

48. Dickerson M, Sun Y, Howerton B, and Glazer EC. Modifying charge and hydrophilicity of simple ru(II) polypyridyl complexes radically alters biological activities: Old complexes, surprising new tricks. *Inorg. Chem.* **53**, 10370–10377 (2014).

49. Svensson FR, Matson M, Li M, and Lincoln P. Lipophilic ruthenium complexes with tuned cell membrane affinity and photoactivated uptake. *Biophys. Chem.* **3**, 102–106 (2010).

50. Mandal S, Poria DK, Ghosh R, Ray PS, and Gupta P. Development of a cyclometalated iridium complex with specific intramolecular hydrogen-bonding that acts as a fluorescent marker for the endoplasmic reticulum and causes photoinduced cell death. *Dalton Trans.* **43**, 17463–17471 (2014).

51. Ross MF, Filipovska A, Smith RA, Gait MJ, and Murphy MP. Cell-penetrating peptides do not cross mitochondrial membranes even when conjugated to a lipophilic cation: Evidence against direct passage through phospholipid bilayers. *Biochem. J.* **383**, 457–468 (2004).

52. Mahon KP, Potocky TB, Blair D, Roy MD, Stewart KM, Chiles TC, and Kelley SO. Deconvolution of the cellular oxidative stress response with organelle-specific peptide conjugates. *Chem. Biol.* **14**, 923–930 (2007).

53. Kostrhunova H, Florian J, Novakova O, Peacock AFA, Sadler PJ, and Brabec V. DNA interactions of monofunctional organometallic osmium(II) antitumor complexes in cell-free media. *J. Med. Chem.* **51**, 3635–3643 (2008).

54. Svensson FR, Matson M, Li M, and Lincoln P. Lipophilic ruthenium complexes with tuned cell membrane affinity and photoactivated uptake. *Biophys. Chem.* **3**, 102–106 (2010).

55. Brooks H, Lebleu B, and Vivès E. Tat peptide-mediated cellular delivery: Back to basics. *Adv. Drug Deliv. Rev.* **4**, 559–577 (2005).

56. Neugebauer U, Pellegrin, Y, Devocelle, M, Forster, RJ Signac, W, Moran N, and Keyes TE. Ruthenium polypyridyl peptide conjugates: Membrane permeable probes for cellular imaging. *Chem. Commun.* **42**, 5307–5309 (2008).

57. Raszeja L, Maghnouj A, Hahn S, and Metzler-Nolte NA. Novel organometallic ReI complex with favourable properties for bioimaging and applicability in solid-phase peptide synthesis. *ChemBioChem.* **12**, 371–376 (2011).

58. Wang T, Zabarska N, Wu YZ, Lamla M, Fischer S, Monczak K, Ng DYW, Rau S, and Weil T. Receptor selective ruthenium-somatostatin photosensitizer for cancer targeted photodynamic applications. *Chem. Commun.* **51**, 12552–12555 (2015).

59. Zhao R, Hammitt R, Thummel RP, Liu Y, Turro C, and Snapka RM. Nuclear targets of photodynamic tridentate ruthenium complexes. *Dalton Trans.* **48**, 10926–10931 (2009).

60. Escriou V, Carrière M, Scherman D, and Wils P. NLS bioconjugates for targeting therapeutic genes to the nucleus. *Adv. Drug Deliv. Rev.* **2**, 295–306 (2003).

61. Ragin AD, Morgan RA, and Chmielewski J. Cellular import mediated by nuclear localization signal peptide sequences. *Chem. Biol.* **8**, 943–948 (2002).

62. Brandén LJ, Mohamed AJ, and Smith CIE. A peptide nucleic acid–nuclear localization signal fusion that mediates nuclear transport of DNA. *Nat. Biotechnol.* **17**, 784–787 (1999).

63. Sturzua A, Regenbogena M, Klosea U, Echnerc H, Gharabaghib A, and Heckla S. Novel dual labelled nucleus-directed conjugates containing correct and mutant nuclear localisation sequences. *European Journal of Pharmaceutical Sciences.* **33**, 207–216 (2008).

64. Blackmore L, Moriarty R, Dolan C, Adamson K, Forster RJ, Devocelle M, and Keyes TE. Peptide directed transmembrane transport and nuclear localization of ru(II) polypyridyl complexes in mammalian cells. *Chem. Commun.* **49**, 2658 (2013).

65. Boulikas T. Putative nuclear localization signals (NLS) in protein transcription factor. *J. Cell Biochem.* **55**, 32–58 (1994).

66. Ragin AD, Morgan RA, and Chmielewski J. Cellular import mediated by nuclear localization signal peptide sequences. *Chem. Biol.* **8**, 943–948 (2002).

67. Duchen MR and Szabadkai G. Roles of mitochondria in human disease. *Essays Biochem.* **47**, 115–137 (2010).

68. Tiede LM, Cook E A, Morsey B, and Fox HS. Oxygen matters: Tissue culture oxygen levels affect mitochondrial function and structure as well as responses to HIV viroproteins. *Cell Death Dis.* **2**, 246 (2011).

69. Galluzzi L, Larochette N, Zamzami N, and Kroemer G. Mitochondria as therapeutic targets for cancer chemotherapy. *Oncogene* **25**, 4812–4830 (2006).

70. Green K, Brand MD, and Murphy MP. Prevention of mitochondrial oxidative damage as a therapeutic strategy in diabetes. *Diabetes* **53**, 110–118 (2004).

71. Tiede LM, Cook EA, Morsey B, and Fox HS. Oxygen matters: Tissue culture oxygen levels affect mitochondrial function and structure as well as responses to HIV viroproteins. *Cell Death Dis.* **3**, 274 (2012).

72. Pisani MJ, Weber DK, Heimann K, Collins JG, and Keene FR. Selective mitochondrial accumulation of cytotoxic dinuclear polypyridyl ruthenium(II) complexes. *Metallomics* **2**, 393–396 (2010).

73. Kondrashina AV et al. A phosphorescent nanoparticle-based probe for sensing and imaging of (intra)cellular oxygen in multiple detection modalities. *Adv. Funct. Mater.* **22**, 4931–4939 (2012).

74. Frantz M and Wipf P. Mitochondria as a target in treatment. *Environ. Mol. Mutagen.* **51**, 462–475 (2010).

75. Heller A, Brockhoff G, and Goepferich A. Targeting drugs to mitochondria. *Eur. J. Pharm. Biopharm.* **82**, 1–18 (2012).

76. Malhi SS and Murthy RSR. Delivery to mitochondria: A narrower approach for broader therapeutics. *Expert Opin. Drug Deliv.* **9**, 909–935 (2012).

77. Horton KL, Stewart KM, Fonseca SB, Guo G, and O'Kelley S. Mitochondriapenetrating peptides. *Chem. Biol.* **15**, 375–382 (2008).

78. Yousif LF, Stewart KM, and Kelley SO. Targeting mitochondria with organelle-specific compounds: Strategies and applications. *ChemBioChem.* **10**, 1939–1950 (2009).

79. Desgrosellier JS and Cheresh DA. Integrins in cancer: Biological implications and therapeutic opportunities. *Nat. Rev. Cancer* **10**, 9–22 (2010).

80. Adamson K, Devocelle M, Moran N, Forster RJ, and Keyes TE. RGD peptide labelled metal complex probes reveal activation status of integrin from fluorescence anisotropy. *Bioconjug. Chem.* **2014**, 25, 928.

81. Wang T, Zabarska N, Wu Y, Lamla M, Fischer S, Monczak K, Ng DK, Rau S, and Weil T. Receptor selective ruthenium-somatostatin photosensitizer for cancer targeted photodynamic applications. *Chem. Commun.* **51**, 12552–12555 (2015).

82. Kikkeri R, García-Rubio I, and Seeberger PH. Ru(II)–carbohydrate dendrimers as photoinduced electron transfer lectin biosensors. *Chem. Commun.* **2**, 235–237 (2009).

83. Louie M, Liu H, Lam, MH, Lam Y, and Lo KK. Luminescent rhenium(I) polypyridine complexes appended with an a-d-glucose moiety as novel biomolecular and cellular Probes. *Chem. Eur. J.* **17**, 8304–8308 (2011).

84. Law WH, Lee LC, Louie M, Liu H, Ang TW, and Lo KK. Phosphorescent cellular probes and uptake indicators derived from cyclometalated iridium(III) bipyridine complexes appended with a glucose or galactose entity. *Inorg Chem.* **52**, 13029–13041 (2013).

85. Lo KK-W, Hui W-K, and Ng DCM. Novel Rhenium(I) polypyridine biotin complexes that show luminescence enhancement and lifetime elongation upon binding to avidin. *J. Am. Chem. Soc.* **124**, 9344–9345 (2002).

86. Langdon-Jones EE et al. Fluorescent rhenium-naphthalimide conjugates as cellular imaging agents. *Inorg. Chem.* **53**, 3788–3797 (2014).

87. Balakrishnan G et al. Interaction of rhenium(I) complex carrying long alkyl chain with calf thymus DNA: Cytotoxic and cell imaging studies. *Inorg. Chim. Acta* **434**, 51–59 (2015).

88. Amoroso AJ, Arthur RJ, Coogan MP, Court JB, Fernandez-Moreira V, Hayes AJ, Lloyd D, and Millet C, Pope SJA. 3-Chloromethylpyridyl bipyridine fac-tricarbonyl rhenium: A thiol-reactive luminophore for fluorescence microscopy accumulates in mitochondria. *New J. Chem.* **32**, 1097–1102 (2008).

89. Marie E, Sagan S, Cribier S, and Tribet CJ. Amphiphilic macromolecules on cell membranes: From protective layers to controlled permeabilization. *Membr. Biol.* **247**, 861–81 (2014).

90. Knop K, Hoogenboom R, Fischer D, and Schubert US. Poly(ethylene glycol) in drug delivery: Pros and cons as well as potential alternatives. *Angew. Chem. Int. Ed.* **49**, 6288–6308 (2010).

91. Samanta D, Kratz K, Zhang, X, and Emrick T. A synthesis of PEG- and phosphorylcholine-substituted pyridines to afford water-soluble ruthenium benzylidene metathesis catalysts. *Macromolecules* **41**, 530–532 (2008).

92. Li SPY, Liu HW, Zhang KY, and Lo KK. Modification of luminescent iridium(III) polypyridine complexes with discrete poly(ethylene glycol) (PEG) pendants: Synthesis, emissive behavior, intracellular uptake, and PEGylation properties. *Chem. Eur. J.* **16**, 8329–8339 (2010).

93. Lo KK. Luminescent rhenium(I) and iridium(III) polypyridine complexes as biological probes, imaging reagents, and photocytotoxic agents. *Acc. Chem. Res.* **48**, 2985–2995 (2015).

94. Li S, LC, Louie M, Lam Y, Cheng S, and Lo KK. Mitochondria-targeting cyclometalated iridium(III)–PEG complexes with tunable photodynamic activity. Biomaterials **34**, 7519–7532 (2013).

95. Li SP, Lau CT, Louie MW, Lam YW, Cheng SH, and Lo KK. Mitochondria-targeting cyclometalated iridium(III)–PEG complexes with tunable photodynamic activity. *Biomaterials* **34** (30), 7519–7532 (2013).

96. Burns A, Owb H, and Wiesner U. Fluorescent core–shell silica nanoparticles: Towards "Lab on a Particle" architectures for nanobiotechnology. *Chem. Soc. Rev.* **35**, 1028–1042 (2006).

97. Yang ZG, Cao JF, He YX, Yang JH, Kim T, Peng XJ, and Kim JS. Macro-/microenvironment-sensitive chemosensing and biological imaging. *Chem. Soc. Rev.* **43**, 4563–4601 (2014).

98. Cao L, Zhang R., ZhangW, Du Z, Liu C, Ye Z, Song B, and Yuan J. A ruthenium(II) complex-based lysosome-targetable multisignal chemosensor for in vivo detection of hypochlorous acid. *Biomaterials* **68**, 21–31, 2015.

99. Lakowicz JR, Szmacinski H, Nowaczyk K, Lederer WJ, Kirby MS, and Johnson ML. Fluorescence lifetime imaging of intracellular calcium in COS cells using quin-2. *Cell Calcium* **15**, 7–27 (1994).

100. Levitt JA, Kuimova MK, Yahioglu G, Chung P, Suhling K, and Phillips D. Membrane-bound molecular rotors measure viscosity in live cells via fluorescence lifetime imaging. *J. Phys. Chem. C* **113**, 11634–11642 (2009).
101. Ryder AG, Glynn TJ, Przyjalgowski M, and Szczupak B. A compact violet diode laser-based fluorescence lifetime microscope. *J. Fluoresc.* **12**, 177–180 (2002).
102. Lakowicz JR. *Principles of Fluorescence Spectroscopy*, Vol. 954, 3rd edn. Springer, 741–755 (2006).
103. Becker W. Fluorescence lifetime imaging–techniques and applications. *J Microsc.* **2**, 119–36 (2012).
104. Levitt JA, Kuimova MK, Yahioglu G, Chung P, Suhling K, and Phillips D. Membrane-bound molecular rotors measure viscosity in live cells via fluorescence lifetime imaging. *J Phys Chem C* **113**, 11634–11642 (2009).
105. Sarder P, Maji D, and Achilefu S. Molecular probes for fluorescence lifetime imaging. *Bioconjug. Chem.* **26**, 963–974 (2015).
106. Adamson K, Dolan C, Moran N, ForsterRJ, and Keyes TE. RGD labeled Ru(II) polypyridyl conjugates for platelet integrin alpha(IIb)beta(3) recognition and as reporters of integrin conformation. *Bioconj. Chem.* **25**, 928–944 (2014).
107. Byrne A, Dolan C, Moriarty RD, Martin A, Neugebauer U, Forster RJ, Davies A, Volkov Y, and Keyes TE. Osmium(II) polypyridyl polyarginine conjugate as a probe for live cell imaging; a comparison of uptake, localization and cytotoxicity with its ruthenium(II) analogue. *Dalton Trans.* **44**, 14323–14332 (2015).
108. Bertout JA, Patel SA, and SimonMC. The impact of O_2 availability on human cancer. *Nat. Rev. Cancer* **8**, 967–975 (2008).
109. Ruggi A, van Leeuwen FWB, and Velders AH. Interaction of dioxygen with the electronic excited state of Ir(III) and Ru(II) complexes: Principles and biomedical applications. *Coord. Chem. Rev.* **255**, 2542–2554 (2011).
110. (a) Demas JN, Diemente D, and Harris EW. Oxygen quenching of charge-transfer excited states of ruthenium(II) complexes. Evidence for singlet oxygen production. *J. Am. Chem. Soc.* **95**, 6864–6865 (1973); (b) Demas JN, Harris EW, Flynn CM, and Diemente D. Luminescent osmium(II) and iridium(III) complexes as photosensitizers. *J. Am. Chem. Soc.* **97**, 3838–3839 (1975). (c) Demas JN, Harris EW, and McBride RP. Energy transfer from luminescent transition metal complexes to oxygen. *J. Am. Chem. Soc.* **99**, 3547–3551 (1977).
111. Zhong W, Urayama P, and Mycek M. Imaging fluorescence lifetime modulation of a ruthenium-based dye in living cells: The potential for oxygen sensing. *J. Phys. D Appl. Phys.* **36**, 1689–1695 (2003).
112. Martin A, Byrne A, Burke CS, Forster RJ, and Keyes TE. Peptide-bridged dinuclear Ru(II) complex for mitochondrial targeted monitoring of dynamic changes to oxygen concentration and ROS generation in live mammalian cells. *J. Am. Chem. Soc.* **136**, 15300–15309 (2014).
113. Cosgrave L, Neugebauer U, Forster RJ, and Keyes TK. Multimodal cell imaging by ruthenium polypyridyl labelled cell penetrating peptides. *Chem. Commun.* **46**, 103–105 (2010).
114. Jamieson ER and Lippard SJ. Structure, recognition, and processing of cisplatin-DNA adducts. *Chem. Rev.* **99**, 2467–2498 (1999).
115. Medici S et al. Noble metals in medicine: Latest advances. *Coord. Chem. Rev.* **284**, 329–350 (2015).
116. Brandt WW, Dwyer FP, and Gyarfas ED. Chelate complexes of 1,10-phenanthroline and related compounds. *Chem. Rev.* **54**, 959–1017 (1954).
117. D'Aurora V, Stern AM, and Sigman DS. Inhibition of *E. coli* DNA polymerase I by 1,10-phenanthroline. *Biochem. Biophys. Res. Commun.* **78**, 170–176 (1977).

118. (a) Yamagishi, A. Evidence for stereospecific binding of tris(1,10-phenanthroline)-ruthenium(II) to DNA is provided by electronic dichroism. *J. Chem. Soc. Chem. Commun.* 572–573 (1983). (b) Yamagishi A. Electric dichroism evidence for stereospecific binding of optically active tris chelated complexes to DNA. *J. Phys. Chem.* **88**, 5709–5713 (1984).

119. (a) Barton JK, Dannenberg JJ, and Raphael AL. Enantiomeric selectivity in binding tris(phenanthroline)zinc(II) to DNA. *J. Am. Chem. Soc.* **104**, 4967–4969 (1982); (b) Barton JK, Danishefsky A, and Goldberg J. Tris(phenanthroline)ruthenium(II): Stereoselectivity in binding to DNA. *J. Am. Chem. Soc.* **106**, 2172–2176 (1984).

120. (a) Barton JK, Goldberg JM, Kumar CV, and Turro NJ. Binding modes and base specificity of tris(phenanthroline)ruthenium(II) enantiomers with nucleic acids: Tuning the stereoselectivity. *J. Am. Chem. Soc.* **108**, 2081–2088 (1986); (b) Hiort C, Norden B, and Rodger A. Enantiopreferential DNA binding of [ruthenium(II) (1,10-phenanthroline)3]2+ studied with linear and circular dichroism. *J. Am. Chem. Soc.* **112**, 1971–1982 (1990); (c) Rehmann JP and Barton JK. Proton NMR studies of tris(phenanthroline) metal complexes bound to oligonucleotides: Structural characterizations via selective paramagnetic relaxation. *Biochemistry* **29**, 1710–1717 (1990); (d) Haworth IS, Elcock AH, Freeman J, Rodger A, and Richards WG. Sequence selective binding to the DNA major groove: Tris(1,10-phenanthroline) metal complexes binding to poly(dG-dC) and poly(dA-dT). *J. Biomol. Struct. Dyn.* **9**, 23–44 (1991); (e) Eriksson M, Leijon M, Hiort C, Norden B, and Graeslund A. Minor groove binding of [Ru(phen)3]2+ to [d(CGCGATCGCG)]2 evidenced by two-dimensional NMR. *J. Am. Chem. Soc.* **114**, 4933–4934 (1992); (f) Satyanarayana S, Dabrowiak JC, and Chaires JB. Neither.DELTA.-nor.LAMBDA.-tris(phenanthroline)ruthenium(II) binds to DNA by classical intercalation. *Biochemistry* **31**, 9319–9324 (1992); (g) Satyanarayana S, Dabrowiak JC, and Chaires JB. Tris(phenanthroline)ruthenium(II) enantiomer interactions with DNA: Mode and specificity of binding. *Biochemistry* **32**, 2573–2584 (1993); (h) Eriksson M, Leijon M, Hiort C, Norden, B, and Graeslund A. Binding of.DELTA.- and LAMBDA.-[Ru(phen)3]2+ to [d(CGCGATCGCG)]2 Studied by NMR. *Biochemistry* **33**, 5031–5040 (1994); (i) Lincoln P and Nordén B. DNA binding geometries of ruthenium(II) complexes with 1,10-Phenanthroline and 2,2'-bipyridine ligands studied with linear dichroism spectroscopy. Borderline cases of intercalation. *J. Phys. Chem. B* **102**, 9583–9594 (1998).

121. (a) Barton JK, Basile LA, Danishefsky A, and Alexandrescu A. Chiral probes for the handedness of DNA helices: Enantiomers of tris(4,7-diphenylphenanthroline) ruthenium(II). *Proc. Natl. Acad. Sci. USA* **81**, 1961–1965 (1984); (b) Barton JK. Metals and DNA: Molecular left-handed complements. *Science* **233**, 727–734 (1986);

122. Amouyal E, Homsi A, Chambron J-C, and Sauvage J-P. Synthesis and study of a mixed-ligand ruthenium(II) complex in its ground and excited states: bis(2,2'-bipyridine) (dipyrido[3,2-a : 2',3'-c]phenazine-N4N5)ruthenium(II). *J. Chem. Soc. Dalton Trans.* **6**, 1841–1845 (1990).

123. Friedman AE, Chambron JC, Sauvage JP, Turro NJ, and Barton JK. A molecular light switch for DNA: Ru (bpy) 2 (dppz) 2+. *J. Am. Chem. Soc.* **112**, 4960–4962 (1990).

124. (a) Sitlani A, Long EC, Pyle AM, and Barton JK. DNA photocleavage by phenanthrenequinone diimine complexes of rhodium(III): Shape-selective recognition and reaction. *J. Am. Chem. Soc.* **114**, 2303–2312 (1992); (b) Sitlani A and Barton JK. Sequence-specific recognition of DNA by phenanthrenequinone diimine complexes of rhodium(III): Importance of steric and van der waals interactions. *Biochemistry* **33**, 12100–12108 (1994); (c) Jackson BA and Barton JK. Recognition of DNA base mismatches by a rhodium intercalator. *J. Am. Chem. Soc.* **119**, 12986–12987 (1997); (d) Junicke H et al. A rhodium(III) complex for high-affinity DNA base-pair mismatch recognition. *Proc. Natl. Acad. Sci. USA* **100**, 3737–3742 (2003).

125. (a) Pascu GI, Hotze ACG, Sanchez-Cano C, Kariuki BM, and Hannon MJ. Dinuclear ruthenium(II) triple-stranded helicates: Luminescent supramolecular cylinders that bind and coil DNA and exhibit activity against cancer cell lines. *Angew. Chem. Int. Ed.* **46**, 4374–4378 (2007); (b) Malina J, Hannon MJ, and Brabec V. Interaction of dinuclear ruthenium(II) supramolecular cylinders with DNA: Sequence-specific binding, unwinding, and photocleavage. *Chem. Eur. J.* **14**, 10408–10414 (2008); (c) McDonnell U, Hicks MR, Hannon MJ, and Rodger A. DNA binding and bending by dinuclear complexes comprising ruthenium polypyridyl centres linked by a bis(pyridylimine) ligand. *J. Inorg. Biochem.* **102**, 2052–2059 (2008).

126. (a) Lecomte JP, Mesmaeker AK-D, Kelly JM, Tossi AB, and Görner H. Photo-induced electron transfer from nucleotides to ruthenium-*tris*-1,4,5,8tetraazaphenanthrene: Model for photosensitized DNA oxidation. *Photochem. Photobiol.* **55**, 681–689 (1992); (b) Feeney MM, Kelly JM, Tossi AB, Mesmaeker AK, and Lecomte JP. Photoaddition of ruthenium(II)-tris-1,4,5,8-tetraazaphenanthrene to DNA and mononucleotides. *J. Photochem. Photobiol. B Biol.* **23**, 69–78 (1994); (c) Jacquet L, Davies RJH, Kirsch-De Mesmaeker A, and Kelly JM. Photoaddition of Ru(tap)2(bpy)2+ to DNA: A new mode of covalent attachment of metal complexes to duplex DNA. *J. Am. Chem. Soc.* **119**, 11763–11768 (1997); (d) Moucheron C, Kirsch-De Mesmaeker A, and Kelly JM. Photoreactions of ruthenium (II) and osmium (II) complexes with deoxyribonucleic acid (DNA). *J. Photochem. Photobiol. B* **40**, 91–106 (1997); (e) Pauly M et al. In vitro inhibition of gene transcription by novel photo-activated polyazaaromatic ruthenium(II) complexes. *Chem. Commun.* **21**, 1086–1087 (2002). (f) Le Gac S et al. A photoreactive ruthenium(II) complex tethered to a guanine-containing oligonucleotide: A biomolecular tool that behaves as a 'seppuku molecule'. *Angew. Chem.* **121**, 1142–1145 (2009); (g) For a review of the work: Elias B and Kirsch-De Mesmaeker A. Photo-reduction of polyazaaromatic Ru(II) complexes by biomolecules and possible applications. *Coord. Chem. Rev.* **250**, 1627–1641 (2006).

127. Gill MR et al. A ruthenium(II) polypyridyl complex for direct imaging of DNA structure in living cells. *Nat. Chem.* **1**, 662–667 (2009).

128. Baggaley E et al. Dinuclear ruthenium(II) complexes as two-photon, time-resolved emission microscopy probes for cellular DNA. *Angew. Chem.* **126**, 3435–3439 (2014).

12 Three-Dimensional Electrical Impedance Tomography for Pulmonary Embolism
A Simulation Study of a 16-Electrode System

D. Trang Nguyen, M.A. Barry, R. Kosobrodov,
A. Thiagalingam, and Alistair L. McEwan

CONTENTS

12.1 Principles of Electrical Impedance Tomography 256
    12.1.1 Forward Problem 256
    12.1.2 Inverse Problem 257
    12.1.3 Advantages and Disadvantages of EIT..258
12.2 EIT for Perfusion Defect Detection 259
12.3 Methods 261
    12.3.1 Finite Element Models of the Human Thorax 261
    12.3.2 Experiment Paradigm...262
    12.3.3 Inverse Solving: Image Reconstruction...262
    12.3.4 Evaluation of Results 265
12.4 Results...265
    12.4.1 Images..265
    12.4.2 Amplitude Difference...266
12.5 Discussion..269
12.6 Conclusion 272
References..273

12.1   PRINCIPLES OF ELECTRICAL IMPEDANCE TOMOGRAPHY

Electrical impedance tomography (EIT) is a medical imaging modality that typically uses a single ring of external electrodes to image the impedance changes within the body. Impedance refers to the complex value that is used in electrical engineering to describe the electrical properties, resistance and reactance, of an object. In a biological system such as the human body, electrolytes such as Na⁺, Ka⁺, and Cl⁻ are abundant. The distribution of these ions in the body results in the different electrical impedances of each tissue and by extension each organ of the body. Physiological processes as fundamental as breathing or circulation of blood change the electrical impedance of the corresponding anatomy, which can be imaged using EIT. The estimation of internal impedance is achieved by the excitation of the electric field in the body using a small electrical current at high frequency, typically less than 5 mA at 50 kHz through two of the 16 or 32 surface electrodes. Changes in voltages at the same frequency (50 kHz) can be measured via the rest of the surface electrodes. Mathematical modeling can then be used to infer the internal distribution of impedance through these voltage measurements (Figure 12.1).

12.1.1   FORWARD PROBLEM

In order to understand the principles of EIT, it is necessary to revisit the Maxwell's theory of electromagnetism, applying to electric potential of a volume. The forward problem deals with the solution to the question: "What is the distribution of the electrical potential when the internal conductivity and applied current are known?"

Considering a homogenous volume of Ohmic conductor (Ω), when a current I is injected to the body, a scalar potential field ϕ is generated, which has a gradient opposite to the electrical field $\mathbf{E}$ [3]:

$$\mathbf{E} = -\nabla\phi \qquad (12.1)$$

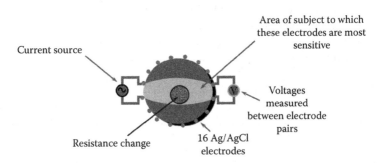

FIGURE 12.1   The 2D electrical impedance tomography concept. Electrical impedance is measured between the current source pair and voltage pair electrodes, which are sensitive to internal impedance or resistance changes. With a 16-channel system, this 4-electrode measurement is switched between 96 different electrode combinations from the 16 surface electrodes.

Furthermore, Ohm's law states that

$$\mathbf{J} = -\sigma\mathbf{E} \tag{12.2}$$

where
$\mathbf{J}$ is the current density, resulting from the injection of current I
σ is the conductivity distribution [3]:

$$\mathbf{J} = -\sigma\mathbf{E} = \sigma\nabla\phi \tag{12.3}$$

If the volume contains no current source, by Kirchhoff's law, the net flux of charge out of the boundary of the volume must be 0. In other words, the divergence of the current density $\mathbf{J}$ is equal to 0 [3]. Thus,

$$\nabla\cdot(\mathbf{J}) = \nabla\cdot(\sigma\nabla\phi) = 0 \tag{12.4}$$

Expanding this equation leads to the well-known Poisson's equation [3]:

$$\nabla^2\phi = -\frac{\nabla\sigma\cdot\nabla\phi}{\sigma} \tag{12.5}$$

This equation describes the relationship between the potential field and the conductivity. The solution for the distribution of ϕ through the domain given the distribution of σ is the solution to the so-called forward problem. Given a simple enough geometry, Equation 12.5 can be solved analytically. However, in reality, with more complicated and irregular geometry, the forward problem is often solved by adaptation of a numerical technique such as finite element modeling (FEM) method. With the FEM method, the volume is discretized into elements of homogeneous conductivity. The forward problem is then solved for each element. Electric potential at the edges of each elements is governed by Maxwell's equations such that the continuity of electric potential ϕ and current density $\mathbf{J}$ is maintained [2].

Forward modeling is an imperative step in most EIT reconstruction algorithms, which first solve for the boundary voltages based on an initial assumptions (a priori) of the conductivity distribution and then gradually improve this solution until the simulated voltages are within a reasonable range of the real measurements. Therefore, if the boundary shape of the forward model is considerably different from the actual object, for example, a cylindrical FEM and the human thorax, it is not probable for even the best algorithm to produce an accurate result [10].

12.1.2 INVERSE PROBLEM

In difference EIT, the vector of conductivity change $\mathbf{x} = \sigma-\sigma_r$ of all of the elements in the FEM is determined from the change on the boundary voltage measurements

$\mathbf{y} = \mathbf{y} - \mathbf{y}_r$, where σ_r and $\mathbf{y}_r$ are the conductivity and boundary voltage of the reference frame, respectively. For sufficiently small changes and if one assumes isotropic conductivity distribution, $\mathbf{x}$ and $\mathbf{y}$ are approximately linearly related:

$$\mathbf{y} = \mathbf{Ax} + \mathbf{n} \qquad (12.6)$$

where $\mathbf{A}$ is the Jacobian matrix, for each element of the FEM, $\mathbf{A}_{ij} = \partial y_i / \partial x_j$, and $\mathbf{n}$ is the measurement (white) noise [2].

The problem is severely ill-posed because $\mathbf{x}$ is a much larger vector than $\mathbf{y}$. For example, with the true thorax shape of the first sheep used in subsequent experiments, there were 2.5×10^4 elements in the FEM where there were only 256 boundary voltage measurements (16 channel system) for each frame. Thus, A is a not a square matrix and therefore noninvertible.

A number of linear reconstruction algorithms, instead, estimate $\mathbf{x}$ using a reconstruction matrix $\mathbf{R}$:

$$\hat{\mathbf{x}} = \mathbf{Ry} \qquad (12.7)$$

For Graz consensus reconstruction algorithm for EIT (GREIT), which performs exceedingly well for ventilation data, the reconstruction matrix $\mathbf{R}$ is calculated as [1]

$$\mathbf{R} = \tilde{\mathbf{X}}\mathbf{Y}\left(\mathbf{J}\Sigma_x\mathbf{A}^T + \lambda\Sigma_n\right)^{-1} \qquad (12.8)$$

where

- $\tilde{\mathbf{X}} = 1/n(\tilde{x}^{(1)} \dots \tilde{x}^{(n)})$ is a matrix calculated from the horizontal concatenation of n desired solution
- $\tilde{\mathbf{Y}} = 1/n(\tilde{y}^{(1)} \dots \tilde{y}^{(n)})$ is the horizontal concatenation matrix of n simulated measurements
- λ is a hyperparameter that controls the amount of regularization
- Σ_x^{-1} is the image covariance matrix
- Σ_n^{-1} is the noise covariance matrix

12.1.3 ADVANTAGES AND DISADVANTAGES OF EIT

EIT has been applied for imaging the lung, heart, brain, gastrointestinal tract, and breast, to name just a few [11]. Among these, EIT has been extensively studied and proved to have the ability to detect both physiologic and pathophysiologic changes of regional pulmonary ventilation [4]. This application of EIT has recently been described as being on the verge of clinical application for pulmonary monitoring with medical device manufactures offering 2D imaging systems for trial [4] (Figure 12.2).

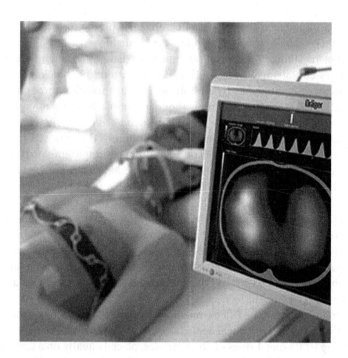

FIGURE 12.2   The PulmoVista500 electrical impedance tomography system from Dräeger (Lübeck, Germany), introduced for bedside continuous lung monitoring.

EIT is a noninvasive, radiation-free, and portable imaging technique, suitable for bedside use, especially in the ICU. Nonetheless, to infer the internal impedance, distribution from limited surface measurements is a *severely ill-posed inverse* problem [2]. Due to the *ill-posed* nature of the inverse problem, EIT has very low spatial resolution compared with other imaging techniques such as MRI or CT. It is also prone to noise, artifacts, and anatomical distortion.

These anatomical errors in EIT images are a result of the mismatch between the FEM used in forward modeling. In [10], it was shown that mismatch in shape between the true anatomy and the FEM *has a strong detrimental effect* on the image quality. In fact, mismatch as small as 5% is observable on the resulting images in the case of ventilation EIT.

12.2 EIT FOR PERFUSION DEFECT DETECTION

From a physiological point of view, pulmonary perfusion is as important as lung ventilation as both elements are necessary to sufficiently describe the regional behavior of lung tissues for applications such as the adjustment of ventilator settings [8]. As the EIT signal of the thorax contains both the ventilation and the perfusion information, it is possible to independently and concurrently detect regional changes due to these two processes. However, the significantly smaller amplitude of the perfusion impedance change signal compared with the respiratory impedance change signal renders perfusion mapping very difficult.

A number of studies have been attempted to detect perfusion defects caused by pulmonary embolism (PE) using EIT. PE one or multiple blockages of the pulmonary artery or its branches by emboli, which have traveled through the bloodstream. Acute PE is life-threatening, requiring immediate attention of medical professionals to begin appropriate treatment such as anticoagulant and thrombolytic therapy [7]. PE can be diagnosed by combining the patient's medical history and ventilation/perfusion scans or CT pulmonary angiography. Both of these imaging techniques expose patients to radiation and are not preferred especially for critically unwell patients. EIT experiments to detect PEs [12,13] showed promising results. Both studies showed that EIT was able to detect regions of the lung where perfusion was disrupted, while ventilation remained normal, indicating PE. However, the images published in their papers showed large emboli. A more recent study in pigs, simulated emboli by inflating or deflating a catheter balloon, showed that with the aid of a contrast agent, NaCl, injected intravenously, perfusion distribution acquired by EIT was comparable to the results obtained from electron beam CT [7].

The current resolution of EIT is very low, rendering its sensitivity for perfusion defects not yet suitable for clinical application. In previous work, we have shown that in performing EIT using 4 rings of 16 electrodes, placed equally on a simulated realistic FEM thorax model, pulmonary perfusion defects larger than 5% lung volume are observable. However, our EIT system, KHU Mark 2.5, like most currently available systems, is currently constrained to 16 electrodes, albeit, with parallel hardware implementations for each channel, 3D EIT [15] is practically available. Thus, pragmatically, we pose the following questions:

1. Is it possible to detect a reasonably small perfusion defect with this system (16 electrodes)?
2. Would arranging electrodes in two rings of 8 increase the sensitivity of EIT in terms of detecting perfusion anomaly? Arranging electrodes in more than one cross-sectional plane would capture more information in the craniocaudal (head-to-toe) direction [15]. For PE detection, since there is no prior knowledge on the location of the affected lung area, we hypothesize that this would be more advantageous than the traditional one-ring arrangement of electrodes.

In this chapter, we present the simulation results obtained numerically on a realistic 2.5D FEM of the human thorax with various sizes of perfusion anomaly. The results for two different electrode configurations, namely, one ring of 16 electrodes and two rings of 8 electrodes, are presented and compared. The study is also extended to study the response of each of the two electrode configurations not only to different sizes of anomalies but also to the distance between the center of these anomalies and the one ring of electrode or to the mean vertical distance between the two rings of electrodes. This is to study the efficacy of each electrode configuration in detecting out-of-plane anomalies.

12.3 METHODS

This research used EIDORS,* an open-source package for MATLAB®† that provides algorithms for forward and inverse modeling for EIT. Details on the usage of EIDORS can be found in [2].

12.3.1 FINITE ELEMENT MODELS OF THE HUMAN THORAX

A protruded model of the human thorax, as segmented from a human lung CT (Figure 12.3), was created using Netgen [17], through EIDORS interface. The lungs were also created in similar manner. The CT image and protruded meshing with

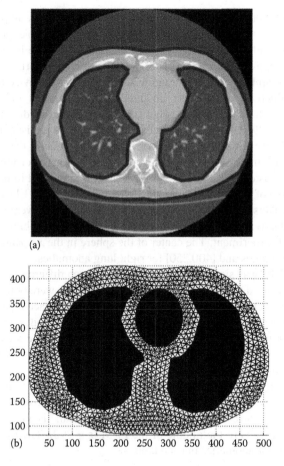

(a)

(b)

FIGURE 12.3   CT lung image used for mesh generation (a) and planar (*xy*) view (b) of the generated finite element modeling mesh for forward solving, showing the difference in conductivity between the heart and the lungs.

---

* http:// eidors3d.sourceforge.net/.
† MathWorks, MA, USA, http://www.mathworks.com.

Netgen are standard functionality within EIDORS. The heart region could not be easily determined from the CT image. Therefore, the heart was represented as an ellipsoid, as shown in Figure 12.4. For the first electrode configuration, one ring of 16 electrodes, all the 16 electrodes were arranged at an equal angular distance from each other. In the second electrode configurations, two rings of 8 electrodes were located at $z = 150$ AU (arbitrary unit) and $z = 250$ AU, approximately 7.2% of the thorax circumference apart. In each ring, electrodes were placed at equal angular distance from each other. Electrodes on the second ring were placed at the same planar (x, y) coordinates as those on the first ring.

12.3.2 EXPERIMENT PARADIGM

The relationship between cardiac activities and the pulmonary impedance change is characterized by the following events. After ventricular systole, pulmonary arteries are filled with blood, resulting in higher conductivity in the lung because blood has higher conductivity than lung tissues. As venous return starts, blood volume in the pulmonary capillary drops, leading to decreasing conductivity in the lung and increasing conductivity in the heart. To simulate these changes, the conductivity of the lungs was increased by 3% of its original value, while the conductivity of the heart was decreased by 3% [6]. In order to make the model more realistic, tissues conductivities at 50 kHz were assigned to each corresponding regions in the model [17] since 50 kHz is the most usually used frequency for lung EIT [5]. The original conductivity of each tissue is listed in Table 12.1. The system was simulated to have a signal-to-noise ratio (SNR) of 85 and 40 dB, which are realistic for recent EIT systems [14].

Perfusion defects such as parts of the lungs affected by PE were simulated by no change in conductivity in the affected volume. The defective volume was defined by a sphere in each experiment. The center of the sphere in the xy-plane was [150,250] for left lung anomalies and [400,250] for right lung anomalies.

Difference EIT was simulated, in which the voltage difference between systole and diastole was used to reconstruct EIT images. Voltages were obtained by simulatively injecting electrical current (1 mA, 50 kHz) to each adjacent pair of electrodes and measuring between the remaining pairs of adjacent electrodes. In the case of two rings, one interring current injection and one interring voltage measurement were performed. Other than that, all other measurements and current injection were the same as in the one-ring case [9].

For each electrode configuration, we perform simulations with different sizes of anomaly (Table 12.2) at a range of vertical (z-axis) distances from the position of the ring of electrode in the case of one ring and from the mean distance of the two rings, which in both cases is in the middle of the FEM ($z = 200$). The same experiments were performed for anomalies in each lung.

12.3.3 INVERSE SOLVING: IMAGE RECONSTRUCTION

In order to avoid *inverse crimes* [2], images were reconstructed on a different mesh from the forward mesh in both cases, one ring of 16 electrodes and two rings of 8 electrodes. In reality, it is rare that the internal lung and heart anatomy is available

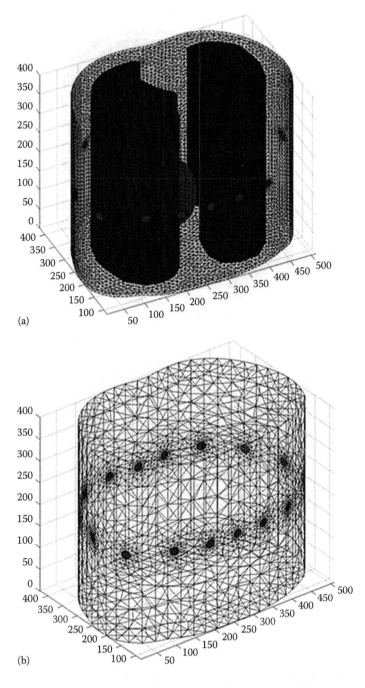

(a)

(b)

FIGURE 12.4 Finite element modeling (FEM) mesh generated from CT image in Figure 6.1, (a) FEM for forward modeling with1 ring of 16 electrodes located at $z = 200$ (256,821 elements), (b) FEM for inverse solving with 1 ring of 16 electrodes located at $z = 200$ (16,667 elements). *(Continued)*

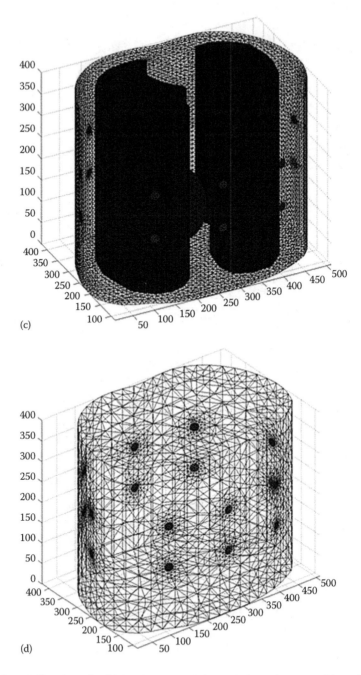

FIGURE 12.4 (*Continued*)   Finite element modeling (FEM) mesh generated from CT image in Figure 6.1, (c) FEM for forward modeling with 2 rings of 8 electrodes located at $z = 150$ and 250 (256,389 elements), (d) FEM for inverse solving with 2 rings of 8 electrodes located at $z = 150$ and 250 (18,184 elements). The lungs and trunk are protruded based on the contours extracted from the CT image in Figure 6.1, while the heart is represented as an ellipsoid.

TABLE 12.1
Conductivities of Each Tissue Type in the Model

Tissue Type	Diastole Conductivity Value (S/m)	Systole Conductivity Value
Lungs	0.1	0.103
Heart	0.19	0.1843
Background	0.02 (for bones) + 0.35 (for muscle)	0.02 (for bones) + 0.35 (for muscle)

TABLE 12.2
Different Sizes of Simulated Anomalies

Anomaly	Radius in the Model (AU)	Volume (AU3)	Ratio to LL (%)	Ratio to RL (%)
1	45	3.8170e+5	3.5	3
2	50	5.2360e+5	4.8	4.1
3	55	6.9691e+5	6.4	5.5
4	60	9.0478e+5	8.3	7.1

Abbreviations: LL, left lung; RL, right lung.

unless CT or MRI is performed on the same subject prior to EIT imaging. For that reason, the FEMs used for inverse solving were based on the extrusion of the trunk only, as shown in Figure 12.4.

Inverse solving is based on EIDORS Gaussian Newton reconstruction algorithm, which is described in detail in [16].

12.3.4 EVALUATION OF RESULTS

In order to quantify the results, a series of test parameters were used, including amplitude, position error, resolution, shape deformation, and ringing effects, which were proposed for the testing of GREIT algorithms [1]. However, due to the highly distorted and low impedance difference, only the amplitude parameter provided meaningful results.

12.4 RESULTS

12.4.1 IMAGES

As depicted in Figures 12.5 and 12.6, using the no-defect case as the reference, in all cases, defects as small as 3% can be discerned. Qualitatively, two rings of 8 electrodes performed better than one ring of 16 electrodes. There are more changes in the reconstructed lungs with two rings of 8 electrodes compared with the same lung when no defect was simulated. The changes are more subtle in the

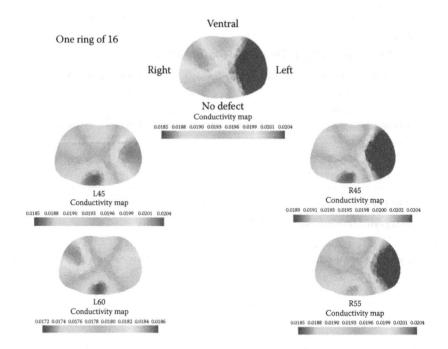

FIGURE 12.5   Image results with one ring of 16 electrodes, L45 and L60 show the reconstructed images when the anomaly is simulated in the left lung (center at [150, 250, 200]) with the radii of 45 AU and 60, respectively, and R45 and R55 show the reconstructed images when the anomaly is simulated in the right lung (center at [400, 250, 200]). The cross-sectional slices are the tomography of the respective reconstructed volume at $z = 200$.

one ring of 16 electrodes case. It is important to note that the results shown in Figures 12.5 and 12.6 were reconstructed results in the cases where the anomalies are on the same z-coordinate ($z = 200$) as the ring of electrodes in the case of one ring and in the middle of the two electrode rings in the case of two rings. The amount of difference between the defective perfusion cases and the normal case worsens as the anomalies are moved further away from the electrode ring(s), as shown in Figure 12.7.

The distortions of the anatomical structures of the heart and lungs are quite high; however, this is expected of EIT. Visually, the reconstructed images in the case of two rings of 8 electrodes showed better distinction between the left lung and the heart, which appears to be blended into one another when using only one ring of electrodes.

12.4.2   AMPLITUDE DIFFERENCE

Figure 12.7 shows the change in GREIT parameter amplitude as the anomaly vertical position changes, as well as amplitude difference when size of the anomaly varies. Notably, the y-axis of the graphs in Figure 12.7 is a relative ratio, normalized

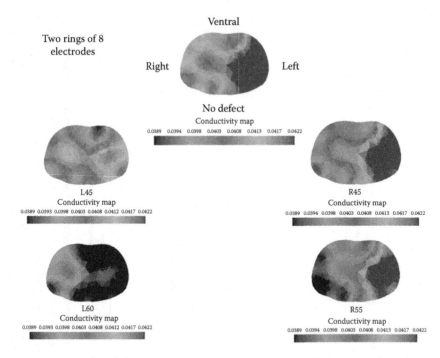

FIGURE 12.6 Image results with two rings of 8 electrodes, L45 and L60 show the reconstructed images when the anomaly is simulated in the left lung (center at [150, 250, 200]) with the radii of 45 AU and 60, respectively, and R45 and R55 show the reconstructed images when the anomaly is simulated in the right lung (center at [400, 250, 200]). The cross-sectional slices are the tomography of the respective reconstructed volume at $z = 200$.

against the amplitude response computed when there is no anomaly, instead of the raw amplitude value. This is because the absolute response is different between the two electrode configurations.

From Figure 12.7, a number of interesting observations can be made:

1. The effect of the anomaly is reduced as it was moved further away from the ring position or the mean of the two rings position.
2. The amplitude response of the same anomaly at different z-locations forms a bell curve, apparently following an upside-down Gaussian function in all cases, except for the largest anomaly in the left heart detected when using 2 rings of the 8 electrodes.
3. The larger anomaly has sharper amplitude response.
4. Using two rings of 8 electrodes, the amplitude response in all cases seems to have a larger full width at half maximum (FWHM), that is, the drop in effect as the anomaly is moved further away from the mean vertical location of the two rings is not as dramatic as it is when only one ring of 16 electrodes is used.

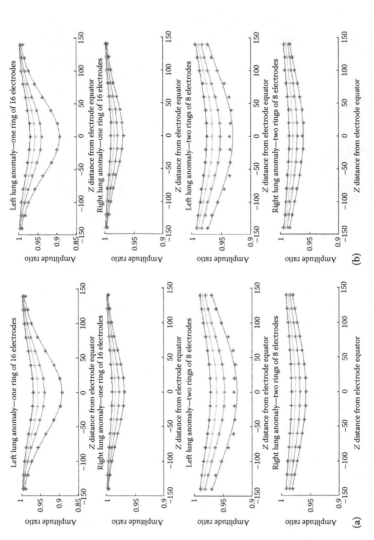

FIGURE 12.7 Changes of the Graz consensus reconstruction algorithm for electrical impedance tomography amplitude parameter with respect to the difference in vertical (z_r) distance between the centers of the anomalies and the center of the electrode ring in the case of one ring and the middle of two electrode rings in the other case, the highest, data and fitted curve correspond to anomaly of radius 45, the next, data and fitted curve correspond to anomaly of radius 50, the next, data and fitted curve correspond to anomaly of radius 55, and finally, the last, data and fitted curve correspond to anomaly of radius 60. (a) 3% conductivity change with a signal-to-noise ratio (SNR) of 85 dB and (b) 3% conductivity change with an SNR of 40 dB.

In order to characterize this amplitude response, we modeled each response with a Gaussian function, utilizing the curve fitting tool in MATLAB. The actual function that was used is a slight variation of the normal Gaussian function:

$$f(z) = 1 - \frac{a}{A} e^{-(z-b)^2/2c^2} \tag{12.9}$$

where

z is the difference in distance between the z-coordinate of the anomaly and $z = 200$, where one ring of the 16 electrodes is located and the mean z-distance between the two rings of 8 electrodes

A is the amplitude difference when there is no defect (this is because of the global amplitude ratio that GREIT calculated; anomalies in the same lung detected with the same electrode configuration have nearly the same A)

a is the amplitude (peak) of the Gaussian curve

b represents the distance between where the peak of the curves occurs and $z = 200$ (it is generally very close to 0)

c represents the sharpness of the curve

FWHM can be calculated from c by

$$\text{FWHM} = 2 \times \sqrt{2 \times \ln(2)c} \tag{12.10}$$

Thus, Figure 12.7 shows the change in amplitude ratios $R = a/A$ as the anomaly moves away from the electrode equator.

Figures 12.8 and 12.9 show the variation in parameters $R = a/A$ for right lung anomalies and FWHM in each case, respectively. From these figures, it can be seen that

- In terms of amplitude response, the configuration with two rings of 8 electrodes produces higher amplitude changes for right lung anomalies but not for left lung anomalies.
- In terms of FWHM, in both left lung and right lung anomalies, the configuration with two rings of 8 electrodes has a higher FWHM than the configuration with one ring of 16 electrodes.
- The difference between systems with SNR of 40 and 85 dB is very small.

12.5 DISCUSSION

The simulation result showed that in reference to the case of no perfusion defect, anomalies as small as 3% can be detected. However, without reference to the no-defect case, defects that are smaller than 5% cannot be easily distinguished. The intrinsic asymmetry of the two lungs and the effect of the heart make it even more challenging to accurately and objectively identify if an embolism has occurred. Furthermore, due to the position of the heart, which is located off center to the left, it is easier to identify a perfusion defect in the right lung than it is to identify one in the left lung.

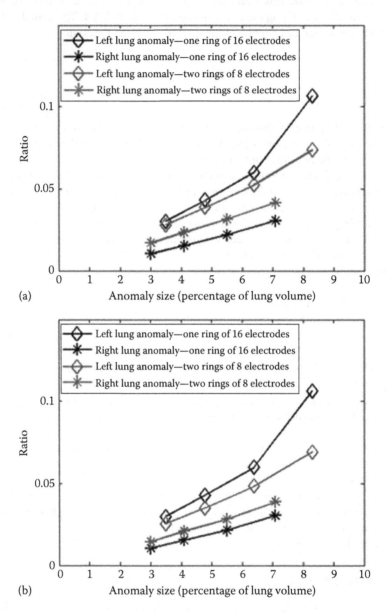

FIGURE 12.8 Amplitude ratio $R = a/A$ of fitted Gaussian functions in four different cases: (1) left lung anomaly with one ring of 16 electrodes, (2) right lung anomaly with one ring of 16 electrodes, (3) left lung anomaly with two rings of 8 electrodes, and (4) right lung anomaly with two rings of electrodes. In each curve, the x-axis is the radius of the anomaly and the y-axis is the ratio $1-R = a/R$ of the function in Equation 12.1. (a) 3% conductivity change with a signal-to-noise ratio (SNR) of 85 dB and (b) 3% conductivity change with an SNR of 40 dB.

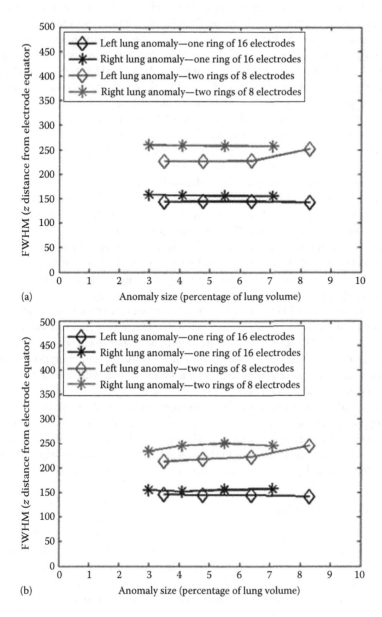

FIGURE 12.9 Full width at half maximum (FWHM) of fitted Gaussian functions in four different cases: (1) left lung anomaly with one ring of 16 electrodes, (2) right lung anomaly with one ring of 16 electrodes, (3) left lung anomaly with two rings of 8 electrodes, and (4) right lung anomaly with two rings of electrodes. In each curve, the x-axis is the radius of the anomaly and the y-axis is the FWHM calculated from Equations 12.1 and 12.2. (a) 3% conductivity change with a signal-to-noise ratio (SNR) of 85 dB and (b) 3% conductivity change with an SNR of 40 dB.

From Figures 12.5 and 12.6, it is also observable that the effect of the anomaly on the conductivities of the lung is global. That is, one cannot, from the reconstructed EIT images, pinpoint the location of the anomaly other than the lung that it affected (left or right lung). Although it can be argued that this is a disadvantage of EIT compared with other techniques, for example, CTPA or SPECT, both of which, to different extents, show the regional effect of PE, clinically, the treatment for PE is an anticoagulant that acts globally, and therefore any indication that PE has occurred is sufficient to start a treatment. Thus, this slight disadvantage should not discourage the development of EIT for PE detection.

As shown in Figure 12.7, the fitted $f(z)$ described the behavior of each anomaly in most cases very well. In fact, with the exception of the anomaly of radius 60 AU in the left lung detected by two rings of 8 electrodes where the root mean square error (RMSE) is 26%, the RMSE in all other cases is less than 0.9%. The reason for the misfit of the case mentioned earlier is because the data in this case do not follow a Gaussian function as it has an inflection point in the center.

While the amplitude responses of both electrode configurations increase more steeply as the size of the anomaly gets larger in both electrode configurations, the steepness of this increase is higher when using two rings of 8 electrodes, indicating this configuration is more sensitive to the size of the anomaly.

In addition, the FWHM of the amplitude response curves stays relatively the same for each electrode configuration. This parameter is therefore independent of the anomaly size. Higher FWHM of the amplitude response curves indicates better amplitude response for out-of-plane anomalies. Thus, the configuration with two rings of 8 electrodes performs better than the configuration with one ring of 16 electrodes for out-of-plane anomalies.

In this study, we simulate the effect of the heart by an equal change in conductivity in the opposite direction to the two lungs. In reality, the conductivity change of the heart between systole and diastole is more complicated as it involves not only the movement of blood but also the contraction of the heart itself. Therefore, one can expect the influence of the heart to be more overwhelming than it was in our simulations.

12.6 CONCLUSION

The configuration with two rings of 8 electrodes performs better than the configuration with one ring of 16 electrodes. The amplitude difference is higher using two rings than just one. Furthermore, two rings are more sensitive to anomaly that are out of plane, evidenced by higher FWHMs. Additionally, the configuration with two rings of 8 electrodes shows better differentiation for different sizes of anomalies and less anatomical distortions.

With either electrode configurations, perfusion defects more than 5% of one lung volume can be observed visually from the reconstructed images. It is easier to recognize anomalies in the right lung than it is in the left lung.

We also show the changes in amplitude difference, as the anomalies move out of the plane or planes of electrodes, follow a Gaussian function with characteristic amplitude a and standard deviation c. These parameters are, in our opinion, of use in the characterization of the performance of 3D EIT algorithms or electrode configurations.

Future *in vivo* experiments on human or animal models should be performed to confirm the results of this study.

REFERENCES

1. A. Adler, J. H. Arnold, R. Bayford, A. Borsic, B. H. Brown, P. Dixon, T. J. Faes et al. Greit: A unified approach to 2d linear EIT reconstruction of lung images. *Physiological Measurement*, 30(6):S35–S55, June 2009.
2. A. Adler and W. R. B Lionheart. Uses and abuses of EIDORS: An extensible software base for EIT. *Physiological Measurement*, 27(5):S25–S42, May 2006.
3. D. C. Barber and B. H. Brown. Recent developments in applied potential tomography. In *Information Processing in Medical Imaging*, (ed) Stephen L. Bacharach. Amsterdam, the Netherlands: Nijho, 1986.
4. M. Bodenstein, M. David, and K. Markstaller. Principles of electrical impedance tomography and its clinical application. *Critical Care Medicine*, 37(2):713–724, February 2009.
5. L. Borcea. Electrical impedance tomography. *Inverse Problems*, 18:R99–R136, 2002.
6. B. M. Eyuboglu, B. H. Brown, and D. C. Barber. *In vivo* imaging of cardiac related impedance changes. *IEEE Engineering in Medicine and Biology Magazine: The Quarterly Magazine of the Engineering in Medicine & Biology Society*, 8(1):39–45, January 1989.
7. I. Frerichs, J. Hinz, P. Herrmann, G. Weisser, G. Hahn, M. Quintel, and G. Hellige. Regional lung perfusion as determined by electrical impedance tomography in comparison with electron beam CT imaging. *IEEE Transactions on Medical Imaging*, 21(6):646–652, June 2002.
8. I. Frerichs, S. Pulletz, G. Elke, F. Reifferscheid, D. Schadler, J. Scholz, and N. Weiler. Assessment of changes in distribution of lung perfusion by electrical impedance tomography. *Respiration; International Review of Thoracic Diseases*, 77(3):282–291, January 2009.
9. B. M. Graham and A. Adler. Electrode placement configurations for 3d EIT. *Physiological Measurement*, 28:S29–S44, 2007.
10. B. Grychtol, W. R. B. Lionheart, M. Bodenstein, G. K. Wolf, and A. Adler. Impact of model shape mismatch on reconstruction quality in electrical impedance tomography. *IEEE Transactions on Medical Imaging*, 31(9):1754–1760, 2012.
11. D. Holder. Introduction to biomedical electrical impedance tomography. In *Electrical Impedance Tomography Methods, History and Applications*, (ed) David S. Holder, pp. 423–449. Bristol, U.K.: Institute of Physics, 2005.
12. A. D. Leathard, B. H. Brown, J. Campbell, F. Zhang, A. H. Morice, and D. Tayler. A comparison of ventilatory and cardiac related changes in EIT images of normal human lungs and of lungs with pulmonary emboli. *Physiological Measurement*, 15(Suppl. 2):A137–A146, May 1994.
13. F. J. McArdle, A. J. Suggett, B. H. Brown, and D. C. Barber. An assessment of dynamic images by applied potential tomography for monitoring pulmonary perfusion: An assessment of dynamic images by applied potential tomography for monitoring pulmonary perfusion. *Clinical Physics and Physiological Measurement*, 9(Suppl. A):87–91, 1988.
14. A. McEwan, G. Cusick, and D. S Holder. A review of errors in multi-frequency EIT instrumentation. *Physiological Measurement*, 28(7):S197–S215, 2007.
15. L. P. Metheral, D. C. Barber, R. H. Smallwood, and B. H. Brown. Three-dimensional electrical impedance tomography. *Nature*, 380:509–512, 1996.
16. N. Polydorides. Image reconstruction for soft-field tomography. PhD thesis, University of Manchester Institute of Science and Technology, Manchester, U.K., 2002.
17. J. Schoberl. Netgen an advancing front 2d/3d-mesh generator based on abstract rules. *Computing and Visualization in Science*, 1(1):41–52, 1997.

Index

A

ABCD, *see* Air-breakdown coherent detection
Acousto-optic (AO) effect, 126
Acousto-optic frequency-shifting technique,
  126–128
Acousto-optic modulator (AOM), 126–128
Active pixel sensor (APS), 144–146
ADC, *see* Analog-to-digital conversion
Air-breakdown coherent detection (ABCD),
  94–95
Analog-to-digital conversion (ADC), 151–152
AOM, *see* Acousto-optic modulator
APS, *see* Active pixel sensor
Avalanche photodiodes (APDs)
  avalanche effect, 147
  Geiger mode, 147–148
  high-performance scientific imaging, 149
  internal multiplication gain, 147
  nonlinear-mode APDs, 147
  quenching, 147–148
  SPAD, 147–149

B

Biosensors, 141; *see also* Optical biosensors
Bragg cells, 126

C

Cancer metastasis
  CTCs detection
    conventional *in vitro* detection methods,
      215
    IVFC, 215–224
    PAFC, 219–220, 222–224
  definition, 213
  illustration diagram, 213–214
Cell imaging
  membrane sensing, 243–245
  molecular oxygen sensing, 241–243
  transition metal complexes
    cell permeability, 231–232
    cell uptake mechanisms, 231
    lipophilicity, 232
    molecular charge, 232–234
  transition metal luminophores
    confocal imaging, 239
    d⁶ transition metal luminophores, 239

FLIM, 240–241
HIV-1 TAT peptide, 234
main classes of, 229–230
mitochondria-targeting peptide
  sequences, 235–237
neuropeptide, 234
NF-κβ peptide sequence, 235–236
NLS peptides, 235
peptide somatostatin, 236–237
polymer conjugates, 238–239
protein-binding peptides, 236
rhenium prolabels, 238–239
sterols, 238
sugar conjugates, 237–238
Chirped photonic crystal fiber (CPCF)
  cornea imaging, 209–210
  cross section of, 202–203
  skin imaging, 208–209
  specifications of, 202–203
  tooth imaging, 210
  ultrashort laser output, spectra, 203–204
CMOS photodetectors
  advantages, 143
  avalanche photodiodes
    avalanche effect, 147
    Geiger mode, 147
    high-performance scientific imaging, 149
    internal multiplication gain, 147
    nonlinear-mode APDs, 147
    quenching, 147–148
    SPAD, 147–149
  linear-mode P–N junction photodetectors
    APS structure, 144–146
    cross-sectional view and operation
      diagram, 143–144
    PPD structure, 145–146
    PPS structure, 143–145
    simplified R-C model, 143–144
    silicon P–N junction diode, inherent optical
      response, 142
Complementary metal-oxide semiconductor
  time-correlated single-photon
  counting (CMOS TCSPC) system
  chip architecture, 187–188
  chip-level TDC integration
    arrayed implementation, 183, 185
    single-detector implementation, 183–184
  column-level integration, 183, 185–186
  pixel-level TDC integration, 185–188

275

Concentric-circular-grating (CCG) THz QCLs,
     62–63
  design
     active region gain, 65
     adjustment of slit locations, 66–67
     azimuthal dependence, 63
     azimuthal modes, 66, 68
     commercial finite element method solver, 63
     effective refractive index, 64
     electromagnetic field distributions, 64–65
     emission loss, 64, 66
     feedback, 65–66
     mode spectrum, 64–65
     optical microscope image, 66, 68
     optical modes with arbitrary azimuthal
        order, 64
     perturbation, three-spoke structure, 68
     refractive index of active region, 64
     standard second-order DFB grating, 64
     superposition of Bessel functions, 63, 66
     surface emitting, 63
     three dimensional schematic
        representation, 63
     2D cross-sectional simulation, 63
     2D PDE mode, 64
  fabrication, experimental results, and analysis
     anisotropic sidewall profiles, wet
        chemical etching, 69, 72
     azimuthal modes, 71
     electric field distribution, 69, 71
     emission spectra, 69–70
     fabrication, 68–69
     light–current–voltage characterization,
        69–70
     simulated far-field pattern, 70, 72
     2D far-field emission pattern, 69, 71
Confocal microscopy, 196
CPCF, see Chirped photonic crystal fiber

D

DCF, see Double-clad fiber
Delta Pro system, 176
Differential LDV
  advantages, 121
  basic structure, 122–123
  conventional scanning techniques, 122
  cross-sectional velocity distribution
     one dimension measurements, 122
     two-dimensional measurements, 129–136
  Doppler frequency shift, 123
  fine spatial resolution, 122
  measurement volume, 122
  mechanical scanning techniques, 121
  nonmechanical scanning technique
     compact and reliable probe, 124
     directional discrimination, 126–129

  dual-axis scanning LDV, 126
  fiber-optic nonmechanical scanning LDV,
     123–124
  in-line polarization beam splitter, 125
  nonmechanical axial scanning LDV
     structure, 124–125
  optical fibers, 123–124
  PMF, 124–125
  reflection-type ruled diffraction
     gratings, 125
  relative measurement positions, 125
  transverse direction, 126
  tunable laser and diffraction gratings, 123
Distributed feedback (DFB) grating, 62
Double-clad fiber (DCF), 199–201
Double-graphene layer (DGL) van der Waals
     heterostructures
  inter-GL RT
     band diagram, 19
     band offset alignment voltage, 20
     band offset vs. gate bias, 20–21
     carrier transit time, 21
     contact resistance, 22
     current–voltage characteristics, 4, 20–21
     cutoff frequency, 21
     degenerated electron and hole systems, 20
     nonlinear electrical properties, 19
     schematic cross section, 19
     static carrier density, 20
     Z-shape NDC, 22
  photon-assisted resonant tunneling
     band diagrams, THz emitter/detector
        structures, 22
     Drude absorption, 23
     electric field distribution, 23
     frequency dependence, 22–23
     simulated DGL-PD responsivity vs.
        photon energy, 24–25
     spontaneous THz emission, 22
     TM photon modes, 22
  plasmon-assisted resonant tunneling
     boundary conditions, 26
     complex frequency, 28
     dynamic system of equations, 26
     in-phase/out-of-phase plasma
        excitation, 24
     inter-GL RT current density, 24
     net ac current, 26
     small-signal admittance, 26–27
     variation of RT current density, 24
  THz modulators
     characteristic plasma frequency, 30
     DGL core shell as optical intensity
        modulator, 28–29
     interband and intraband absorption, 28
     modulation characteristics, 28
     momentum relaxation time, 30

normalized modulation depth *vs.*
modulation frequency, 28–29
plasma effects, 28
THz photomixers (PMs), 28

E

Electrical impedance tomography (EIT)
advantages and disadvantages, 258–259
description, 256
finite element models, human thorax, 261–264
fitted Gaussian functions
amplitude ratio, 269–270
full width at half maximum of, 269, 271
forward problem, 256–257
Graz consensus reconstruction algorithm, 266–269
image reconstruction, 262, 265
inverse problem, 257–258
one ring of 16 electrodes, image results, 265–266
for perfusion defect detection, 259–260
simulated anomalies size, 262, 265
tissue types, conductivities, 262, 265
2D EIT, 256
two rings of 8 electrodes, image results, 265, 267
Electro-optic (EO) detection
balanced detection scheme, 92–93
disadvantage, 92
GaP crystal, 93
Pockels effect, 91
real-time near-field THz imaging, 99
ZnTe crystals, 92–93
Electro-optic (EO) effect, 128

F

Fabry–Pérot resonators, 8
Fermi–Dirac distribution function, 7
Fluorescence lifetime correlated spectroscopy (FLCS), 175–176
Fluorescence lifetime imaging microscopy (FLIM), 153–154, 175, 240–241
Förster resonance energy transfer (FRET), 174–175
Fourier transform infrared (FTIR) spectroscopy, 90, 92

G

Gas plasma technique
ambient air/pure nitrogen, gas target, 90
applications, 91
BBO crystal, 89
broadband single-cycle THz pulses, 89
ionized gas, 88

laser pulses, 90
one-color and two-color schemes, 88–89
recorded terahertz energy, 91–92
R-THz-ABCD *vs.* traditional TDS *vs.* FTIR
spectroscopy, 90, 92
schematic setup, 89–90
subwavelength resolution THz imaging
techniques, 91
terahertz temporal pulse and spectrum, 90–91
Geiger mode, 147–148
Graphene-based terahertz (THz) devices
electrons and holes, massless Dirac fermions, 3–4
graphene-based van der Waals
heterostructures, 4 (*see also* Double-graphene layer van der Waals heterostructures)
groundbreaking discovery, 3
physics of plasmons, 4
SPPs, 4
Graphene-channel field-effect transistors, 18
Graphene layers (GLs), 4
Graz consensus reconstruction algorithm, 266–269

H

Hartree–Fock semiconductor Bloch equations, 38
Human immunodeficiency virus type-1 (HIV-1)
TAT peptide, 234

I

Integrated CMOS optical biosensor systems
CMOS imagers, luminescence spectroscopy
intensity-based luminescence imaging, 150–156
postsilicon processing, 150
time-resolved luminescence imaging, 150, 153–160
CMOS photodetectors
advantages, 143
avalanche photodiodes, 147–149
linear-mode P–N junction photodetectors, 143–146
silicon P–N junction diode, inherent optical response, 142
commercial off-the-shelf CMOS camera, 162–164
diffuse optical tomography, 163
IR spectroscopy, 160, 163
pulse oximeter, 160
Intense THz pulses
detection techniques
air-breakdown coherent detection, 94–95
electro-optic sampling, 91–93
spectral domain interferometry, 93–94

generation techniques
  gas plasma technique, 88–92
  optical rectification technique, 86–89
nonlinear terahertz spectroscopy
  biomedical-induced nonlinearity, 97–98
  classification, 95
  nonlinear THz response, 95–97
  real-time near-field THz imaging,
    98–100
Intensity-based luminescence imaging
  ADC, 151–152
  chip package, glass culture chamber, and
    mounted emission filter, 153
  DNA detection device, 150–151
  DNA microarray, 150
  optical band-pass/longpass filter, 153
  packaging and encapsulation, CMOS chips,
    153
  sampling schemes, 151–153
Intracellular dioxygen (ico$_2$), 241
Intrinsic graphene, 11
In vivo flow cytometer (IVFC)
  basic principle of, 216–217
  fluorescence-based IVFC
    electronic module, 219
    hepatocellular carcinoma study,
      223–224
    optical module, 217–219
    software module, 219
  history of, 215–216
  signal processing, 220–222
  two-color two-channel, 217–218

L

Lanthanides, 228–229
Laser Doppler velocimeter (LDV), 121; *see also*
    Differential LDV
LEDs, *see* Light-emitting diodes
LEF, *see* Linewidth enhancement factor
Lensless ultrawide-field-of-view cell monitoring
  CD4 cell counting, HIV monitoring, 162
  digital image processing, 162–163
  holographic imaging, 162–163
  incoherent LED light source, 163–164
  lens-free cell phone microscope, 163–164
Light-emitting diodes (LEDs), 156–157
LiNbO$_3$ (LN) phase shifter, 128–129
Linewidth enhancement factor (LEF)
  definition, 53
  as a function of applied bias, 55–56
  interband semiconductor diode lasers, 54
  intersubband transition characteristics, 54
  Kramers–Kronig relation, 54
  many-body Coulomb interactions, 54–55
  nonparabolicity, 54–55
  nonzero LEF, 54

optical emission linewidth, 53
resonant-tunneling transport coherence,
    54–55
Schawlow–Townes limit, 53
self-heating effect, 54
zero LEFs, 54
Luminescence-based imaging methods, 228
Luminescent coordination compounds, 229
Luminescent inorganic probes, 228

M

Michelson interferometer, 107
Monoexponential fluorescence decays, 170
Multiexponential decays, 170
Multiphoton microscopy (MPM) endoscopy
  cancer diagnosis and long-term disease
    monitoring, 198
  for clinical applications, 198
  CPCF
    cornea imaging, 209–210
    cross section of, 202–203
    skin imaging, 208–209
    specifications of, 202–203
    tooth imaging, 210
    ultrashort laser output, spectra, 203–204
  DCF, 199–201
  dispersion management, 205–207
  free-space components, 198
  imaging system setup, 204–205
  PCF, 201–202
  principle of, 196–198
  SMF, 200
  SPM, 199–200
  two-photon microendoscope, 199

N

Negative differential conductance (NDC), 4,
    20–22
Nonlinear terahertz spectroscopy
  biomedical-induced nonlinearity, 97–98
  classification, 95
  nonlinear THz response
    monolayer graphene, 95–96
    silicon waveguide, 96–97
  real-time near-field THz imaging
    EO sampling, 99
    linear spatial resolution, 99
    LN tilted-pulse-front setup, 98
    metallic dipole antenna, schematic and
      visible representations, 99–100
    SRR, 100
    terahertz near-field microscope
      experimental setup, 98
    ultrafast electric-/magnetic-field
      switching, 86

Nonmechanical scanning LDV
  compact and reliable probe, 124
  directional discrimination
    acousto-optic frequency-shifting
      technique, 126–128
    optical serrodyne frequency-shifting
      technique, 128–129
  dual-axis scanning LDV, 126
  fiber-optic nonmechanical scanning LDV,
    123–124
  in-line polarization beam splitter, 125
  nonmechanical axial scanning LDV
    structure, 124–125
  optical fibers, 123–124
  PMF, 124–125
  reflection-type ruled diffraction gratings,
    125
  relative measurement positions, 125
  transverse direction, 126
  tunable laser and diffraction gratings, 123
  two-dimensional nonmechanical scanning
    LDV
    commercial 24-channel PMF array, 132
    measurement point, 131–132
    measurement position, 131–133
    model of transmitting optics, 130–131
    structure, 129–130
    transverse measurement position, 129
    unsteady/pulsatile flow, 132
Nuclear localizing signalling (NLS) peptides,
    235

O

Ohm's law, 257
Optical biosensors
  CMOS technology (see also Integrated
    CMOS optical biosensor systems)
    applications, 150–164
    CMOS image sensors, 142
    mixed-signal readout circuits, 142
    photodetector array, 142
    photodetectors, 142–149
    instrumentation, 142
    luminescence, 141
Optical coherence tomography, 196
Optical gain, THz QCLs
  as a function of bias, 49–50
  as a function of doping density, 50–53
  as a function of injection and extraction
    coupling strength, 50–52
  many-body interaction and nonparabolicity
    effects
    computed gain spectra, 47–48
    density-matrix model, 47, 49
    vs. free-carrier model, 47–48
Optical imaging, 195

Optical rectification (OR) technique
  excitation pulse bandwidth, 87
  femtosecond pulse propagation, 86–87
  lithium niobate (LN) crystals
    calculated THz-generation efficiency vs.
      pump pulse duration, 88–89
    THz pulse energy output vs. pump pulse
      energy, 88–89
    tilted-pulse-front technique, 87–88
  phase matching, 87
  ZnTe crystals
    disadvantages, 87
    THz pulse energy output vs. pump pulse
      energy, 88–89
Optical serrodyne frequency-shifting technique,
    128–129
Optoelectronic properties, graphene
  current-injection pumping
    carrier–carrier scattering, 9
    dual-gate field-effect transistor structure
      and band diagram, 9–10
    electron–hole plasma, 9
    lasing gain profiles, 11
    optical phonon decay time, 9
    waveguiding THz-emitted waves, 11
  optical conductivity
    Drude conductivity, 5
    gapless and linear energy spectra,
      electrons and holes, 5
    interband carrier transitions, 5
    interband optical conductivity, 5–6
    intraband carrier transport, 5
    Kubo formula, 4–5
    optical pumping, 6
  ultrafast carrier dynamics and THz gain
    carrier relaxation and recombination
      dynamics, 7
    defect-originated symmetry breaking, 7
    Dirac points, 7
    dynamic conductivity, temporal evolution,
      8–9
    honeycomb lattice, 6
    interband carrier–carrier scatterings, 7
    lattice deformation, 7
    nonequilibrium relaxation dynamics, 7
    optical phonon energies, 6
    quasi-Fermi level, time evolution, 7–8
Organic fluorophores, 228
OR technique, see Optical rectification technique

P

PAFC, see Photoacoustic flow cytometry
PA-GPMC, see Planar periodic array of graphene
    plasmonic microcavities
Partial differential equation (PDE) mode, 64
Passive pixel sensor (PPS), 143–145

PCF, *see* Photonic crystal fiber
Perfusion defect detection, 259–260
Photoacoustic flow cytometry (PAFC)
    melanoma study, 224
    setup of, 219–220
    signal processing, 222–223
Photoacoustic imaging, 195–196
Photoexcited graphene, 11
Photomultiplier tube (PMT), 178–179
Photonic crystal fiber (PCF), 201–202
Pinned photodiode (PPD) pixel, 145–146
Planar periodic array of graphene plasmonic
        microcavities (PA-GPMC), 16–18
Plasmonic properties, graphene
    dispersions and modes, 11–12
    giant THz gain, SPP
        absorption coefficient, 13
        effective refractive index, 12–13
        excitation and propagation, 11
        Fourier-transformed gain spectra, 15
        gain coefficient, monolayer graphene, 13
        gain enhancement factor could, 16
        nonequilibrium plasmons, 12
        optical-pump, THz-probe, and optical-
            prove measurement, 14
        population-inverted graphene, 11–12
        quasi-Fermi energies, 14
        resonant plasmon absorption, 11
        spatial charge-density modulation, 16
        spatial field distribution, THz probe pulse
            intensities, 15–16
        superradiant THz emission, 11
        temporal responses, THz photon-echo
            probe pulse, 14–15
    plasmonic THz intensity modulation, 18–19
    plasmonic THz photodetection and
        photomixing, 18
    superradiant THz plasmonic lasing, graphene
            metasurfaces
        amplification coefficient, 18
        graphene micro-/nanocavity array, 16–17
        wave amplification, 17
Plasmonic waveguide structures
    fabrication, experimental results, and analysis
        fabrication, 75, 77
        groove widths of devices, 77
        horizontal and vertical line scans, 77, 79
        LIV characteristics, 77
        near-field light intensity distributions,
            78–80
        pulsed light–current–voltage
            characteristics, 78
        two dimensional far-field patterns, 77
    in-plane integration of SSP structure
        calculated far-field intensity profile, 74, 76
        electric field distribution, tapered THz
            QCL, 73, 75

fabricated device, scanning electron
    microscope image, 73–74
geometrical parameters, SSP collimator,
    74, 76
simulated 2D light intensity distribution,
    74, 76
simulated 2D light intensity of laser
    without SSP structure, 75, 77
tapered structure and curved front facet
    design, 73
3D schematic cross-section view, 73–74
uncollimated beams, 74–75
penetration depth of THz field, 72
surface plasmons (SPs), 71
PMT, *see* Photomultiplier tube
Polarization-maintaining fiber (PMF), 124–125
PPD pixel, *see* Pinned photodiode pixel
PPS, *see* Passive pixel sensor
PulmoVista500 electrical impedance tomography
    system, 258–259

Q

Quantum cascade lasers (QCLs), 61; *see also*
            Terahertz quantum cascade lasers

R

Reactive ion etching (RIE), 77
Reflection-mode time-domain THz pulse
        measurement
    absorption loss and Fresnel reflections, 112
    PET monolayer bottle, 113
    PET multilayer preform, 112–113
    refractive index, 113
    retroreflector delay line, 114
    thickness measurement, TeTechS PlasThick™
        unit, 111–112
Reflective THz ABCD, 90, 92
Resonant tunneling (RT), 4

S

SDI, *see* Spectral domain interferometry
Self-phase modulation (SPM), 199–200
Semiclassical Boltzmann equations, 11
Single-mode fiber (SMF), 200
Single-photon avalanche detector (SPADs), 179
Single-photon avalanche diodes (SPADs)
    guard ring configurations, 147
    implementation, 147–148
    pixel schematic with passive quenching and
        passive recharge circuit, 147–149
    timing diagram, 148–149
    transient response, 149
Spectral domain interferometry (SDI), 93–94
Split-ring resonator (SRR), 100

SPM, *see* Self-phase modulation
Spoof surface plasmons (SSPs), *see* Plasmonic
   waveguide structures
Steady-state luminescence imaging, *see*
   Intensity-based luminescence imaging
Surface plasmon polaritons (SPPs), 4

T

TCSPC, *see* Time-correlated single-photon
   counting
TDC, *see* Time-to-digital converter
TDS, *see* Time-domain spectroscopy
Terahertz photoconductive antennas
   (THz-PCAs), 107–108
Terahertz quantum cascade lasers (THz QCLs)
   Bloch equation and Maxwell's equations
      coupling
      amplitude gain, 46
      induced polarization, 46
      intensity gain, 46–47
      laser field, 46
      LEF, 47
      macroscopic polarization, 47
      microscopic electric dipole moment, 45
      wave equation, 45
   CCG, 62–63
      design, 63–68
      fabrication, experimental results, and
         analysis, 68–72
   DFB grating, 62
   edge-emitting and surface-emitting
      devices, 62
   frequency range, 61
   intersubband semiconductor Bloch
      equations
      chemical potentials and electron
         temperatures, 43–45
      conduction band diagram, 39
      Coulomb interactions representation, 43
      density matrix elements, 41–42
      dynamical equations of motion, 40
      electron–electron scattering and electron–
         phonon scattering, 44–45
      electron occupation, 41–42
      equations of motion, 41
      Fermi–Dirac distribution with lattice
         temperature, 45
      Hamiltonian of system, 40
      Hartree–Fock contributions, 41
      lifetime of upper laser level, 43
      magnitude squared envelope wave
         functions, 39
      near-resonant screening, 41
      particle and energy conservation, 44
      relaxation-rate approximation, 44
      screening Coulomb matrix element, 40

LEF, microscopic analysis
   definition, 53
   as a function of applied bias, 55–56
   interband semiconductor diode
      lasers, 54
   intersubband transition characteristics, 54
   Kramers–Kronig relation, 54
   many-body Coulomb interactions, 54–55
   nonparabolicity, 54–55
   nonzero LEF, 54
   optical emission linewidth, 53
   resonant-tunneling transport coherence,
      54–55
   Schawlow–Townes limit, 53
   self-heating effect, 54
   zero LEFs, 54
macroscopic model, 38
many-body Coulomb interaction
   bandstructure and Rabi frequency
      renormalizations, 38
   Coulomb-induced subband coupling, 38
   Direct numerical treatment, 38
   gain peak frequency, 39
   Hartree–Fock semiconductor Bloch
      equations, 38
   nonparabolicity effect, 38–39
metal–metal waveguide, 62
Monte Carlo and nonequilibrium Green's
   functions, 37–38
optical gain, microscopic analysis
   as a function of bias, 49–50
   as a function of doping density, 50–53
   as a function of injection and extraction
      coupling strength, 50–52
   many-body interaction and
      nonparabolicity effects, 47–49
optical properties and electron transport, 38
plasmonic waveguide structures
   fabrication, experimental results, and
      analysis, 75, 77–80
   in-plane integration of SSP structure,
      73–77
   penetration depth of THz field, 72
   surface plasmons (SPs), 71
resonant-phonon (RP) design, 37
simplified density-matrix model, 38
2D photonic crystals, 62
wavelengths, 37
Terahertz (THz) technology
   black plastic polymers identification and
      sorting
      absorption coefficient, 109–111
      cost-effective recovery, 108
      float sink process, 109
      near-infrared (NIR) separation
         technology, 109
      refractive index, 109

resin recovery from electronic waste, 108
TeTechS' Rigel system, 109–110
TeTechS sensor system, 108
THz-TDS, 109
THz waves, 109
hose and tubes wall thickness measurement,
    114–115
industrial applications
    compact THz sources and detectors, 107
    photonic-based THz sources and
      detectors, 107
    THz-PCAs, 107–108
    THz-TDS systems, 108
IV bags inspection, 116–117
Michelson interferometer, 107
multilayer plastic bottles and preforms
    measurements
    Hall effect (Magna-Mike) measurement
      probes, 111
    infrared interferometry, 110–111
    reflection-mode time-domain THz pulse
      measurement, 111–114
    TeTechS' PlasThick™ THz sensor system,
      110
    THz-TDS setup, 110
    transparent PET preform, 112–113
noninvasive and nonionizing imaging, 107
Terahertz time-domain spectroscopy (THz-TDS)
    systems, 108
TeTechS
    PlasThick™ THz sensor system
    hose and tubes wall thickness
      measurement, 114–115
    PET multilayer preform, 112–113
    plastic bottles wall thickness
      measurement, 110–111
    Rigel system, 109–110
    sensor system, 108
    THz-based plastic sorting system, 108
    Vega™ THz vision sensor system, 116–117
THz QCLs, *see* Terahertz quantum cascade
    lasers
TIA, *see* Transimpedance amplifier
Time-correlated single-photon counting
    (TCSPC)
    APD array, 156–157
    CMOS TCSPC system
      chip-level TDC integration, 183–185
      column-level integration, 183, 185–186
      pixel-level TDC integration, 185–188
    constant fraction discriminator, 180
    Delta Pro system, 176
    detectors
      afterpulsing, 180
      dark counts, 180
      dead time, 180
      photon-counting detectors, 179

PMT-based detection system, 178–179
    single-photon detector, 178
    SPADs, 179
    excitation sources, 177
    fluorescence lifetime
      bleaching, 171
      FLCS, 175–176
      FLIM, 175
      FRET, 174–175
      Jablonski diagram, 171–172
      natural lifetime, 171
      quantum yield, 171
    frame rate, 156
    implementations, 156
    limitation, 156
    operation principles, 154–155, 172–174
    temporal resolution, 170
    timing electronics
      TAC/ADC implementation, 181
      TDC implementation, 181–182
Time-domain spectroscopy (TDS), 90, 92, 100
Time-gated method, 170
Time-resolved luminescence imaging, 150
    advantages, 153
    digital phase imager architecture, 158, 161
    direct phase-to-digital conversion, 158
    FLIM, 153–154
    frequency-domain phase-modulation
      techniques, 154–155
    LEDs, 156–157
    lifetime dependency, 158
    lock-in amplifier, 155–156
    sensitivity, 158
    TCSPC, 154–157
    TDC, 155, 160
    TIA, 159–160
    xerogel-based oxygen sensor, 158–160
    ZCD, 158, 160
Time-to-digital converter (TDC), 155, 160
Transimpedance amplifier (TIA), 159–160
Transition metal luminophores, cell imaging
    confocal imaging, 239
    d^6 transition metal luminophores, 239
    FLIM, 240–241
    HIV-1 TAT peptide, 234
    main classes of, 229–230
    mitochondria-targeting peptide sequences,
      235–237
    neuropeptide, 234
    NF-κβ peptide sequence, 235–236
    NLS peptides, 235
    peptide somatostatin, 236–237
    polymer conjugates, 238–239
    protein-binding peptides, 236
    rhenium prolabels, 238–239
    sterols, 238
    sugar conjugates, 237–238

Two-dimensional cross-sectional velocity
      distribution measurement
   nonmechanical scanning and spatial
      encoding combination
      LN phase-shifter array, 134
      measurement points, 134, 136
      measurement positions, 134, 136
      six-channel spatially encoded beam array,
         134
      structure, 134–135
   spatial encoding method, 133
   two-dimensional nonmechanical scanning LDV
      commercial 24-channel PMF array, 132
      measurement point, 131–132

   measurement position, 131–133
   model of transmitting optics, 130–131
   structure, 129–130
   transverse measurement position, 129
   unsteady/pulsatile flow, 132

X

Xerogel-based oxygen sensor, 158–160

Z

Zero-crossing detection (ZCD), 158, 160

Printed in the United States
by Baker & Taylor Publisher Services